T0207248

Lecture Notes in Computer Science

Lecture Notes in Artificial Intelligence 13938

Founding Editor

Jörg Siekmann

Series Editors

Randy Goebel, *University of Alberta, Edmonton, Canada*
Wolfgang Wahlster, *DFKI, Berlin, Germany*
Zhi-Hua Zhou, *Nanjing University, Nanjing, China*

The series Lecture Notes in Artificial Intelligence (LNAI) was established in 1988 as a topical subseries of LNCS devoted to artificial intelligence.

The series publishes state-of-the-art research results at a high level. As with the LNCS mother series, the mission of the series is to serve the international R & D community by providing an invaluable service, mainly focused on the publication of conference and workshop proceedings and postproceedings.

Hisashi Kashima · Tsuyoshi Ide · Wen-Chih Peng
Editors

Advances in Knowledge Discovery and Data Mining

27th Pacific-Asia Conference on Knowledge Discovery and Data Mining, PAKDD 2023 Osaka, Japan, May 25–28, 2023 Proceedings, Part IV

Editors
Hisashi Kashima
Kyoto University
Kyoto, Japan

Tsuyoshi Ide
IBM Research, Thomas J. Watson Research Center
Yorktown Heights, NY, USA

Wen-Chih Peng
National Chiao Tung University
Hsinchu, Taiwan

ISSN 0302-9743 ISSN 1611-3349 (electronic)
Lecture Notes in Artificial Intelligence
ISBN 978-3-031-33382-8 ISBN 978-3-031-33383-5 (eBook)
https://doi.org/10.1007/978-3-031-33383-5

LNCS Sublibrary: SL7 – Artificial Intelligence

© The Editor(s) (if applicable) and The Author(s), under exclusive license
to Springer Nature Switzerland AG 2023
Chapters "Road Network Representation Learning with Vehicle Trajectories", "MetaCitta: Deep Meta-Learning for Spatio-Temporal Prediction Across Cities and Tasks", "Words Can Be Confusing: Stereotype Bias Removal in Text Classification at the Word Level" and "cPNN: Continuous Progressive Neural Networks for Evolving Streaming Time Series" are licensed under the terms of the Creative Commons Attribution 4.0 International License (http://creativecommons.org/licenses/by/4.0/). For further details see license information in the chapters.
This work is subject to copyright. All rights are reserved by the Publisher, whether the whole or part of the material is concerned, specifically the rights of translation, reprinting, reuse of illustrations, recitation, broadcasting, reproduction on microfilms or in any other physical way, and transmission or information storage and retrieval, electronic adaptation, computer software, or by similar or dissimilar methodology now known or hereafter developed.
The use of general descriptive names, registered names, trademarks, service marks, etc. in this publication does not imply, even in the absence of a specific statement, that such names are exempt from the relevant protective laws and regulations and therefore free for general use.
The publisher, the authors, and the editors are safe to assume that the advice and information in this book are believed to be true and accurate at the date of publication. Neither the publisher nor the authors or the editors give a warranty, expressed or implied, with respect to the material contained herein or for any errors or omissions that may have been made. The publisher remains neutral with regard to jurisdictional claims in published maps and institutional affiliations.

This Springer imprint is published by the registered company Springer Nature Switzerland AG
The registered company address is: Gewerbestrasse 11, 6330 Cham, Switzerland

General Chairs' Preface

On behalf of the Organizing Committee, we were delighted to welcome attendees to the 27th Pacific-Asia Conference on Knowledge Discovery and Data Mining (PAKDD 2023), held in Osaka, Japan, on May 25–28, 2023. Since its inception in 1997, PAKDD has long established itself as one of the leading international conferences on data mining and knowledge discovery. PAKDD provides an international forum for researchers and industry practitioners to share their new ideas, original research results, and practical development experiences across all areas of Knowledge Discovery and Data Mining (KDD). PAKDD 2023 was held as a hybrid conference for both online and on-site attendees.

We extend our sincere gratitude to the researchers who submitted their work to the PAKDD 2023 main conference, high-quality tutorials, and workshops on cutting-edge topics. We would like to deliver our sincere thanks for their efforts in research, as well as in preparing high-quality presentations. We also express our appreciation to all the collaborators and sponsors for their trust and cooperation.

We were honored to have three distinguished keynote speakers joining the conference: Edward Y. Chang (Ailly Corp), Takashi Washio (Osaka University), and Wei Wang (University of California, Los Angeles, USA), each with high reputations in their respective areas. We enjoyed their participation and talks, which made the conference one of the best academic platforms for knowledge discovery and data mining. We would like to express our sincere gratitude for the contributions of the Steering Committee members, Organizing Committee members, Program Committee members, and anonymous reviewers, led by Program Committee Co-chairs: Hisashi Kashima (Kyoto University), Wen-Chih Peng (National Chiao Tung University), and Tsuyoshi Ide (IBM Thomas J. Watson Research Center, USA). We feel beholden to the PAKDD Steering Committees for their constant guidance and sponsorship of manuscripts.

Finally, our sincere thanks go to all the participants and volunteers. We hope all of you enjoyed PAKDD 2023 and your time in Osaka, Japan.

April 2023

Naonori Ueda
Yasushi Sakurai

PC Chairs' Preface

It is our great pleasure to present the 27th Pacific-Asia Conference on Knowledge Discovery and Data Mining (PAKDD 2023) as the Program Committee Chairs. PAKDD is one of the longest-established and leading international conferences in the areas of data mining and knowledge discovery. It provides an international forum for researchers and industry practitioners to share their new ideas, original research results, and practical development experiences from all KDD-related areas, including data mining, data warehousing, machine learning, artificial intelligence, databases, statistics, knowledge engineering, big data technologies, and foundations.

This year, PAKDD received a record number of 869 submissions, among which 56 submissions were rejected at a preliminary stage due to policy violations. There were 318 Program Committee members and 42 Senior Program Committee members involved in the reviewing process. More than 90% of the submissions were reviewed by at least three different reviewers. As a result of the highly competitive selection process, 143 submissions were accepted and recommended to be published, resulting in an acceptance rate of 16.5%. Out of these, 85 papers were primarily about methods and algorithms and 58 were about applications. We would like to thank all PC members and reviewers, whose diligence produced a high-quality program for PAKDD 2023. The conference program featured keynote speeches from distinguished researchers in the community, most influential paper talks, cutting-edge workshops, and comprehensive tutorials.

We wish to sincerely thank all PC members and reviewers for their invaluable efforts in ensuring a timely, fair, and highly effective PAKDD 2023 program.

April 2023

Hisashi Kashima
Wen-Chih Peng
Tsuyoshi Ide

Organization

General Co-chairs

Naonori Ueda	NTT and RIKEN Center for AIP, Japan
Yasushi Sakurai	Osaka University, Japan

Program Committee Co-chairs

Hisashi Kashima	Kyoto University, Japan
Wen-Chih Peng	National Chiao Tung University, Taiwan
Tsuyoshi Ide	IBM Thomas J. Watson Research Center, USA

Workshop Co-chairs

Yukino Baba	University of Tokyo, Japan
Jill-Jênn Vie	Inria, France

Tutorial Co-chairs

Koji Maruhashi	Fujitsu, Japan
Bin Cui	Peking University, China

Local Arrangement Co-chairs

Yasue Kishino	NTT, Japan
Koh Takeuchi	Kyoto University, Japan
Tasuku Kimura	Osaka University, Japan

Publicity Co-chairs

Hiromi Arai	RIKEN Center for AIP, Japan
Miao Xu	University of Queensland, Australia
Ulrich Aivodji	ÉTS Montréal, Canada

Proceedings Co-chairs

Yasuo Tabei	RIKEN Center for AIP, Japan
Rossano Venturini	University of Pisa, Italy

Web and Content Chair

Marie Katsurai	Doshisha University, Japan

Registration Co-chairs

Machiko Toyoda	NTT, Japan
Yasutoshi Ida	NTT, Japan

Treasury Committee

Akihiro Tanabe	Osaka University, Japan
Aya Imura	Osaka University, Japan

Steering Committee

Vincent S. Tseng	National Yang Ming Chiao Tung University, Taiwan
Longbing Cao	University of Technology Sydney, Australia
Ramesh Agrawal	Jawaharlal Nehru University, India
Ming-Syan Chen	National Taiwan University, Taiwan
David Cheung	University of Hong Kong, China
Gill Dobbie	University of Auckland, New Zealand
Joao Gama	University of Porto, Portugal
Zhiguo Gong	University of Macau, Macau
Tu Bao Ho	Japan Advanced Institute of Science and Technology, Japan
Joshua Z. Huang	Shenzhen Institutes of Advanced Technology, Chinese Academy of Sciences, China
Masaru Kitsuregawa	University of Tokyo, Japan
Rao Kotagiri	University of Melbourne, Australia
Jae-Gil Lee	Korea Advanced Institute of Science & Technology, Korea

Tianrui Li	Southwest Jiaotong University, China
Ee-Peng Lim	Singapore Management University, Singapore
Huan Liu	Arizona State University, USA
Hady W. Lauw	Singapore Management University, Singapore
Hiroshi Motoda	AFOSR/AOARD and Osaka University, Japan
Jian Pei	Duke University, USA
Dinh Phung	Monash University, Australia
P. Krishna Reddy	International Institute of Information Technology, Hyderabad (IIIT-H), India
Kyuseok Shim	Seoul National University, Korea
Jaideep Srivastava	University of Minnesota, USA
Thanaruk Theeramunkong	Thammasat University, Thailand
Takashi Washio	Osaka University, Japan
Geoff Webb	Monash University, Australia
Kyu-Young Whang	Korea Advanced Institute of Science & Technology, Korea
Graham Williams	Australian National University, Australia
Raymond Chi-Wing Wong	Hong Kong University of Science and Technology, Hong Kong
Min-Ling Zhang	Southeast University, China
Chengqi Zhang	University of Technology Sydney, Australia
Ning Zhong	Maebashi Institute of Technology, Japan
Zhi-Hua Zhou	Nanjing University, China

Contents – Part IV

Texts, Web, Social Media

Time-Series and Streaming Data

Scientific Data

Inline Citation Classification Using Peripheral Context and Time-Evolving Augmentation

Priyanshi Gupta[1], Yash Kumar Atri[1(✉)], Apurva Nagvenkar[2], Sourish Dasgupta[2], and Tanmoy Chakraborty[3]

[1] IIIT-Delhi, Delhi, India
{priyanshig,yashk}@iiitd.ac.in
[2] Crimson AI, Mumbai, India
{apurva.nagvenkar,sourish.dasgupta}@crimsoni.ai
[3] IIT Delhi, Delhi, India
tanchak@iitd.ac.in

Abstract. Citation plays a pivotal role in determining the associations among research articles. It portrays essential information in indicative, supportive, or contrastive studies. The task of inline citation classification aids in extrapolating these relationships; However, existing studies are still immature and demand further scrutiny. Current datasets and methods used for inline citation classification only use citation-marked sentences constraining the model to turn a blind eye to domain knowledge and neighboring contextual sentences. In this paper, we propose a new dataset, named 3Cext, which along with the cited sentences, provides discourse information using the vicinal sentences to analyze the contrasting and entailing relationships as well as domain information. We propose PeriCite, a Transformer-based deep neural network that fuses peripheral sentences and domain knowledge. Our model achieves the state-of-the-art on the 3Cext dataset by +0.09 F1 against the best baseline. We conduct extensive ablations to analyze the efficacy of the proposed dataset and model fusion methods.

Keywords: citation classification · bibliometrics · transformer

1 Introduction

For the past several decades, there has been an interest in citation analysis for research evaluation. Researchers have emphasized the necessity for new methodologies that take into account various components of citing sentences. A well-known qualitative technique for assessing the scientific influence is to analyze the sentence in which the research article is mentioned to ascertain the purpose behind the citation. The context of the citation, or the text in which the cited document is mentioned, has proven to be an effective indicator of the citation's intent [25]. Measuring the scientific impact of research articles requires a fundamental understanding of citation intent. A great way to gauge the significance of a scientific publication is to determine why citations are made in one's work and how significant they are.

© The Author(s), under exclusive license to Springer Nature Switzerland AG 2023
H. Kashima et al. (Eds.): PAKDD 2023, LNAI 13938, pp. 3–14, 2023.
https://doi.org/10.1007/978-3-031-33383-5_1

Previous methods for citation context categorization used a range of annotation techniques with low-to-high granularity. Comparing the earlier systems is extremely difficult due to the absence of standardized methodologies and annotation schemes. The 3C shared task [12,13] used a piece of the Academic Citation Typing (ACT) dataset to categorize the reference anchor into 'function' or 'purpose' by looking at the citing sentence or the text that contains the citation [19]. Only quantitative elements are considered in traditional citation analysis based solely on the citation count. One of the biggest obstacles to citation context analysis for citation identification is that there is no multidisciplinary dataset and that there isn't any medium to fine-grained schemes that adequately represent the function and its influence [8]. To address this challenge, Kunnath et al. [12] provided a unified task, called the 3C Shared Task, to compare several citation classification approaches on the same dataset to address the shortcomings of citation context categorization. The main distinction in the second iteration of this task [13] was that the subtasks contained full-text datasets. However, even with the full text, the metadata associated with the citation sentence was not adequate to understand the reasoning for the citation.

To alleviate the above limitations, we propose a new dataset, named `3Cext`, and a new model, named `PeriCite` that combines the advantages of augmentation and peripheral context. Experiments show that the cited sentences heavily rely on the peripheral context to strengthen an argument by contrasting or entailing information. Our main contributions are as follows

1. We extend the 3C dataset [13] – `3Cext`, which, along with the cited sentence, adds more discourse information by providing contrasting and entailing information using the peripheral sentences.
2. We propose a novel model, `PeriCite`, which uses spatial fusion and cross-text attention to attend to contextual information for the peripheral sentences and time-evolving augmentation to counter class imbalance during the training time.
3. We also compare our proposed model against various baselines and show the efficacy of the module along with ablation studies and error analysis.

2 Related Work

Citations are important for persuasion since they provide a source of support for the assertions made by authors. Understanding whether the writers agree or disagree with the assertions made in the cited publication is crucial because not all citations are used with a similar purpose. In order to classify citations according to their context, a sizable corpus of research has previously examined the language used in scientific discourse. Several frameworks have been devised to categorize the intent of citations [16]. Many strategies were used in the early efforts for automatic citation intent categorization; they included rule-based systems [7,18], machine learning techniques based on language patterns [9], and manually-constructed features from the citation context. Teufel et al. [25] introduced how to annotate citations in scholarly articles for 12 classes and used machine learning techniques to replicate annotation. These classes were split into four top-level groups, namely neural class, citations that expressly address weaknesses, citations that contrast or compare, and citations that concur with, use, or are compatible with the citing work. Abu-Jbara et al. [1] utilized a linear SVM and lexical, structural, and syntactic characteristics for categorization. Additionally, feature-based techniques [4,5] for locating quoted spans in the mentioned publications have been studied. Improvements were shown by a joint prediction of cited spans and citation function using a CNN-based model [24].

Most of these initiatives offer far too fine-grained citation categories, some of which are infrequently used in articles. They, therefore, serve little purpose in automated analyses of scientific articles. Jurgens et al. [10] developed a six-category technique to incorporate earlier research and suggested a more precise categorization scheme expanding all previous feature-based work on citation intent classification. The authors also added six categories and 1,941 samples from computational linguistics studies in addition to the three original features-pattern-based, topic-based, and prototype argument-based. They also used structural topology, lexical semantics, grammatical, field positions and values, and usage characteristics.

All of the methods listed above, classified data using hand-engineered features. Cohan et al. [3] proposed a neural multi-task learning technique for classifying citation intent using non-contextualized (GloVe [17]) and contextualized embeddings (ELMo [23], Bidirectional LSTM, and attention method). The authors used two auxiliary tasks to support the primary classification task in order to accomplish multi-task learning. Their recent research [3] included only three citation categories and 11,020 instances from the Computer Science and Medical domains to make up their new dataset (SciCite). Beltagy et al. [2] released SciBERT, a model pre-trained over 1.14 million papers from Semantic Scholar. To support the study in this area, a recent analysis by [11] evaluated 60 research articles on this topic, the difficulties the researchers had while conducting their work, and the knowledge gaps that still need to be addressed.

3 Methodology

In this section, we discuss our proposed model, `PeriCite`. It comprises two stacked Transformers, each with four blocks. It uses Cross-Text Attention to capture the discourse between the cited and peripheral sentences. `PeriCite` also houses Time-evolving Augmentation to synthetically generate data as per label loss and Spatial Fusion to fuse the final representations of stacked Transformers. Figure 1 shows the schematic diagram of `PeriCite` .

3.1 Cross-Text Attention

Attention formulation over a single input text may not provide adequate information to the model. However, when fused with the peripheral context attention, the model can learn important excerpts relative to the label. To fuse peripheral context attention to the main input, we propose Cross-Text Attention (CTA). It computes pairwise weights between the main text and a peripheral context.

Given the self-attention as

$$Attn_{\text{self}}(\mathbf{x}) = softmax\left(\frac{QK^T}{\sqrt{d_k}}\right)V,$$

initially, queries $Q \in \mathbb{R}^{N \times d_k}$, keys $K \in \mathbb{R}^{N \times d_k}$, and values $V \in \mathbb{R}^{N \times d_v}$ are generated for the main input text with d_k and d_v as their dimensions, respectively. Next, to compute the CTA score, pairwise weights between the main input and peripheral context are computed using

$$Attn_{bidir}(x_s, x_t) = Attn_{cross}(x_t, x_s) + Attn_{cross}(x_s, x_t),$$

$$Attn_{cross}(x_t, x_s) = softmax\left(\frac{Q_t K_s^T}{\sqrt{d_k}}\right)V_s,$$

$$Attn_{cross}(x_s, x_t) = softmax\left(\frac{Q_s K_t^T}{\sqrt{d_k}}\right)V_t$$

Here x_s denotes the contextual representation of the main input text, x_t denotes the context of the peripheral text, and Q, K, and V with s and t represents queries, keys, and values based on the main and peripheral text, respectively. We then perform a linear projection of attention heads to capture language comprehension. Finally, the computed attention weights are passed through a feed-forward layer.

3.2 Spatial Fusion

Since the fusion module combines the generated attention vectors from both peripheral encoders, it is crucial to determine which vector is more significant, how it contributes to the main context and interacts with each other. Keeping this intuition in mind, we utilise a fusion strategy based on Spatial Fusion (SF). SF was first introduced by Li et al. [15] to fuse multiple images to decipher deep features. We extend it in our setting to fuse texts together to form deep contextual features.

In spatial fusion as suggested in [27], global features are modeled by a self-attention layer, which is obtained from a convolutional and a dimensionality reduction layer. The combination of these two captures the local as well as the global relations between the feature set. Further, we pass the obtained features to a range of convolutional blocks to enhance the contexts during fusion. Finally, an ordered weighting average map is created to merge features with the source text.

The multi-scale feature matrix is denoted by a, where a is the list of tensors $a \in \{1, 2, \cdots, a\}, a = 4$. The fusion mechanism takes two inputs – one from the cited sentence as P_1^a, and the other from the peripheral sentence as P_2^a. The spatial attention [15] C_f^a is created by the fusion and fed to the final decoding layer.

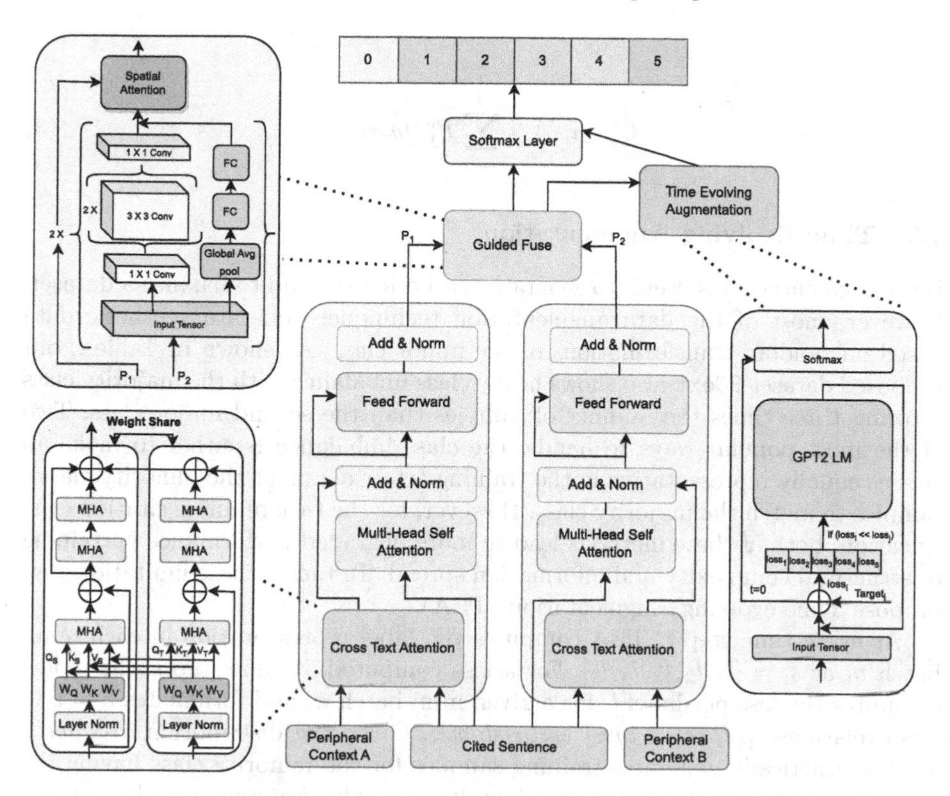

Fig. 1. Illustration of PeriCite: It comprises two parallel stacked Transformer blocks. The cross-text attention is computed between the main context and peripheral sentence. The output from the Transformer blocks is fused using the Guided Fuse. The time-evolving augmentation based on the label representation in the mini-batch generates synthetic training samples.

The Ordered Weighing maps are computed using the L1 normalization and softmax over P_1^a and P_2^a resulting in weight matrix W_1^a and W_2^a, respectively. The final Ordered Weighing maps are computed using Eq. 1 as follows,

$$S_j^a(m,n) = \frac{P_j^a(m,n)}{\sum_{i=1}^{j} ||P_i^a(m,n)} \tag{1}$$

Here L1 norm is computed for both P_j^a and P_i^a with j ranging in $[0,2]$ set. The position in the feature set P_j^a and P_i^a are indicated by (m,n) with a fixed vector size V. The $P_j^a(m,n)$ also outputs a vector of size V.

The feature vectors P_1^a and P_2^a over the weight matrix W_1^a and W_2^a are further enhanced by weighing them using α_k^m using Eq. 2,

$$P_j^a(m,n) = \alpha_j^a(m,n) \times P_j^a(m,n) \tag{2}$$

Finally, the fused vector C_f^a is computed by projecting it against the enhanced feature set using Eq. 3

$$C_f^a(m,n) = \sum_{i=1}^{j} P_j^a(m,n) \tag{3}$$

3.3 Time Evolving Augmentation

Data augmentation is useful to generate synthetic data and to balance a dataset. However, most of the data augmentation techniques generate synthetic data based on random transformations of the minor class. As shown in Table 2, our proposed dataset `3Cext` also shows heavy class imbalance with the majority class showing three times the number of samples than the second major class. Two of the most popular ways to handle the class imbalance is either to make all classes equally representative in the training set or augment the minority class's samples to match the majority class. However, for the task of inline citation classification, both of these methods lead to more degraded performance pertaining to structural complexity and information spread. To tackle these limitations, we propose Time-evolving Augmentation (TEA).

At every time step t, TEA computes the label representation in each mini-batch m as $s_i = [l_1, l_2, l_3, l_4, l_5]$. For a loss computed at time step t, the model computes the loss per label l_i for a given mini batch m and formulates a loss to label relationship $loss \rightarrow label$ as $losslabel_i$. Given the distribution $losslabel_i$, TEA synthetically generates training samples for the minority class having the highest loss in a given mini-batch. The data samples are augmented using the GPT2 language model [20]. The $loss \rightarrow label$ representation is independent of the global representation of a number of samples per class and only takes into account the representation of the given mini-batch. This method helps the model keep the loss in check for each label at every step. The loss of representation per label is a guiding factor for the TEA to evolve at every time step, helping to model to learn equal representation during the training phase.

4 Experiments

4.1 Dataset

In this section, we discuss our proposed `3Cext` dataset in detail. Kunnath et al. [12] introduced the ACT dataset, with annotations for 11,233 citations annotated by 883 authors. The cited label was masked with "#AUTHOR TAG" denoting the position of the cited object. Additionally, the 3C dataset contained full text and the label denoting the class of a particle citation (c.f. Table 2).

In our work, we extend the 3C dataset to house more discourse information to explain better why a citation is present in a sentence. Our intuition is that the cited sentences mostly either entail or contrast the adjoining sentences. To capture the peripheral sentences, we extract the full-text files corresponding to the COREIDs (unique paper ID) in our dataset to follow through on this discovery.

Table 1. Instance of `3Cext` dataset. First Sentence represents the prefixed sentence, Cited Sentence represents the main cited sentence, and Second Sentence is the suffix sentence. We mark first or second sentence as EOF if the cited sentence is either first or last.

First Sentence	Cited Sentence	Second Sentence
[CITE] describe a hybrid recommender system that exploits ontologies to increase the accuracy of the profiling process and hence the usefulness of the recommendations	#AUTHOR_TAG use a different strategy by representing user profiles as bags-of-words and weighing each term according to the user interests derived from a domain ontology	Razmerita et al. [CITE] describe OntobUM, an ontology-based recommender that integrates three ontologies: i) the user ontology, which structures the characteristics of users and their relationships, ii) the domain ontology, which defines the domain
Content-based recommender systems [CITE] rely on pre-existing domain knowledge to suggest items more similar to the ones that the user seems to like. They usually generate user models that describe user interests according to features [CITE]	This API supports a number of applications, including Smart Book Recommender, Smart Topic Miner [CITE], the Technology-Topic Framework #AUTHOR_TAG, a system that forecasts the propagation of technologies across research communities, and the Pragmatic Ontology Evolution Framework [CITE]	<EOF>

Table 2. Number of instances in each class. The classes represents Background, COM-CONT: Compare-Contrast, EXTENSN: Extensions, FUTR: Future Works, MOTIVN: Motivation and Uses.

Dataset	Classes					
	BACKGROUND	COM-CONT	EXTENSN	FUTR	MOTIVN	USES
`3Cext`	1318	380	294	221	137	50

For a given document, we map the location of the main cited sentence and find the prefixed and the suffixed sentence. We use heuristic methods like regex, Levenshtein distance, and hard rules like full-stop identification and author name identification to mark three sentences. In our dataset, the first sentence indicates being the prefix sentence before the citation and second, the suffix sentence after the citation. The six categories of the classes are distributed in labels between 0 and 5, as 0 - BACKGROUND, 1 - COMPARES CONTRASTS, 2 - EXTENSION, 3 - FUTURE, 4 - MOTIVATION, 5 - USES (as suggested in [12]). Table 1 illustrates a sample instance from `3Cext`, and Table 2 represents the number of instances per class.

4.2 Implementation Details

Dataset Setting: On the dataset part, we first preprocess the data by removing stopwords, lowering cases, and removing all special characters. We clip the sentences at 256 token lengths for train and test instances. We use 2400 instances as a train set and 600 as a test set.

Table 3. Performance benchmarks over the 3C and `3Cext` datasets. We provide six classification baselines along different ablation versions of `PeriCite`.

Baseline System	Dataset					
	3C			`3Cext`		
	Precision	Recall	F1 Score	Precision	Recall	F1 Score
Multi-layer Perceptron	0.30	0.38	0.30	0.39	0.32	0.39
LSTM-Attention-Scaffold	0.27	0.48	0.46	0.30	0.55	0.39
Transformer	0.38	0.37	0.34	0.36	0.38	0.37
DistilBERT	0.43	0.48	0.37	0.46	0.55	0.40
BART	0.41	0.46	0.34	0.47	0.52	0.41
T5	0.41	0.43	0.35	0.43	0.48	0.38
SciBERT	0.45	0.51	0.46	0.52	0.53	0.51
`PeriCite`						
`PeriCite` w/ SF	0.38	0.37	0.34	0.36	0.38	0.37
`PeriCite` w/ CTA	0.34	0.36	0.32	0.38	0.41	0.37
`PeriCite` w/ TEA	0.36	0.43	0.41	0.38	0.42	0.39
`PeriCite` w/ TEA, SA, CTA	0.46	0.44	0.42	0.60	0.63	0.60

5 Baselines

We discuss the baseline systems in detail. For the language model (LM)-based baselines, we fine-tune LM with the training samples of 3C and `3Cext` till the convergence.

1. **Multilayer Perceptron:** We use three stacked dense layers [6] with softmax activation. We use Glove embeddings as input representation with cross-entropy as loss.
2. **BiLSTM with Attention and Scaffolding** [3]: This baseline uses BiLSTM with Attention with Glove as input embedding and Elmo as contextual representation. A 20-node MLP is used for the scaffolding task. We preserve all the original hyperparameters for the baseline.
3. **Transformer:** We use the vanilla Transformer [26] architecture to run as baseline. We use 4 stacked layers each in Encoder and Decoder with a max sequence length of 32, with softmax as the activation function and cross-entropy as loss.
4. **DistilBERT** [22]: It follows the same architecture of BERT but reduces the number of parameters by making use of knowledge distillation during pretraining. We use the huggingface ported model for the baseline.
5. **SciBERT:** SciBERT [2] uses the standard BERT architecture with pretraining performed on the scientific documents. The hyperparameters are similar to the Transformer baseline.
6. **BART:** Similar to SciBERT, BART [14] uses a bidirectional encoder along with an autoregressive decoder. It is pre-trained over the Books and Wikipedia data. We use similar hyperparameters to the SciBERT baseline.

7. **T5:** It is a pretrained Transformer based encoder-decoder language model [21]. It is pretrained as a text-to-text Transformer over various supervised and unsupervised downstream tasks.

6 Analysis

We perform ablation studies on our proposed `PeriCite` model to showcase the efficacy of each module. We show that the peripheral context alone can significantly improve the model's performance by providing contextual information. The addition of TEA pushes the performance for each class, concluding that controlled synthetic generation of training data improves the system's overall performance. Table 3 shows that our model attains an improvement of +0.09 F1 points with CTA, SA, and TEA against the best baseline. The improvements are seen in every module. With the introduction of CTA, our model attains an improvement of +0.02 Recall against the base Transformer network. TAdding TEA shows an improvement of +0.02 F1 and +0.04 Recall against the Transformer.

Table 4. Class-wise performance metric of the proposed model. We report Precision, Recall and F1 score of each class.

Model	Class	Precision	Recall	F1 Score
`PeriCite`	BACKGROUND	0.67	0.83	0.74
	COMPARE CONTRAST	0.49	0.28	0.36
	EXTENSION	0.30	0.09	0.14
	FUTURE	0.33	0.17	0.22
	MOTIVATION	0.55	0.31	0.40
	USES	0.61	0.61	0.62

Table 4 shows the class-wise performance of `PeriCite`. It shows that our model is able to capture contextual information for all classes. When compared to the best baseline's (SciBERT) confusion matrix in Table 6, we see that the baseline leans heavily towards the majority class and predicts 0 for almost all other classes. However, our model was able to predict all classes uniformly. We also analyze the model's prediction errors in Table 5. For the sentence in the second row, the model might be distracted by the phrase "we assess the similarity" giving it the impression that it is a "use" category. The third row is also likely to be distracted by the phrase "Following the process of reflection". The mislabelling is probably due to very low number of training samples for these classes. Providing more contextual information and large number of qualitative training samples can help improve the performance of the model.

Table 5. Sample of PeriCite classification error on the 3Cext dataset.

Main Text	True	Pred
What has been termed episodic foresight (#AUTHOR_TAG, 2010), along with autobiographic memory and theory of mind, also makes up much of our mind wandering (Spreng and Grady, 2009), as we preview some future activity or consider possible future options in order to select appropriate action	1	2
We assess the similarity of two semantic vectors using the cosine similarity #AUTHOR_TAG, since this measure relies on the orientation but not the magnitude of the topic weights in the vector space, allowing us to compare editorial products associated with a different number of chapters	0	5
Following the process of reflection presented by #AUTHOR_TAG (1996), in the new version, the first word of 4 questions was added as a visual prompt	0	5
Some more recent models, though, have also included domain experts to define the learning content of the educational game (#AUTHOR_TAG et al., 2017)	4	1

Table 6. Confusion matrix over PeriCite/SciBERT (best baseline). The classes represents Background, Com-Cast: Compare-Contrast, Extensions, Future, Motivation and Uses.

		Predicted					
		Background	Com-Cast	Extensions	Future	Motivation	Uses
Actual	**Background**	280/320	13/0	5/0	1/0	9/0	27/0
	Com-Cast	46/73	21/0	1/0	0/0	3/0	3/1
	Extensions	25/34	0/0	3/0	1/0	2/0	3/0
	Future	9/12	1/0	0/0	2/0	0/0	0/0
	Motivation	28/55	3/0	1/0	1/0	17/0	5/0
	Uses	30/79	5/0	0/0	1/0	0/0	59/16

7 Conclusion

In this paper, we proposed a new dataset 3Cext, where we extended the 3C dataset by introducing peripheral contextual information to analyze the contrasting and entailing information. We also introduced a novel model, PeriCite that uses cross-text attention to attend to the contextual information present in citation input and the peripheral sentences. We also introduced time-evolving augmentation to generate synthetic data for the minority classes during each time step and spatial fusion to attend to the critical information in the input space. Our proposed model achieves a new state-of-the-art on 3Cext by +0.09 F1 score against the best baseline. We also conducted extensive ablations to analyze the efficacy of the proposed dataset and model fusion methods. For future works, an exciting line of work can be to utilize the discourse information of the sections to provide more context to the inline citations. The contrasting or entailing information in the neighbouring sentence is crucial in understanding the reasoning's of citation intent. Additional tasks like baseline recommendations, scientific paper recommendations, etc., can be greatly improved with the performance improvement over this task.

Acknowledgment. The research reported in this paper is funded by Crimson AI Pvt. Ltd.

References

1. Abu-Jbara, A., Ezra, J., Radev, D.: Purpose and polarity of citation: towards NLP-based bibliometrics. In: NAACL, pp. 596–606 (2013)
2. Beltagy, I., Lo, K., Cohan, A.: Scibert: a pretrained language model for scientific text. arXiv preprint arXiv:1903.10676 (2019)
3. Cohan, A., Ammar, W., van Zuylen, M., Cady, F.: Structural scaffolds for citation intent classification in scientific publications. In: NAACL, Minneapolis, Minnesota, pp. 3586–3596. ACL (2019)
4. Cohan, A., Goharian, N.: Contextualizing citations for scientific summarization using word embeddings and domain knowledge. In: ACM SIGIR, pp. 1133–1136 (2017)
5. Cohan, A., Soldaini, L., Goharian, N.: Matching citation text and cited spans in biomedical literature: a search-oriented approach. In: NAACL, pp. 1042–1048 (2015)
6. Gardner, M.W., Dorling, S.: Artificial neural networks (the multilayer perceptron)-a review of applications in the atmospheric sciences. Atmos. Environ. **32**(14–15), 2627–2636 (1998)
7. Garzone, M., Mercer, R.E.: Towards an automated citation classifier. In: Hamilton, H.J. (ed.) AI 2000. LNCS (LNAI), vol. 1822, pp. 337–346. Springer, Heidelberg (2000). https://doi.org/10.1007/3-540-45486-1_28
8. Hernández-Alvarez, M., Gómez, J.M.: Citation impact categorization: for scientific literature. In: 2015 IEEE ICCSE18th International Conference on Computational Science and Engineering, pp. 307–313. IEEE (2015)
9. Ikram, M.T., Afzal, M.T.: Aspect based citation sentiment analysis using linguistic patterns for better comprehension of scientific knowledge. Scientometrics **119**(1), 73–95 (2019). https://doi.org/10.1007/s11192-019-03028-
10. Jurgens, D., Kumar, S., Hoover, R., McFarland, D., Jurafsky, D.: Measuring the evolution of a scientific field through citation frames. Trans. Assoc. Comput. Linguist. **6**, 391–406 (2018)
11. Kunnath, S.N., Herrmannova, D., Pride, D., Knoth, P.: A meta-analysis of semantic classification of citations. Quant. Sci. Stud. **2**(4), 1170–1215 (2022)
12. Kunnath, S.N., Pride, D., Gyawali, B., Knoth, P.: Overview of the 2020 WOSP 3C citation context classification task. In: Proceedings of the 8th International Workshop on Mining Scientific Publications. pp. 75–83. Association for Computational Linguistics (2020)

13. Kunnath, S.N., Pride, D., Herrmannova, D., Knoth, P.: Overview of the 2021 SDP 3C citation context classification shared task. Association for Computational Linguistics (2021)
14. Lewis, M., et al.: Bart: denoising sequence-to-sequence pre-training for natural language generation, translation, and comprehension. arXiv preprint arXiv:1910.13461 (2019)
15. Li, H., Wu, X.J., Durrani, T.: NestFuse: an infrared and visible image fusion architecture based on nest connection and spatial/channel attention models. IEEE Trans. Instrum. Meas. **69**(12), 9645–9656 (2020)
16. Moravcsik, M.J., Murugesan, P.: Some results on the function and quality of citations. Soc. Stud. Sci. **5**(1), 86–92 (1975)
17. Pennington, J., Socher, R., Manning, C.: GloVe: global vectors for word representation. In: EMNLP, Doha, Qatar, pp. 1532–1543. ACL (2014). https://doi.org/10.3115/v1/D14-1162. https://aclanthology.org/D14-1162/
18. Pham, S.B., Hoffmann, A.: A new approach for scientific citation classification using cue phrases. In: Gedeon, T.T.D., Fung, L.C.C. (eds.) AI 2003. LNCS (LNAI), vol. 2903, pp. 759–771. Springer, Heidelberg (2003). https://doi.org/10.1007/978-3-540-24581-0_65
19. Pride, D., Knoth, P., Harag, J.: Act: an annotation platform for citation typing at scale. In: ACM/IEEE JCDL, pp. 329–330. IEEE (2019)
20. Radford, A., Wu, J., Child, R., Luan, D., Amodei, D., Sutskever, I.: Language models are unsupervised multitask learners (2019)
21. Raffel, C., et al.: Exploring the limits of transfer learning with a unified text-to-text transformer. J. Mach. Learn. Res. **21**(140), 1–67 (2020)
22. Sanh, V., Debut, L., Chaumond, J., Wolf, T.: Distilbert, a distilled version of BERT: smaller, faster, cheaper and lighter. CoRR abs/1910.01108 (2019)
23. Sarzynska-Wawer, J., et al.: Detecting formal thought disorder by deep contextualized word representations. Psychiatry Res. **304**, 114135 (2021)
24. Su, X., Prasad, A., Kan, M.Y., Sugiyama, K.: Neural multi-task learning for citation function and provenance. In: 2019 ACM/IEEE Joint Conference on Digital Libraries (JCDL), pp. 394–395. IEEE (2019)
25. Teufel, S., Siddharthan, A., Tidhar, D.: Automatic classification of citation function. In: EMNLP Proceedings of the 2006 Conference on Empirical Methods in Natural Language Processing, pp. 103–110 (2006)
26. Vaswani, A., et al.: Attention is all you need. In: Advances in Neural Information Processing Systems, vol. 30 (2017)
27. Vs, V., Valanarasu, J.M.J., Oza, P., Patel, V.M.: Image fusion transformer. In: 2022 IEEE International Conference on Image Processing (ICIP), pp. 3566–3570. IEEE (2022)

Social Network Analysis

Post-it: Augmented Reality Based Group Recommendation with Item Replacement

Wei-Pin Wang[1(✉)], Hsi-Wen Chen[1], De-Nian Yang[2], and Ming-Syan Chen[1]

[1] Department of Electrical Engineering, National Taiwan University, Taipei, Taiwan
{wpwang,hwchen}@arbor.ee.ntu.edu.tw, mschen@ntu.edu.tw

[2] Institute of Information Science, Research Center for Information Technology Innovation, Academia Sinica, Taipei, Taiwan
dnyang@iis.sinica.edu.tw

Abstract. AR shopping has attracted significant attention with the emergence of popular AR applications, e.g., IKEA Place and Nike AR Fit. However, most AR stores merely provide auxiliary information for each user, instead of group recommendations that leverage new AR features. In this paper, we make the first attempt to explore AR shopping to simultaneously: i) leverage flexible display and item replacement to maximize preference, ii) construct an immersive environment with virtual item haptic feedback, and iii) stimulate social interactions with common user interests. We formulate the Tangible AR Group Shopping (TARGS) problem to partition users and recommend virtual substitutes. We then develop the Social-aware Tangible AR Replacement and Recommendation (STAR3) system, including a connected graph neural network for modeling graph features from different domains, a virtual-physical item mapping model to enable haptic experience by leveraging on-shelf items' passive haptic feedback, and a multi-goal recommender to dynamically split user groups and recommend substitutes for satisfaction maximization. Experimental results manifest that STAR3 outperforms baselines by 23% to 63% in three real-world datasets and a user study.

Keywords: Augmented reality · Replacement-based recommendation

1 Introduction

Augmented reality (AR) has attracted significant attention from industry and academia. According to Valuates,[1] the AR market size is 1.85 billion USD, which will continuously grow to 6.73 billion USD by 2028. Nike and AISLE have prioritized and developed AR-driven retail applications.[2] Compared to conventional shopping, AR shopping enhances user experience and increases 50% of the purchase rate [5]. Despite the promising progress, most AR applications only render auxiliary information for physical items without leveraging new AR features to enhance retail functionality.

AR shopping provides the following features i) **Personalized interface.** An AR system can recommend and display virtual products in the AR views according to user

[1] https://reports.valuates.com/market-reports/QYRE-Othe-2Z282/ar-in-retail.

[2] https://time.com/6138147/augmented-reality-shopping/; http://aisle411.com/walgreens/.

© The Author(s), under exclusive license to Springer Nature Switzerland AG 2023
H. Kashima et al. (Eds.): PAKDD 2023, LNAI 13938, pp. 17–29, 2023.
https://doi.org/10.1007/978-3-031-33383-5_2

preference. Replacement-based recommendations can improve user satisfaction by displaying preferred virtual products to occlude disliked physical products. According to Alter Agents,[3] 73% of AR users agree personalized item display improves shopping experiences. In other words, users would be more satisfied when the AR views overlay personal preferred virtual items on top of uninterested physical ones (please see Appendix A.1). However, existing AR applications [7] merely render digital information of on-shelf items for all users, instead of providing personalized recommendations. ii) **Tangible item replacement.** Tangible AR (TAR) systems [18] ensure *haptic consistency* by allowing users to touch and feel AR-rendered virtual objects that share similar shapes and touches with the replaced physical objects. According to [10], tangible experience improves user immersion and purchase willingness compared to online shopping. Also, Retail Dive[4] reports that 62% of users prefer in-store shopping because they can feel the products. However, most tangible AR research dedicates to developing haptic devices [14] without considering the passive tactile feedback of on-shelf products for tangible item replacement [18]. iii) **Social-aware co-display.** Mora et al. [12] observe that 68% of customers prefer to shop with companions. Existing personal and group recommendations can not support dynamic AR displays to recommend substitutes for multiple subgroups. The social-aware co-display technique [8] can foster social discussions and enhance purchasing intention [22] by flexibly co-displaying common virtual items for multiple users with similar interests, while enabling each user to view additional personally preferred items at the same time.

For TAR group shopping, we formulate a new replacement-based recommendation problem, called *Tangible AR Group Shopping (TARGS)*. Specifically, while typical recommendations utilize a user-item *preference graph* from history to predict preferred items, it is important for TARGS to find the replaced on-shelf items for personalized interface and tangible item replacement. Therefore, we additionally consider a user-item *interaction graph* according to in-store users' real-time behaviors in their views of AR devices (e.g., mobile phones). An illustrated example is presented in Appendix A.1.

Indeed, TARGS faces new research challenges compared with existing recommendations. 1) How to correctly model the heterogeneous user activities in AR shopping, including online transactions and in-store behaviors? 2) How to find the optimal replacement pairs, i.e., virtual and on-shelf physical items, for each group member to maximize satisfaction and ensure haptic consistency. 3) How to dynamically group users and ensure each member's co-display consistency (i.e., the co-displayed items overlay identical physical items) to benefit discussion. 4) How to maximize user satisfaction with all implicit factors with limited ground truth? Therefore, we propose a three-phase graph learning framework, *Social-aware TAR Replacement and Recommendation (STAR3)* in Fig. 1 with three components: i) *Interaction and Preference-aware Graph Attention Network (IPGAN)*, ii) *Haptic-aware Virtual Candidate Item Generator (HVCIG)*, and iii) *Social- and Haptic-aware Recommender (SHaRe)*.

Specifically, for challenge 1, a naive way is to model user preferences and interactions from history (*preference graph*) and in-store behaviors (*interaction graph*) by different models and concatenate the extracted features. However, the learned

[3] https://tinyurl.com/2hhyh8ad.

[4] https://tinyurl.com/2z7hzdud.

distribution from different models may lie in different spaces and reduce performance, i.e., distribution shift [3]. Moreover, existing feature fusion [13] requires bijective feature mapping. However, for TARGS, the numbers of viewed items and candidate substitutes are different, and node mapping is not bijective, i.e., candidate virtual objects in the preference graph may not exist in the user view (interaction graph). Therefore, we propose IPGAN to extract node intra-relations of each graph by GNNs while exploring node inter-relations across the two graphs. We design the identifying encoding and attention-based fusion model to address distribution shift and non-bijective mapping.

For challenge 2, the optimal replacement pair of the virtual and on-shelf items may differ for each user since uninterested items are diverse. A simple way is to pre-define a fixed display location for every user [7], but virtual items may occlude preferred physical products of some users since it does not consider user preferences. In contrast, we introduce HVCIG to predict a replaced item for each virtual object by jointly considering user preferences and haptic features. Specifically, we first design a personalized substitute grouping network to infer the virtual-to-physical item replacing pairs for each user, by modeling user interest and item haptic correlation. Afterward, to find common interests, a pooling model is developed to examine different replacing combinations and predict the virtual item display locations (i.e., replacement pairs) for each group.

For challenge 3, we introduce SHaRe to solve TARGS by flexibly grouping users to foster social interactions and recommending substitutes for preference maximization. For substitute recommendations, SHaRe recommends an item to users to encourage social interactions. Moreover, we ensure co-display consistency by splitting the users into several discussion groups. In contrast, previous works [16] merely split the group by dense graph discovery without considering personalized item replacement for the co-display interface. Last, if we only consider group interests, it may sacrifice individual preferences [1,21] since minority members have diverse preferences. Similar to personal and group interests, there is a trade-off (i.e., the seesaw problem [6]) between preference and item haptic similarity to achieve better immersion. For instance, the virtual and physical items' haptic clues would be mismatched if the desired item always replaces the most uninterested item due to diverse item haptic feedback [2,20]. In contrast, prioritizing haptic correlations reduces preference since preferred items may be occluded [9]. Moreover, due to the lack of ground truth on user partitioning and item replacing, we proposed a reward-based objective to overcome the data sparsity problem. Detailed comparisons with related work are presented in Appendix A.2 and A.3.

- We make the first attempt to propose a replacement-based recommendation and formulate the TARGS problem to carefully examine tactile AR feedback, personal interests, and social interactions in AR shopping.
- We propose a new deep framework, STAR3, for TARGS. We design IPGAN to address the distribution shift issue and infer the multi-relational interplay between users and items. HVCIG is presented to predict item replacing pairs for different groups. SHaRe is introduced to organize user groups and recommend substitutes.
- Extensive experiments are conducted on three real-world datasets with user studies. The results show that STAR3 outperforms all baselines by 37.69%.

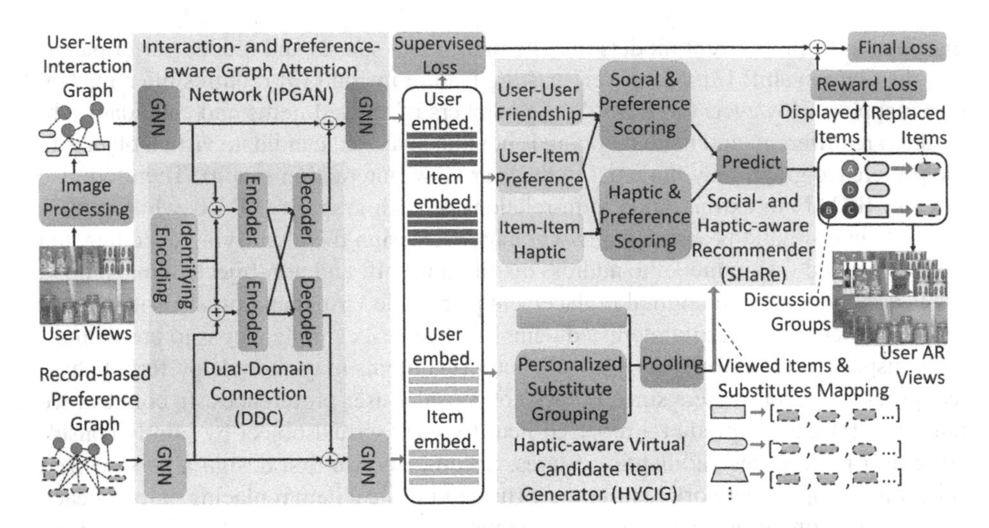

Fig. 1. STAR3 framework.

2 Problem Formulation

In this section, we formally introduce the Tangible Augmented Reality Group Shopping (TARGS) problem. Detailed descriptions and extensional scenarios are presented in Appendix A.4 and A.5. Given a universal item set $c_o \in \mathcal{C}$ and a user-viewed item set $c_k \in \mathcal{C}'$, the TARGS aims to recommend the substitute pair (c_k, c_o) of each user $u_i \in \mathcal{U}'$ in the AR store that maximizes their overall satisfaction. TARGS needs to model the preference gain of displaying item c_o while occluding item c_k. In contrast, conventional recommendation systems [20] merely have to rank and select the user-preferred items.

Moreover, there are two crucial factors for a TARGS problem in AR shopping. i) *co-display consistency* and ii) *haptic consistency*. In AR retail stores, to foster social interactions, the shopping users naturally form subgroups in correspondence with the co-displayed items to be discussed [6]. Also, [22] demonstrates that discussions and interactions in a shopping group are critical to trigger more purchases of an item. Therefore, TARGS partitions the $\mathcal{U}'$ into g disjoint discussion groups $\{\mathcal{U}'_1, \cdots, \mathcal{U}'_g\}$. When TARGS renders the AR items to each user subgroup, it is essential to ensure display consistency for a better group discussion [6]. In other words, group members will see the same object in the same positions during social discussions.

Definition 1 (Co-display consistency). *The display consistency ensures that the substitute location for the AR-rendered item c_o is the same when replacing the same physical item c_k for each user u_i in the same discussion group $\mathcal{U}'_g$.*

For virtual item display in AR systems, haptic feedback (similar to the original physical item) significantly improves user experience and purchasing willingness [10, 18]. To ensure users touch the virtual objects with similar feelings, we first find the haptic similarity of the in-store products by following the tangible item replacement technology [18], in order to sustain the haptic consistency as follows.

Definition 2 (Haptic consistency). *The haptic consistency ensures the haptic similarity $h(c_k, c_o)$ between the AR-rendered item c_o and the physical item c_k is greater than a threshold ϵ_h [18].*

To retain the co-display consistency and haptic consistency, we consider utilities for each user u_i in the following three aspects, including personal preferences, social interactions, and haptic feedback. Given a pair of friends u_i and u_j with a tuple of replaced item $c_k \in \mathcal{C}'$ and substitute $c_o \in \mathcal{C}$, $p(u_i, c_k, c_o)$ represents the *preference utility* of user u_i for replaced item c_k with AR-rendered item c_o. If $c_o = c_k$, $p(u_i, c_k, c_o)$ is the u_i's preference score of item c_o, which can be acquired by following [6]. Otherwise, when $c_o \neq c_k$, we subtract the u_i's preference score of item c_k from the score of item c_o. This is because the virtual displayed item c_o will block the original on-shelf item c_k, i.e., users will not see c_k. $s(u_i, u_j, c_o)$ denotes the *social utility* of user u_i from viewing item c_o together with user u_j, which can be acquired by following [6]. Note that we analyze each user u_j in the same discussion subgroup who also see c_o with u_i for $s(u_i, u_j, c_o)$. Last, $h(c_k, c_o)$ denotes the *haptic utility* of the tactile feedback similarity between items c_k and c_o, which can be acquired by following [18].

Following [6], we capture the overall satisfaction by a weighted combination of the aggregated personal preferences, the retailing benefits from facilitating social interactions, and the user immersion from tangible experience and define the TARGS utility.

$$w(u_i, u_j, c_k, c_o) = (1 - \lambda_1 - \lambda_2)p(u_i, c_k, c_o) + \lambda_1 \cdot h(c_k, c_o) + \lambda_2 \cdot s(u_i, u_j, c_o), \quad (1)$$

where $\lambda_1, \lambda_2 \in [0, 1]$ and $\lambda_1 + \lambda_2 \leq 1$. Note that the preference, social, and haptic utility values, as well as the weights λ, can be directly given by the users or obtained from social-aware personalized recommendation models [19] and TAR research [18]. As a learning framework, we can inductively learn the utilities from historical user data to predict the unseen replacement pair (c_k, c_o) of each user u_i.

Finally, we formally present TARGS recommender as follows.

Definition 3 (TARGS recommender). *A TARGS recommender is a function $A(\cdot) : \mathcal{U}' \rightarrow (\mathcal{C}', \mathcal{C})$ mapping a user to a tuple of a replaced on-shelf product and a recommended virtual item. $A(u_i) = (c_k, c_o)$ means that the recommender displays item c_o on the location of the physical item c_k to user u_i. A TARGS recommender aims to maximize the total TARGS utility w for all AR shoppers $\mathcal{U}'$, i.e., $\sum_{u_i, u_j \in \mathcal{U}'} w(u_i, u_j, c_k, c_o)$. Note that the system can predict multiple items for each user by ranking the total utility.*

Note that the user u_i and u_j are co-displayed an item c_o at the location of item c_k, indicating that they belong to the same group $\mathcal{U}'_g$. To ensure co-display consistency, a naive way is to pre-define the display location of each virtual item for all users. In contrast, after finding the target shopping group, HVCIG predicts the display location of each virtual item. Hence, STAR3 does not require users to specify the new item replacing pairs when adding new users or items to the dataset.

3 STAR3

To solve TARGS, we propose a *Social-aware TAR Replacement and Recommendation (STAR3)* system, including three components: *Interaction and Preference-aware Graph*

Attention Network (IPGAN), *Haptic-aware Virtual Candidate Item Generator (HVCIG)*, and *Social- and Haptic-aware Recommender (SHaRe)*. Figure 1 illustrates the overall architecture of STAR3. IPGAN learns the multi-relational interplay between user and item nodes in the visual-based and knowledge-based graphs (left part). Through an attention-based fusion model to cope with the non-bijective node mapping and distribution shift issues, *Dual-Domain Connection (DDC)* explores node inter-relations and fuses the embeddings across the two graphs. Afterward, HVCIG predicts the optimal item replacing pairs for the shopping group to maximize user satisfaction (bottom right part). Last, SHaRe jointly considers all the factors to flexibly group users and recommends substitutes to maximize personal preference, social interactions, and tangible experiences (upper right part). Time complexity analysis is presented in Appendix A.6.

3.1 Interaction- and Preference-Aware Graph Attention Network

While most web-based recommendation systems implicitly learn user preference from history, in an AR store, modeling users' interactions with on-shelf items is also essential in AR recommendation. To model all AR shopping factors, we first construct a *preference graph* $\mathcal{G}_\Phi$, which consists of three types of edges, i.e., haptic consistency (item-item), transactions (user-item), and friendships (user-user) from all users $\mathcal{U}$ and items $\mathcal{C}$. In addition, we formulate an *interactions graph* $\mathcal{G}_\Psi$, which includes in-store users $\mathcal{U}' \subset \mathcal{U}$ and viewed items $\mathcal{C}' \subset \mathcal{C}$ as nodes and three types of edges, i.e., haptic consistency (item-item), interested products (user-item), and social interactions (user-user). Then, we propose *Interaction and Preference-aware Graph Attention Network (IPGAN)* with two components (see Fig. 1 left part). i) GNN branches extract the features in the *preference graph* (from history) and *interaction graph* (from AR views). ii) *Dual-Domain Connection* with *identifying encoding* integrates multi-modal features with dynamic node mappings and graph sizes across the two graphs. In contrast, prior works cannot learn cross-domain features for TARGS because node embeddings are extracted from different models (i.e., inducing distribution shift), and the node mapping between the two graphs is not a bijection [13], i.e., $|\mathcal{U}| \gg |\mathcal{U}'|$.

To effectively learn the topological and semantic information from the preference and interaction graphs, IPGAN utilizes two independent GNNs that iterative aggregate the neighborhood embeddings. In the lth layer, the embedding $\mathbf{X}^{l+1}$ is aggregated from the previous layer $\mathbf{x}^l \in \mathbf{X}^l$, i.e., $\mathbf{x}_i^{l+1} = \sigma(\mathbf{x}_i^l + \sum_{j \in \mathcal{N}_i} \alpha_{ij} \mathbf{W}^l \mathbf{x}_j^l)$. α_{ij} is the weight for aggregating the neighborhood [19]. $\sigma(\cdot)$ is an activation function. $\mathbf{W}^l$ is a projection matrix. While the embeddings from different graphs may lie in different latent spaces, it may lead to an inaccurate recommendation during preference measurements, e.g., the inner product of the embeddings. To enable the subsequent recommender to analyze preference in a shared latent space, we propose *Dual-Domain Connection (DDC)*. DDC aggregates and fuses the embedding of each layer of GNNs to fully discover user and item correlations between the preference and interaction graphs. Specifically, DDC consists of two phases: the encoding phase to discover node intra-correlation of each graph and the decoding phase to fuse and learn the inter-correlation across two graphs.

Specifically, we first refine the embeddings by a self-attention-based encoder $\mathtt{Enc}$.

$$\mathtt{Enc}(\mathbf{X}) = \mathtt{Norm}(\mathtt{FF}(\mathtt{Att}(\mathbf{X}^l, \mathbf{X}^l, \mathbf{X}^l)) + \mathtt{Att}(\mathbf{X}^l, \mathbf{X}^l, \mathbf{X}^l)). \quad (2)$$

$\texttt{Att}(\cdot)$ is an attention model with residual links as three trainable matrices $\mathbf{W}^q$, $\mathbf{W}^k$, and $\mathbf{W}^v$, i.e., $\texttt{Att}(\mathbf{Q},\mathbf{K},\mathbf{V}) = \texttt{Norm}(\texttt{Softmax}(\frac{(\mathbf{QW}^q)(\mathbf{KW}^k)^\top}{\sqrt{d_h}})\mathbf{VW}^v + \mathbf{QW}^q)$. $\texttt{FF}(\cdot)$ is a feed-forward layer and $\texttt{Norm}(\cdot)$ is a layer-wise normalization layer. We adopt two encoder networks $\texttt{Enc}_\Phi$ and $\texttt{Enc}_\Psi$ for the preference graph features $\mathbf{X}^l_\Phi$ and the interaction graph features $\mathbf{X}^l_\Psi$, respectively. Then, we design a fusion decoder to fuse target domain information to source domain embedding, defined as follows.

$$\texttt{Dec}(\mathbf{X}^l_{src}, \mathbf{X}^l_{ref}) = \texttt{Norm}(\texttt{MLP}(\texttt{Att}(\mathbf{X}^l_{src}, \mathbf{X}^l_{ref}, \mathbf{X}^l_{ref})) + \texttt{Att}(\mathbf{X}^l_{src}, \mathbf{X}^l_{ref}, \mathbf{X}^l_{ref})), \tag{3}$$

where $\mathbf{X}^l_{src}$ and $\mathbf{X}^l_{ref}$ are the encoded information. In the lth-layer, the fused embeddings of the preference and interaction graphs are calculated as follows.

$$\begin{aligned} \mathbf{X}^{l+1}_\Psi &= \mathbf{X}^l_\Psi + \texttt{Dec}_\Psi(\texttt{Enc}_\Psi(\mathbf{X}^l_\Psi), \texttt{Enc}_\Phi(\mathbf{X}^l_\Phi), \texttt{Enc}_\Phi(\mathbf{X}^l_\Phi)), \\ \mathbf{X}^{l+1}_\Phi &= \mathbf{X}^l_\Phi + \texttt{Dec}_\Phi(\texttt{Enc}_\Phi(\mathbf{X}^l_\Phi), \texttt{Enc}_\Psi(\mathbf{X}^l_\Psi), \texttt{Enc}_\Psi(\mathbf{X}^l_\Psi)). \end{aligned} \tag{4}$$

Equipped with the encoder and decoder, DDC can fuse features with different sizes compared to prior fusion methods [13]. Distinct from typical Transformer structures, there are two decoders in DDC to fuse features and transform the embedding distribution for the corresponding GNN.

To identify the nodes across the two graphs, we propose an *identifying encoding* (IE) in the encoding phase to recognize the nodes in both graphs.

$$\texttt{IE}(idx, i) = (i+1)\%2 \times \texttt{sin}(\frac{idx}{10^{8i/d_h}}) + i\%2 \times \texttt{cos}(\frac{idx}{10^{8i/d_h}}), \tag{5}$$

where idx is the node index from the union set of the interaction and preference graphs. d_h is the embedding dimension and i is the position index of the IE vector. Then, we concatenate the identical encoding to the corresponding node embeddings at each layer. The model can distinguish user and item nodes through identical encoding because the same node in different graphs has identical IE. Existing fusion methods [13] cannot properly distinguish the node embeddings because the order and the number of nodes in the interaction and preference graphs are different.

Through DDC, the preference graph embeddings can be updated according to item visual features and user instant behaviors in the interaction graph. Meanwhile, the interaction graph embeddings can exploit virtual items and user substitute interests based on the preference graph. Equipped with DDC, IPGAN can model the complex correlation between the two graphs. Finally, we derive the preference and interaction graph embeddings ($\boldsymbol{\Phi} = \mathbf{X}^L_\Phi$ and $\boldsymbol{\Psi} = \mathbf{X}^L_\Psi$) from the last layer L of IPGAN.

3.2 Haptic-Aware Virtual Candidate Item Generator

For TARGS recommendations, the system needs to model item-replacing pairs for tangible item replacement and social-aware co-display. A method is to adopt a rule-based scheme to predefine the location of virtual objects [7]. However, it is not feasible when the quantities and relations of users and items become large and complex. To generate the virtual-to-physical item mapping for virtual display, a simple way is to calculate the haptic similarity between the preference and interaction graphs by $\boldsymbol{\Phi}_C \boldsymbol{\Psi}^\top_{C'}$ [9].

However, this approach does not consider i) the individual preference for replacement schemes and ii) potential social interactions for social-aware co-display.

Given the in-store user set $\mathcal{U}'$, viewed item set $\mathcal{C}'$, and substitute item set $\mathcal{C}$, we develop a key generator for user $u_i \in \mathcal{U}'$ to integrate multiple factors for physical and virtual item matching. We extend the user embedding ϕ_{u_i} in the preference graph to $\boldsymbol{\Phi}_{u_i} \in R^{|\mathcal{C}| \times d_h}$, where d_h is the feature dimension. The product of group user embeddings $\boldsymbol{\Phi}_{\mathcal{U}'}$ and virtual item embeddings $\boldsymbol{\Phi}_{\mathcal{C}}$ represents the virtual item preferences of the group members. The haptic similarity matrix is estimated by the product of viewed item embeddings $\boldsymbol{\Psi}_{\mathcal{C}'}$ and virtual item embeddings $\boldsymbol{\Phi}_{\mathcal{C}}$.

$$\mathbf{Key}_{u_i} = \mathtt{FF}([\boldsymbol{\Phi}_{\mathcal{C}} || \boldsymbol{\Phi}_{\mathcal{C}} \boldsymbol{\Phi}_{\mathcal{U}'}^{\top} || \boldsymbol{\Phi}_{\mathcal{C}} \boldsymbol{\Psi}_{\mathcal{C}'}^{\top} || \boldsymbol{\Phi}_{u_i}]), \tag{6}$$

where $[\boldsymbol{\Phi}_{\mathcal{C}} || \boldsymbol{\Phi}_{\mathcal{C}} \boldsymbol{\Phi}_{\mathcal{U}'}^{\top} || \boldsymbol{\Phi}_{\mathcal{C}} \boldsymbol{\Psi}_{\mathcal{C}'}^{\top} || \boldsymbol{\Phi}_{u_i}] \in \mathcal{R}^{|\mathcal{C}| \times (2d_h + |\mathcal{U}'| + |\mathcal{C}'|)}$ represents the virtual item embeddings by user preferences, haptic correlations, item features, and user embedding.

Then, we build a virtual-physical item mapping matrix $\mathbf{M}_{u_i} = \mathtt{Att}(\boldsymbol{\Psi}, \mathbf{Key}_{u_i}, \boldsymbol{\Phi})$ by self-attention networks. Compared with regular self-attention networks [13], which only consider item features, the key generator $\mathbf{Key}_{u_i}$ analyzes user information and group members' preferences. HVCIG collects each user's mapping matrix $\mathbf{M}_{u_i}$ and predict the virtual-physical item mapping matrix $\mathbf{M}$ for each group by a pooling model.

$$\mathbf{M} = \mathtt{Softmax}(\mathtt{MLP}(\mathtt{Pool}(\mathbf{M}_{u_1}, \cdots, \mathbf{M}_{u_i}))), \tag{7}$$

where $u_i \in \mathcal{U}'$. $\mathtt{Pool}(\cdot)$ aggregates all features from group members with a 1×1 convolution layer. Then, we predict the mapping matrix $\mathbf{M} \in R^{|\mathcal{C}| \times |\mathcal{C}'|}$ for viewed items $\mathcal{C}'$ to the substitutes $\mathcal{C}$. Then, we discretize the mapping matrix $\mathbf{M} \in [0, 1]$ to a one-hot mapping $\mathbf{M} \in \{0, 1\}$. The main advantage of HVCIG is that it can predict the optimized item replacing pairs for different groups. Compared to normal clustering methods (e.g., k-means), the following recommender SHaRe is designed to jointly optimize the three goals of TARGS, because HVCIG splits the substitutes according to the social relation of group members, their individual preferences, and haptic information.

3.3 Social- and Haptic-Aware Recommender

Finally, *Social- and Haptic-aware Recommender (SHaRe)* is introduced to jointly recommend items, predict locations, and arrange a subgroup for each user. Compared to hybrid recommendations [7] optimizing multiple goals in order, SHaRe jointly considers all the TARGS factors to solve the seesaw problem [6]. While regular group recommendations [1,21] learn group embeddings to capture common interests, the embeddings fail to represent subgroup or individual interests. In contrast, SHaRe employs an information matrix (including members' preferences for personal interests, social interactions, and haptic immersion) to explore shared interests for subgroup partition.

When rendering a virtual item above an actual item, the preference utility needs to subtract the benefit from the physical item, because the AR-render items will block the products on the shelves. Therefore, we consider the preference gain as follows:

$$\mathbf{P} = \boldsymbol{\Phi}_{\mathcal{U}'} \boldsymbol{\Phi}_{\mathcal{C}}^{\top} - \boldsymbol{\Psi}_{\mathcal{U}'} \boldsymbol{\Psi}_{\mathcal{C}'}^{\top} \mathbf{M}^{\top}, \tag{8}$$

where the value $p_{ij} \in \mathbf{P}$ matrix denotes the preference gain for user i to display virtual item j on the corresponding physical item location, which can be inferred by the item mapping matrix $\mathbf{M}$ from HVCIG. Then, we define social preference as follows.

$$\mathbf{S} = \boldsymbol{\Psi}_{\mathcal{U}'}\boldsymbol{\Psi}_{\mathcal{U}'}^{\top} + \boldsymbol{\Phi}_{\mathcal{U}'}\boldsymbol{\Phi}_{\mathcal{U}'}^{\top}, \tag{9}$$

where $\mathbf{S}$ jointly considers frequently interacted shoppers in the AR store by $\boldsymbol{\Psi}_{\mathcal{U}'}\boldsymbol{\Psi}_{\mathcal{U}'}^{\top}$ and user friendships by $\boldsymbol{\Psi}_{\mathcal{U}'}\boldsymbol{\Phi}_{\mathcal{U}'}^{\top}$. Therefore, $s_{ij} \in \mathbf{S}$ represents the utility of co-displaying the same item to user i and j. Next, the item haptic similarity is defined as follows.

$$\mathbf{H} = \boldsymbol{\Psi}_{\mathcal{C}'}\boldsymbol{\Phi}_{\mathcal{C}}^{\top}. \tag{10}$$

With a higher value of $h_{ij} \in \mathbf{H}$, users are envisaged to have a better haptic experience by replacing the actual item c_i with the virtual item c_j. Then, SHaRe collects all factors and examines each substitute with the social and haptic effects of the preference gain.

$$\mathbf{Y} = \texttt{Softmax}(\texttt{MLP}([\boldsymbol{\Omega}_s || \boldsymbol{\Omega}_h])), \text{ where } \boldsymbol{\Omega}_s = \texttt{MLP}([\mathbf{P}||\mathbf{S}]), \text{ and } \boldsymbol{\Omega}_h = \texttt{MLP}([\mathbf{P}||\mathbf{H}]). \tag{11}$$

$\texttt{MLP}(\cdot)$ learns to balance the needs of the input factors. SHaRe simultaneously analyzes the social and haptic effects, i.e., $\boldsymbol{\Omega}_s$ and $\boldsymbol{\Omega}_h$ matrix, to avoid encountering the seesaw problem. For each row of the output matrix $\mathbf{Y} \in R^{|\mathcal{U}'| \times |\mathcal{C}|}$, the virtual item with the largest value is recommended to the corresponding user. To ensure co-display consistency, users with the same recommended item form a discussion subgroup. For each recommended item, SHaRe infers the replaced item by the viewed-substitute item mapping $\mathbf{M}$. For example, if $m_{ij} \in \mathbf{M}$ has the largest value in the i row, item i will be rendered on item j. Unlike using multiple predictors to predict user partitions, substitutes, and display locations, SHaRe recommends substitutes and infers the replaced items and discussion groups to avoid contradictory recommendation results. For instance, two people are grouped together but see different items.

3.4 Overall Objective

A simple way to train STAR3 is to collect the ground truth labels of recommended items, display locations, and user partitions with supervised learning. However, finding the optimal results of item replacement and user partition for a shopping group by brute force search requires massive computing resources ($((|\mathcal{C}'| \times |\mathcal{C}|)^{|\mathcal{U}'|}$ possible combinations). Moreover, once a new item or user is added to the dataset, it is computationally intensive to retrain the whole. As a result, we introduce a hybrid loss function that contains two parts: i) supervised and ii) unsupervised parts for STAR3, to support end-to-end training without the real-time ground truth of item replacement and user partition. First, we apply the max-margin loss [20] to pretrain the node embedding in IPGAN to maximize the similarity between the positive node pairs and minimize the similarity between negative node pairs.

$$L(\mathbf{x}, \mathbf{x}_p, \mathbf{x}_n) = max(0, \mathbf{x}\mathbf{x}_p - \mathbf{x}_p\mathbf{x}_n + \Delta), \tag{12}$$

where Δ denotes the margin hyper-parameter, and $\mathbf{x}_p$ and $\mathbf{x}_n$ are the positive and negative sample to node embedding $\mathbf{x}$. We accumulate the max-margin loss for five relations,

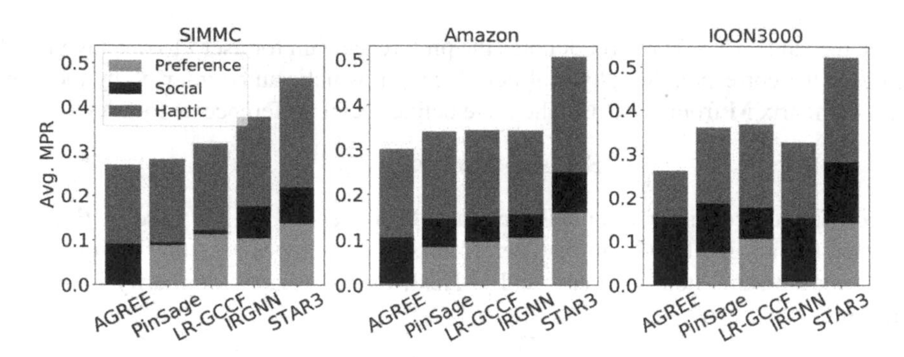

Fig. 2. Quantitative results.

i.e., the virtual item preferences ($\boldsymbol{\Phi}_{\mathcal{U}'}\boldsymbol{\Phi}_{\mathcal{C}}^{\top}$), physical item preferences ($\boldsymbol{\Psi}_{\mathcal{U}'}\boldsymbol{\Psi}_{\mathcal{C}'}^{\top}$), in-store user interactions ($\boldsymbol{\Phi}_{\mathcal{U}'}\boldsymbol{\Phi}_{\mathcal{U}'}^{\top}$), friendships ($\boldsymbol{\Psi}_{\mathcal{U}'}\boldsymbol{\Psi}_{\mathcal{U}'}^{\top}$), and haptic correlations ($\boldsymbol{\Psi}_{\mathcal{C}'}\boldsymbol{\Phi}_{\mathcal{C}}^{\top}$), as our supervised loss $L_{supervised}$.

Due to the high computing costs, we cannot train HVCIG and SHaRe by exploring every combination to find the best user partitioning and item replacing pairs. Inspired by reinforcement learning, we generate the rewarding score for STAR3. With the item substitutes and user subgroups prediction results, we can infer the rewarding score by the ground truth of TARGS utilities as follows.

$$L_{reward} = \gamma_1 p(\cdot) + \gamma_2 s(\cdot) + \gamma_3 h(\cdot), \tag{13}$$

where $p(\cdot)$, $s(\cdot)$, and $h(\cdot)$ are the preference, social, and haptic utility in Sect. 2. γ_1 to γ_3 are the weights of the three factors, which can be set by following [6, 18].

4 Experiments

Setup. Detailed experiment setup, inference time, sensitivity tests, case studies, and user studies are presented in Appendix A.7 to A.11. We evaluate STAR3 on three real datasets. *i) SIMMC* [11] is provided by Meta for AR/VR shopping. It contains $68k$ interaction logs, $5k$ users, and 179 items. *ii) Amazon (home and kitchen)* [4] is a product review dataset with $103k$ users, $98k$ items, and $105k$ reviews. *iii) IQON3000* [17] is a retail dataset with $308k$ outfits, $3k$ users, and $672k$ fashion items. For performance evaluation, STAR3 is compared to four state-of-the-art baselines: *AGREE* [1] for group recommendations, *PinSage* [20] and *LR-GCCF* [2] for personalized recommendations, and *IRGNN* [9] for hybrid recommendations. Following [6, 15], we adopt Mean Percentile Rank $MPR = \frac{1}{|\mathcal{T}|}\sum_{t\in\mathcal{T}}\frac{r}{|rank_t|}$ to evaluate the model performance, where $\mathcal{T}$ is the set of test instances and r is the ranking of the candidate set $rank_t$.

For fair comparisons, the pre-trained node features are used as the initial features for all models following the settings of [20]. The embedding dimension is 256 and the hyperparameters γ_0 to γ_3 are set as 1. We use the Adam gradient optimizer with a 10^{-5} learning rate to train the models on one RTX 2080 Ti GPU. We randomly split 80% and 20% of the data for training and testing.

Table 1. Ablation study.

Model	w/o DDC	w/o IE	w/o haptic	HVCIG w/o haptic	w/o social	HVCIG w/o social	STAR3
Preference	0.4064	0.4072	0.3987	0.4127	0.4299	0.4179	0.4107
Social	0.0831	0.1957	0.2629	0.2148	0.1024	0.1512	0.2464
Haptic	0.6972	0.6949	0.5841	0.6728	0.6575	0.6831	0.7399
Average	0.3995	0.4326	0.4152	0.4335	0.3966	0.4174	0.4657

Quantitative Results. Figure 2 presents the quantitative results, which show that STAR3 significantly outperforms the baselines by at least 24% (SIMMC), 47% (Amazon), and 42% (IQON3000). STAR3 achieves better improvement on Amazon and IQON3000, because it effectively leverages the information from the interaction and preference graphs, especially on sparse datasets. Compared with the group recommendation, the average performance of STAR3 significantly outperforms AGREE by 90%. Group recommendation methods learn group representations to capture users' co-interests for all shopping members. In contrast, IPGAN embeds multi-type interplays between users and items in the preference and interaction graphs to model each user's preferences and social behaviors. For personalized recommendations, STAR3 outperforms PinSage and LR-GCCF by at least 42%, because HVCIG and SHaRe ensure co-display consistency by grouping users and predicting substitutes with the group members' features. In contrast, both personalized approaches need to modify the replaced items to ensure co-display consistency, undermining the overall performance. Finally, STAR3 outperforms the hybrid recommendation IRGNN. Although hybrid approaches can model user preferences and haptic relevances by sequentially adopting preference and item-relation models, the performance is limited by the replaced item selection for each user, since the preference model does not consider haptic relations. In contrast, SHaRe jointly exploits all the factors for discussion groups and substitute recommendations. Therefore, STAR3 can maximize preference, social, and haptic utilities compared to all baselines.

Ablation Study. In Table 1, we conduct the ablation study on the SIMMC dataset to evaluate the importance of each module in STAR3. The six STAR3 variants are *i) STAR3 without connection module (DDC), ii) STAR3 w/o identifying encoding (IE), iii) STAR3 with SHaRe without preference-haptic branch and HVCIG without haptic information, iv) STAR3 without giving haptic information to HVCIG, v) STAR3 with SHaRe without preference-social branch and HVCIG without social information*, and *vi) STAR3 without giving social information to HVCIG*. First, DDC enables GNNs to exchange content and structure information and generate meaningful node embeddings. Moreover, the interaction and preference graph embedders can co-work and encode the haptic information to the item embeddings via the connections. Therefore, DDC in STAR3 improves the average performance by 17.7%. Without IE, it is more difficult for DDC to learn node mapping across different graphs. Thus, the overall performance drops by 7% after removing IE. Without considering the haptic information for substitute grouping (HVCIG) and item recommendation (SHaRe), the performance drops by 10.8%. While adding the parallel branch in SHaRe to handle preference-haptic influence increases the average score by 4.4%, it still has a 7.4% improvement gap between STAR3. This indicates that the substitute grouping will limit the final performance if HVCIG only considers user preference and social relations without haptic feedback. Similar to the experiments of STAR3 without haptic feedback, we also test STAR3 without social information. According to the last three columns in Table 1, the social utility and final performance drop without considering the user relationships. Again, this implies that social interactions are important for group and substitute recommendations.

5 Conclusion

In this paper, we make the first attempt to formulate the TARGS problem and propose a novel STAR3 framework to embed intricate relations and content features, map haptic correlated items, and recommend substitutes and discussion groups to improve user experience. Experimental results show that STAR3 significantly outperforms the baselines on three real-world datasets and a user study by 23% to 47% and 63%, respectively.

Acknowledgement. This work was supported in part by the Ministry of Science and Technology, Taiwan, under grant MOST Project No. 111-2221-E-002-135-MY3 and 111-2223-E-002-006.

References

1. Cao, D., He, X., Miao, L., Xiao, G., Chen, H., Xu, J.: Social-enhanced attentive group recommendation. TKDE **33**(3), 1195–1209 (2019)
2. Chen, L., Wu, L., Hong, R., Zhang, K., Wang, M.: Revisiting graph based collaborative filtering: a linear residual graph convolutional network approach. In: AAAI (2020)
3. Chen, L., Gan, Z., Cheng, Y., Li, L., Carin, L., Liu, J.: Graph optimal transport for cross-domain alignment. In: ICML (2020)
4. He, R., McAuley, J.: Ups and downs: modeling the visual evolution of fashion trends with one-class collaborative filtering. In: WebConf (2016)
5. Javaheri, H., Mirzaei, M., Lukowicz, P.: How far can wearable augmented reality influence customer shopping behavior. In: MobiQuitous (2020)
6. Ko, S.H., Lai, H.C., Shuai, H.H., Lee, W.C., Yu, P.S., Yang, D.N.: Optimizing item and subgroup configurations for social-aware VR shopping. VLDB **13**(8), 1275–1289 (2020)
7. Lam, K.Y., Lee, L.H., Hui, P.: A2W: context-aware recommendation system for mobile augmented reality web browser. In: MM (2021)
8. Lissermann, R., Huber, J., Schmitz, M., Steimle, J., Mühlhäuser, M.: Permulin: mixed-focus collaboration on multi-view tabletops. In: CHI (2014)
9. Liu, W., et al.: Item relationship graph neural networks for e-commerce. IEEE Trans. Neural Netw. Learn. Syst. **33**(9), 4785–4799 (2021)
10. Luo, C., Shen, Y., Liu, Y., Jiang, Z.J.: Look and feel: the importance of sensory feedback in virtual product experience. In: ICIS (2019)
11. Moon, S., et al.: Situated and interactive multimodal conversations. arXiv preprint arXiv:2006.01460 (2020)
12. Mora, J.D., González, E.M.: Do companions really enhance shopping? Assessing social lift over forms of shopper value in Mexico. J. Retail. Consum. Serv. **28**, 228–239 (2016)

13. Prakash, A., Chitta, K., Geiger, A.: Multi-modal fusion transformer for end-to-end autonomous driving. In: CVPR (2021)
14. Salazar, S.V., Pacchierotti, C., de Tinguy, X., Maciel, A., Marchal, M.: Altering the stiffness, friction, and shape perception of tangible objects in virtual reality using wearable haptics. ToH **13**(1), 167–174 (2020)
15. Schedl, M., Zamani, H., Chen, C.W., Deldjoo, Y., Elahi, M.: Current challenges and visions in music recommender systems research. IJMIR **7**, 95–116 (2018)
16. Shen, C.Y., Yang, D.N., Huang, L.H., Lee, W.C., Chen, M.S.: Socio-spatial group queries for impromptu activity planning. TKDE **28**(1), 196–210 (2015)
17. Song, X., Han, X., Li, Y., Chen, J., Xu, X.S., Nie, L.: GP-BPR: personalized compatibility modeling for clothing matching. In: MM (2019)
18. de Tinguy, X., et al.: How different tangible and virtual objects can be while still feeling the same? In: WHC (2019)
19. Yang, L., Liu, Z., Wang, Y., Wang, C., Fan, Z., Yu, P.S.: Large-scale personalized video game recommendation via social-aware contextualized graph neural network. In: WebConf (2022)
20. Ying, R., He, R., Chen, K., Eksombatchai, P., Hamilton, W.L., Leskovec, J.: Graph convolutional neural networks for web-scale recommender systems. In: SIGKDD (2018)
21. Zhang, J., Gao, C., Jin, D., Li, Y.: Group-buying recommendation for social e-commerce. In: ICDE (2021)
22. Zhang, X., Li, S., Burke, R.R.: Modeling the effects of dynamic group influence on shopper zone choice, purchase conversion, and spending. JAMS **46**, 1089–1107 (2018)

Proactive Rumor Control: When Impression Counts

Pengfei Xu, Zhiyong Peng(✉), and Liwei Wang

School of Computer Science, Wuhan University, Hubei, China
{xupengfei,peng,liwei.wang}@whu.edu.cn

Abstract. The spread of rumors in online networks threatens public safety and results in economic losses. To overcome this problem, a lot of work studies the problem of rumor control which aims at limiting the spread of rumors. However, all previous work ignores the relationship between the influence block effect and counts of impressions on the user. In this paper, we study the problem of minimizing the spread of rumors when impression counts. Given a graph $G(V, E)$, a rumor set $R \in V$, and a budget k, it aims to find a protector set $P \in V \backslash R$ to minimize the spread of the rumor set R under the budget k. Due to the impression counts, two following challenges of our problem need to be overcome: (1) our problem is NP-hard; (2) the influence block is non-submodular, which means a straightforward greedy approach is not applicable. Hence, we devise a branch-and-bound framework for this problem with a $(1 - 1/e - \epsilon)$ approximation ratio. To further improve the efficiency, we speed up our framework with a progressive upper bound estimation method, which achieves a $(1 - 1/e - \epsilon - \rho)$ approximation ratio. We conduct experiments on real-world datasets to verify the efficiency, effectiveness, and scalability of our methods.

Keywords: social network · rumor control · random walk · non-submodularity

1 Introduction

World Wide Web and social networks have become the most commonly utilized vehicles for information propagation and changed people's lifestyles greatly due to the increasing popularity of online networks. However, the ease of information propagation is a double-edged sword. But rumors and misinformation could be quickly spread on social networks, which results in undesirable social effects and even leads to economic losses [4]. Therefore, minimizing the spread of rumors in online networks is a crucial problem.

To solve this problem, a lot of work studies the problem of rumor control which aims to minimize the spread of rumor on social network [1–3,5,7,11].

Supplementary Information The online version contains supplementary material available at https://doi.org/10.1007/978-3-031-33383-5_3.

© The Author(s), under exclusive license to Springer Nature Switzerland AG 2023
H. Kashima et al. (Eds.): PAKDD 2023, LNAI 13938, pp. 30–42, 2023.
https://doi.org/10.1007/978-3-031-33383-5_3

However, they only assume that users are passive receivers of rumors even if the users can browse the rumors on their own. Therefore, in this study, we assume that users will actively encounter/contact the rumors via their browsing behaviors, i.e., keyword search, social browsing, etc., which can be modelled by random walk model [10,16]. Unfortunately, existing work [10,16] does not consider the relationship between the influence block and counts of impressions on one user because the model assumes one-time impression is enough. But in the real world, studies in consumer behavior report that users are unlikely to take meaningful action when they receive a message only one time [6,8]. Meanwhile, there is evidence showing that the effect of message repetition should be measured as an S-shaped function (logistic function) [12,14].

To this end, we study the problem of minimizing the spread of rumor when impression counts and call it Rumor Control when Impression Counts (RCIC). Suppose that an online network is represented by a graph $G(V, E)$. Given a rumor set $R \in V$ and a budget k, RCIC aims to find a protector set $P \in V \backslash R$ to minimize the spread of the rumor set R as much as possible under the budget k. To the best of our knowledge, this is the first problem for rumor control when impression counts are considered. As a result, the following challenges are important to be addressed.

The first challenge is the NP-hardness of RCIC as we analyze in Theorem 1. Then, we resort to developing approximate algorithms to solve it efficiently. The second challenge is posed by the property of the logistic function. The influence block model based on the logistic function is non-submodular, which means any straightforward greedy-based the approach is not applicable to address the RCIC problem as shown in Example 1. To overcome this challenge, we proposed a sampling-based greedy method to estimate the upper bounds of the logistic function value. Based on this upper bound estimating method, we devise a branch-and-bound framework for RCIC, with a $(1 - 1/e - \epsilon)$ approximation ratio. Furthermore, we speed up our framework with a progressive upper-bound estimation method. In summary, we make the following contributions.

- We propose and study the RCIC problem, and analyze the monotonicity and non-submodularity of the objective function of RCIC. We show that RCIC is NP-hard.
- To solve the RCIC problem, we present a Monte Carlo based greedy algorithm (Greedy) as the baseline solution. Moreover, we devise an upper-bound estimation method by adaptively solving submodular optimization problems. Based on the upper bound function, we propose a branch-and-bound framework for RCIC, with a $(1 - 1/e - \epsilon)$ approximation ratio.
- To further improve the efficiency, we speed up our framework with a progressive sampling-based greedy method for upper bound estimation, which achieves a $(1 - 1/e - \epsilon - \rho)$ approximation ratio and a significant reduction in running time.
- We conduct extensive experiments on three real-world datasets. The results validate the effectiveness, efficiency, and scalability of our solutions.

2 Related Work

In the following, we discuss the most relevant literature to our problem.

Two proactive rumor control problems in online networks are close to our work [10,16], which also study proactive rumor control problem to minimize the spread of the rumor set under the budget. The core difference lies in the influence block model. In particular, the existing work assumes that one anti-rumor node before the rumor node can block the total influence of the rumor set to the user in one browsing process. It does not consider the relationship between the influence block effect and counts of impressions on one user because the model assumes one time impression is enough. On the contrary, RCIC is built upon a logistic influence block model, which has been widely adopted in consumer behavior studies. To minimize the spread of the rumor set, we need to control the overlap to some extent by impressing the same users several times.

Two other problems close to our problem are influence block and competitive influence maximization. Influence block aims to limit the influence of rumors by blocking some nodes or links in a network [1,7,11]. Their strategies of the seed selection are mainly based on their connectivity, such as degree [1,11], pagerank [7], and betweenness [7]. Different from the first problem, competitive influence maximization tries to identify a set of target seed nodes (or protectors) who will spread an 'anti-rumor' to limit the scale of rumor propagation [2,3,5,18]. Carnes et al. [5], and Bharathi et al. [2] study competitive influence diffusion under the extension of the Independent Cascade model and show that the problem of maximizing the influence of one campaign is NP-hard and submodular, while Borodin et al. [3] studies the similar problem under the Linear Threshold model. Our problem is essentially different from the above work for the following reason. Both influence block and competitive influence maximization assume that the information (or rumors) propagations are driven by the effect of word-of-mouth, and they use the Independent Cascade model and Linear Threshold model to simulate the spread of rumors. However, our problem assumes that rumors spread via browsing behaviors and uses a random walk model to describe the influence spread of rumors.

3 Problem Formulation

In this section, we first formally define the influence model and influence block. In the following, we give the formulation definition of RCIC. In the end, we show the non-submodularity of the objective function of RCIC and prove that RCIC is NP-hard.

3.1 Influence Model

Let $G = (V, E)$ be an online network with $n = |V|$ nodes and $m = |E|$ edges. The random walk process can be used to model the user's browsing process on G as follows [9,10,13,16]. Given a node $u \in V$, a browsing process starting from

u can be represented by a random walk w_u. In particular, w_u picks a neighbor v of u by the probability of $p_{uv} = 1/|\text{neighbors of } v|$ and moves to this neighbor and then follows this way recursively. We say that u hits v at step t, if w_u first visits v after t walk steps.

Similarly, we say that u hits (or is influenced by) set S at the time step t, if w_u first visits set S by a t-hop jump. It is worth noting that t should not be very large in the real world, as most social media users only browse a small number of pages each day. Therefore, we can use a threshold T to bound the hitting time t for any nodes and sets.

3.2 Influence Block

Based on the influence model, we introduce the concept of influence block when impression counts as follows.

Before that, we first introduce the conception of impression. For nodes n_1 and n_2 in a random walk w_u, we define that n_1 have a impression of block the influence of n_2 to u, if w_u visits n_1 before n_2. Therefore, we use the Bernoulli random variable $C_{w_u}(n_1|n_2)$ denoting the states whether n_1 have a impression of block the influence of n_2 to u, where $C_{w_u}(n_1|n_2) = 1$ denotes that n_1 have a impression of block the influence of n_2 to u, otherwise $C_{w_u}(n_1|n_2) = 0$. Then the total impressions of P ($P \subset V$ is a protector set) to block the influence of R ($R \subset V$ is a rumor set) to u in w_u can be computed by $C_{w_u}(P|R) = \sum_{v \in P} C_{w_u}(v|n_r)$. Here n_r is the first node in w_u, which is contained in the set R.

Our influence block when impression counts is based on the logistic function. We use the following equation to compute the influence block of a protector set P to a rumor set R in w_u:

$$I_{w_u}(P|R) = \begin{cases} \frac{1}{1+exp\{\alpha-\beta\cdot C_{w_u}(P|R)\}} & \text{if } C_{w_u}(P|R) > 0 \\ 0 & \text{otherwise} \end{cases} \tag{1}$$

Here α and β are the parameters that control turning point of the user u for being influenced by the protect information, where α controls the overall effectiveness of influence block of P to R and β controls the incremental effectiveness of influence block of one node in P to R in w_u. Then, let $I_u(P|R) = E[I_{w_u}(P|R)]$ for any w_u denote the expected value of possibility that P blocks the influence of R to u.

3.3 Problem Definition

Based on $I_u(P|R)$, the problem of Rumor Control when Impression Counts (RCIC) can be described as follows.

Definition 1 (Problem Definition). *Given a graph $G = (V, E)$, an initial set $R \subset V$ and a budget k, RCIC is dedicated to finding a k-size set $P \subset V \backslash R$, which can maximize the influence block $\mathcal{G}(P|R) = \sum_{u \in V \backslash R} I_u(P|R)$.*

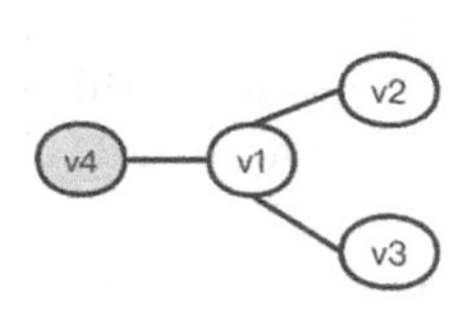

Fig. 1. An example of RCIC

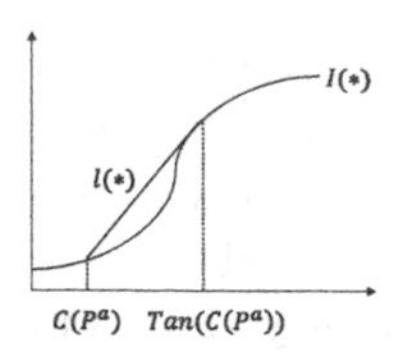

Fig. 2. The upper-bound influence block function

Next, we analyze the monotonicity and submodularity of $\mathcal{G}(P|R)$ and the hardness of RCIC.

Definition 2. *We say that $\mathcal{G}(P|R)$ is monotone iff, for any two assignment protector sets P^a and P^b such that $P^a \subseteq P^b$, it holds that $\mathcal{G}(P^a|R) \leq \mathcal{G}(P^b|R)$. We say that $\mathcal{G}(P|R)$ is submodular iff, for any two such protector sets and any P, it has $\mathcal{G}(P^a \cup P|R) - \mathcal{G}(P^a|R) \geq \mathcal{G}(P^b \cup P|R) - \mathcal{G}(P^b|R)$.*

It is trivial to show that $\mathcal{G}(P|R)$ is monotone. However, as the following counterexample shows, $\mathcal{G}(P|R)$ is not submodular.

Example 1. As shown in Fig. 1, the rumor set $R = \{v4\}$. We choose $P^a = \{\}$, $P^b = \{v1\}$, $P = \{v2\}$, $\alpha = 3$, $\beta = 1$ and $T = 2$. Then we have $\mathcal{G}(P^a|R) = 0$, $\mathcal{G}(P^b|R) = 1.372$ and $\mathcal{G}(P|R) = 0.358$. Furthermore, we have $\mathcal{G}(P^a \cup P|R) - \mathcal{G}(P^a|R) = 0.358$ and $\mathcal{G}(P^b \cup P|R) - \mathcal{G}(P^b|R) = 2.433 - 1.372 = 1.061$. Since $P^a \subseteq P^b$ and $\mathcal{G}(P^a \cup P|R) - \mathcal{G}(P^a|R) \leq \mathcal{G}(P^b \cup P|R) - \mathcal{G}(P^b|R)$. We thus conclude $\mathcal{G}(P|R)$ is not submodular.

Theorem 1. *The RCIC problem is NP-hard.*

Proof. Please see our full version [15] for more details. ■

4 Our Framework

In this section, we first present a Monte Carlo based greedy method (Greedy) as a baseline. Unfortunately, the effectiveness of this method is poor, and Greedy cannot obtain any theoretical guarantees because the objective function of RCIC is non-submodular. Then we devise a Branch-and-Bound framework to solve this problem effectively. The core of this framework is how to estimate the upper bound of each candidate solution. In particular, we propose sampling-based bound estimation techniques for each branch under exploration by setting a submodular function to a tightly upper bound of $I_{w_u}(P|R)$.

4.1 A Baseline

The core idea of Greedy is to select the node u which maximizes the unit marginal gain, i.e., $(\mathcal{G}(P \cup \{u\}|R) - \mathcal{G}(P|R))$, to a candidate solution set P, until the budget k is exhausted. The pseudo-code of Greedy is presented in Algorithm 1. It first initializes P as an empty set and $V \leftarrow V \backslash R$. Next, it finds a set P according to the greedy heuristic (Lines 1.6 to 1.10). In the end, it outputs set P as a result.

Algorithm 1: Greedy($G, R, , k$)

1.1 **Input:** a graph G, a rumor set R and a budget k
1.2 **Output:** a protector set P
1.3 Run X random walks for each node in $V \backslash R$; $Is(R) \leftarrow$ all the random walks influenced by R; Initialize P as an empty set and $V \leftarrow V \backslash R$.
1.4 **repeat**
1.5   Select $u \leftarrow \arg\max_{v \in V}((\mathcal{G}(P \cup \{v\}|R) - \mathcal{G}(P|R)))$
1.6 $V \leftarrow V \backslash \{u\}$ and $P \leftarrow P \cup \{u\}$; $k \leftarrow k - 1$
1.7 **until** $k = 0$
1.8 **return** P

Algorithm 2: BranchAndBound(G, R, k)

2.1 **Input:** a graph G, a rumor set R and a budget k
2.2 **Output:** a protector set P
2.3 Initialize P and P' as an empty set and $V \leftarrow V \backslash R$; $L_G \leftarrow 0$ and $U_G \leftarrow \infty$; Initialize max heap $H \leftarrow \{P', V, U\}$
2.4 **repeat**
2.5 $\{P', V, U\} \leftarrow$ top of H; Select $u \in V$
2.6   **if** $|P'| < k$ **then**
2.7 $V \leftarrow V \backslash u$
2.8 $P^a \leftarrow P' \cup u$ and $P^b \leftarrow P'$
2.9 $\{P^c, L^a, U^a\} \leftarrow SamComputeBound(P^a, V)$
2.10   **if** $L^a > L_G$ **then**
2.11 $L_G \leftarrow L^a$ and $P \leftarrow P^c$
2.12   **if** $U^a > L_G$ **then**
2.13 $H \leftarrow H \cup \{P^a, V, U^a\}$
2.14   Repeat line 2.9 to 2.13 for P^b
2.15 **until** $L_G \geq U_G$
2.16 **return** P

4.2 Branch-and-Bound Framework

As we analyzed above, Greedy cannot obtain any theoretical guarantees because the objective function of RCIC is non-submodular. Then inspired by [17], we introduce a branch and bound framework to solve this problem effectively, and this solution can achieve a theoretical guarantee.

Algorithm 2 shows the pseudo-code of the branch-and-bound framework. We first initialize the global upper bound U_G and global lower bound L_G, and a max heap H with each entry denoted as $\{P', V, U\}$, where P' is the current node set that has been selected as a protector set, V is the set of a node that has not been considered yet, and U is the upper bound influence block of the corresponding search space. H is ordered by the upper bound value of each P'. While $L_G < U_G$, H will pop the top entry that has the maximum upper bound influence block. For each entry, if it matches the budget k constraint, it will generate two new candidate sets (P^a and P^b) by adding a new node $u \in V$ or not. Then it computes the upper bound for each candidate set and updates L_G, P and H when $L^a > L_G$ and $U^a > L_G$, respectively.

Algorithm 3: SamComputeBound(P^a, V)

3.1 **Input:** Protector set P^a and candidate node set V
3.2 **Output:** $\{P, L^a, U^a\}$
3.3 Run X T-random walks for each node in V; $Is(R) \leftarrow$ all the random walks influenced by R; Initialize P as P^a and $k \leftarrow k - |P^a|$
3.4 **repeat**
3.5 Select $u \leftarrow \arg\max_{v \in V}((\overline{\mathcal{G}}(P \cup \{v\}|R) - \overline{\mathcal{G}}(P|R)))$
3.6 $V \leftarrow V \backslash \{u\}$ and $P \leftarrow P \cup \{u\}$; $k \leftarrow k - 1$
3.7 **until** $k = 0$
3.8 **return** P, $L^a \leftarrow \mathcal{G}(P|R)$, $U^a \leftarrow \overline{\mathcal{G}}(P|R)$

4.3 Computing Upper Bound

To estimate the upper bound of the current protector set P^a, we devise a submodular function ($\overline{I}_{w_u}(P|R)$ and $P = P^a \cup P^*$) as shown in Fig. 2 to compute the upper bound of P^a:

$$\overline{I}_{w_u}(P|R) = \begin{cases} l(C(P)) & \text{if } l(x) \text{ exists and} \\ & C(P^a) < C(P) < Tan(C(P^a)) \\ I_{w_u}(P|R) & \text{otherwise} \end{cases} \quad (2)$$

Here, $C(P) = C_{w_u}(P|R)$ for simplicity. $l(x)$ is the tangent through point $(C(P^a), I_{w_u}(P^a|R))$ to function $I_{w_u}(P|R)$ and $Tan(C(P^a))$ is the x-coordinate of the tangent point. It is easy to see that $\overline{I}_{w_u}(P|R)$ is submodular as it concatenates two submodular functions for different domains.

Furthermore, we have the following submodular function $\overline{\mathcal{G}}(P|R) = \sum_{u \in V \backslash R} \overline{I}_u(P|R)$ (here $\overline{I}_u(P|R = E[\overline{I}_{w_u}(P|R)]$ for any w_u) that upper bounds the influence block function $\overline{\mathcal{G}}(P|R)$. It is also easy to see that $\overline{\mathcal{G}}(P|R)$ is submodular as it is a sum of submodular functions.

Due to the submodularity of $\overline{\mathcal{G}}(P|R)$, we turn to devise a greedy-based heuristic algorithm to find the upper bound for a given protector set P^a. In particular, we propose a sampling-based upper bound estimation algorithm to compute the upper bound for a given protector set.

Sampling-Based ComputeBound. As shown in Algorithm 3, it selects the node u which maximizes the unit marginal gain to a candidate solution set P, until the budget k is exhausted. In the end, it outputs set P, $\mathcal{G}(P|R)$ as L^a and $\overline{\mathcal{G}}(P|R)$ as U^a.

4.4 Analysis of Solutions

In this section, we show the proposed branch and bound framework with sampling-based computeBound can achieve a $(1 - 1/e - \epsilon)$-approximation factor through setting an appropriate sampling time X.

Theorem 2. *The branch and bound framework with sampling-based computeBound achieves an approximation factor of $(1 - 1/e - \epsilon)$ for the RCIC through setting an appropriate parameter X.*

Proof. Please see our full version [15] for more details. ∎

5 Progressive Branch-and-Bound

Although Algorithm 2 improves the effectiveness of basic greedy by conducting the branch-and-bound framework, it still suffers from a high computational cost due to heavily invoking Algorithm 3 for bound estimations. To be mores specific, in each greedy search iteration of Algorithm 3, it has to recalculate the marginal gain $(\overline{\mathcal{G}}(P \cup \{v\}|R) - \overline{\mathcal{G}}(P|R))$ for all candidate nodes.

Motivated by this observation, we propose a progressive sampling-based upper bound estimation method (ProSamComputeBound). It selects multiple, but not only one, nodes in each greedy search iteration to cut down the total number of iterations required and hence the computation cost. Meanwhile, we will prove that it can achieve an approximation ratio of $(1 - 1/e - \epsilon - \rho)$ for the upper bound estimation, where ρ is a tunable parameter that provide a trade-off between efficiency and accuracy.

The pseudo-code of ProSamComputeBound is shown in Algorithm 4. ProSamComputeBound first sorts $v \in V$ based on descending order of $\overline{\mathcal{G}}_v(P|R)$ and initializes the threshold h to the value of $\max_{v \in V} \overline{\mathcal{G}}_v(P|R)$. Then, it iteratively fetches all the nodes with their marginal gains not smaller than h into P and meanwhile lowers the threshold h by a factor of $(1+\rho)$ for next iteration (Lines 4.5–4.14). The iteration continues until there are k nodes in P. Unlike the basic greedy method that has to check all the potential nodes in candidate node set V in each iteration, it is not necessary for ProSamComputeBound as it implements an early termination (Lines 4.10–4.11). Since nodes are sorted by $\overline{\mathcal{G}}_v(P|R)$ values, if $\overline{\mathcal{G}}_v(P|R)$ of the current node is smaller than h, all the nodes v‘ pending for evaluation will have their $\overline{\mathcal{G}}_{v‘}(P|R)$ values smaller than h and hence could be skipped from evaluation.

In the following, we show the approximation ratio of the branch-and-bound framework invoking Algorithm 4 for RCIC by Theorem 3.

Theorem 3. *The branch and bound framework with sampling-based computeBound achieves an approximation factor of $(1 - 1/e - \epsilon - \rho)$ for the RCIC through setting an appropriate parameter X.*

Proof. Please see our full version [15] for more details. ■

6 Experiments

In this section, we present our experimental results on the effectiveness, efficiency, memory consumption, and scalability of our proposed methods.

6.1 Experimental Settings

DataSets. We use three real-world datasets in the experiments: Gnutella, Email-Enron, and Gowalla. All the datasets are obtained from an open-source website[1], and their statistics are shown in Table 1. The Gnutella dataset is a peer-to-peer file-sharing network, the Email-Enron dataset is an email communication network, and the Gowalla dataset is a location-based social networking website where users share their locations by checking in.

[1] http://snap.stanford.edu/data/.

Algorithm 4: ProSamComputeBound(P^a, V)

4.1 **Input:** Protector set P^a and candidate node set V
4.2 **Output:** $\{P, L^a, U^a\}$
4.3 Run X T-random walks for each node in V; $Is(R) \leftarrow$ all the random walks influenced by R; Initialize P as P^a; Sort $v \in V$ based on descending order of $\overline{\mathcal{G}}_v(P|R)$; Initialize $h \leftarrow \max_{v \in V} \overline{\mathcal{G}}_v(P|R)$
4.4 **while** $|P| \leq k$ **do**
4.5   **for** *each* $v \in V$ **do**
4.6   **if** $|P| \leq k$ **then**
4.7 $\overline{\mathcal{G}}_v(P|R) \leftarrow (\overline{\mathcal{G}}(P \cup \{v\}|R) - \overline{\mathcal{G}}(P|R))$
4.8   **if** $\overline{\mathcal{G}}_v(P|R) \geq h$ **then**
4.9 $P \leftarrow P \cup v$, $V \leftarrow V \backslash v$
4.10   **if** $\overline{\mathcal{G}}_v(P|R) < h$ **then**
4.11   **break**
4.12   **else**
4.13   **break**
4.14 $h \leftarrow \frac{h}{1+\rho}$
4.15 **return** P, $L^a \leftarrow \mathcal{G}(P|R)$, $U^a \leftarrow \overline{\mathcal{G}}(P|R)$

Table 1. Summary of the datasets.

	n	m	AvgD	MaxD
Gnutella	8.8k	63k	7.2	88
Email-Enron	37k	184k	5.01	1383
Gowalla	197k	950k	4.83	14730

Table 2. Parameter setting.

Parameters	Values
k	50, 100, **150**, 200, 250
$\|R\|$	50, 100, **150**, 200, 250
T	3, 6, **9**, 12, 15

Algorithms. To the best of our knowledge, this is the first work to study RCIC, and thus there exists no previous work for direct comparison. In particular, we compare the four following methods. (1) TopK: It is to select the top-k high block degree nodes in the sampling random walk set as the targeted nodes. (2) Greedy: A basic sampling-based greedy algorithm (Algorithm 1). (3) BranchAndBound (BAB): The branch-and-bound framework (Algorithm 2) with Algorithm 3 for bound estimations. (4) Progressive BranchAndBound (ProBAB): The branch-and-bound framework (Algorithm 2) with Algorithm 4 for bound estimations.

Evaluation Metrics. We evaluate the performance of all methods by the runtime and the blocking percentage of the selected nodes. In particular, the percentage is computed by $\mathcal{G}(P|R)/Is(R)$, where $Is(R)$ denote the random walk set influenced by rumor set R.

Parameter. Table 2 shows the settings of all parameters, such as the budget k, the size of the rumor set R, and the (random walk) length threshold T. Here the default one is highlighted in bold. In this experiment, we set the number of samples $X = 1000$, $\alpha = 7$, $\beta = 3$, and $\rho = 0.1$ as default, see our full version [15] for more details. To simulate the rumor set R, we select nodes uniformly at random from the nodes whose degrees are in the top 10% of G.

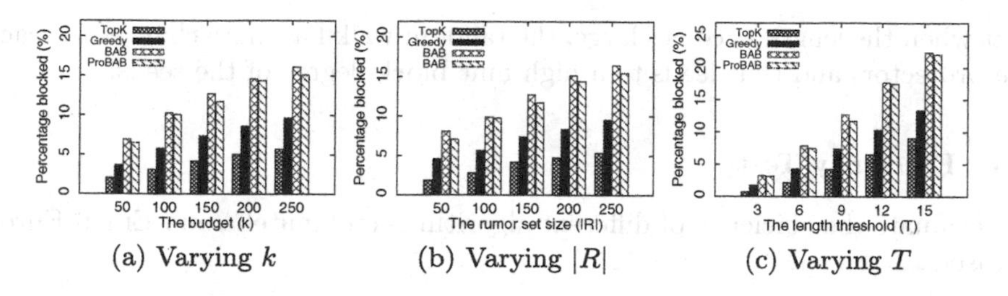

(a) Varying k (b) Varying $|R|$ (c) Varying T

Fig. 3. Effectiveness test on Gnutella

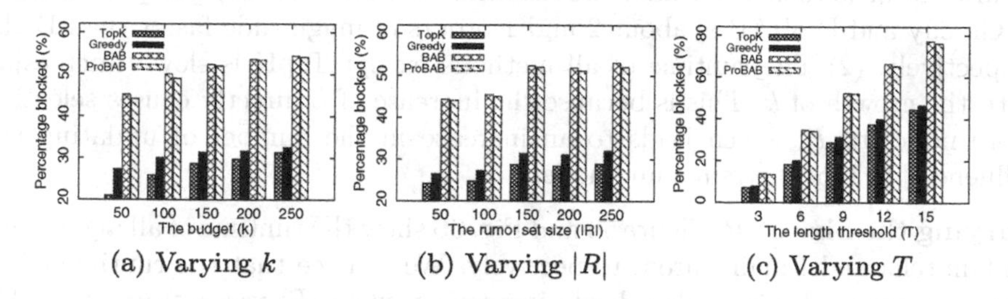

(a) Varying k (b) Varying $|R|$ (c) Varying T

Fig. 4. Effectiveness test on Email-Enron

Setup. All codes are implemented in Java, and experiments are conducted on a server with 2.1 GHz Intel Xeon 8 Core CPU and 32GB memory running CentOS/6.8 OS.

6.2 Effectiveness Test

This section studies how the block degree is affected by varying the budget k, the size of the rumor set R, and the length threshold T of a random walk.

Varying the Budget k. The block degrees of all algorithms on Gnutella and Email-Enron by varying the k are shown in Fig. 3a and Fig. 4a, respectively, and we find that when the budget raises from 50 to 250, BAB outperforms Greedy and TopK by up to 115% in the Email-Enron.

Varying the Size of R. Figure 3b and Fig. 4b show the result by varying the size of R. We find: (1) with the growth of $|R|$, the blocking percentages of all methods are increasing because the increasing influence of R leads to more nodes with higher unit block degrees. (2) ProBAB and BAB are consistently better than that of the rest baselines.

Varying the Random Walk Length Threshold T. Figure 3c and Fig. 4c show the results by varying the threshold T, which determines the length of a random walk starting from a node. We observe that: (1) The rumors on Gnutella dataset are much harder to be controlled than Email-Enron dataset. It implies that the network structure is an important variable for RCIC. (2) With the increase of T, the performance of all algorithms becomes better. The reason is that when the length becomes large, the random walk has more chances to reach the protectors and thus leads to a high unit block degree of the seeds.

6.3 Efficiency Test

We evaluate the efficiency of different algorithms on Gnutella and Email-Enron datasets.

Varying the Budget k. Figure 5a and Fig. 6a present the efficiency result when k varies from 50 to 250. We have the following observations. (1) The performance of Greedy and ProBAB is about 2 and 1 orders of magnitude faster than BAB, respectively. (2) The runtime of all methods except TopK is slowly increasing with the growth of k. This is because the increase of k directly causes selecting more nodes to P, which leads to an increase in the number of updating the influence block of the remaining node.

Varying the Size of R. Figure 5b and Fig. 6b show the runtime of all algorithms on Gnutella and Email-Enron, respectively. We can see that the runtime of all methods except TopK is also slowly increasing when $|R|$ varies from 50 to 250 on all datasets. This is because the influence set $Is(R)$ of R is increasing with the growth of $|R|$.

Varying the Random Walk Length Threshold T. We evaluate the efficiencies of algorithms by varying T from 3 to 15. The result is shown in Fig. 5c and Fig. 6c. We can see that all the algorithms except for TopK scale linearly with

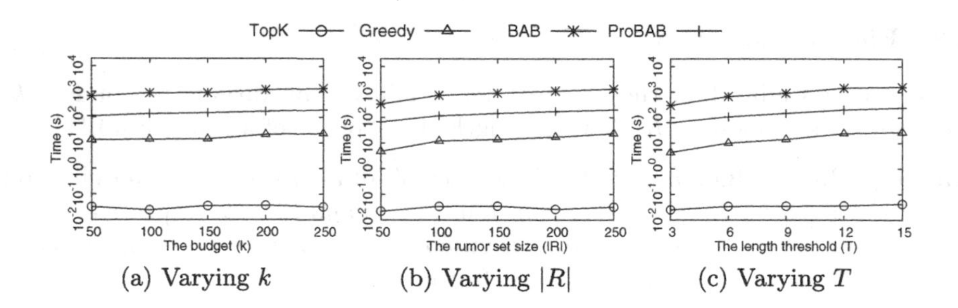

(a) Varying k (b) Varying $|R|$ (c) Varying T

Fig. 5. Efficiency test on Gnutella

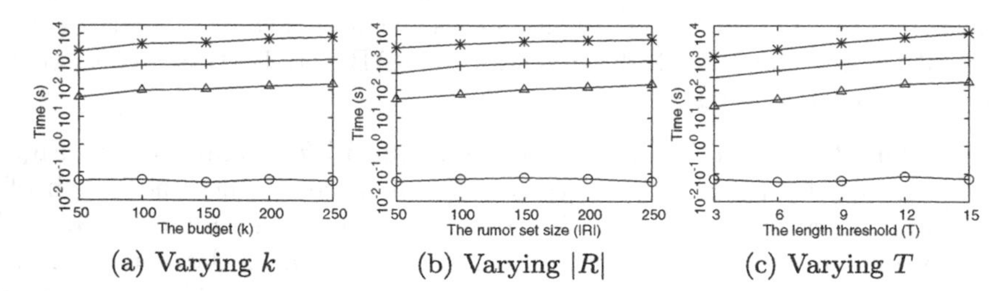

(a) Varying k (b) Varying $|R|$ (c) Varying T

Fig. 6. Efficiency test on Email-Enron

respect to T, which is because them need to scan more nodes to compute the influence block in each random walk.

6.4 Scalability Test

This experiment is to evaluate the scalability of Greedy and BAB when we increase the network size. To vary the network size, we partition Gowalla dataset into five subgraphs, and each of them covers 20% nodes of the dataset. To avoid smashing the network into pieces, each subgraph is generated by a breadth-first traversal process. Figure 7 shows the result, and we have the following observations. (1) The performance of Greedy and ProBAB is about 2 and 1 orders of magnitude faster than BAB, respectively. (2) When the graph size is increasing, the memory consumption of Greedy, BAB and ProBAB is increasing slowly but no more than 25 GB.

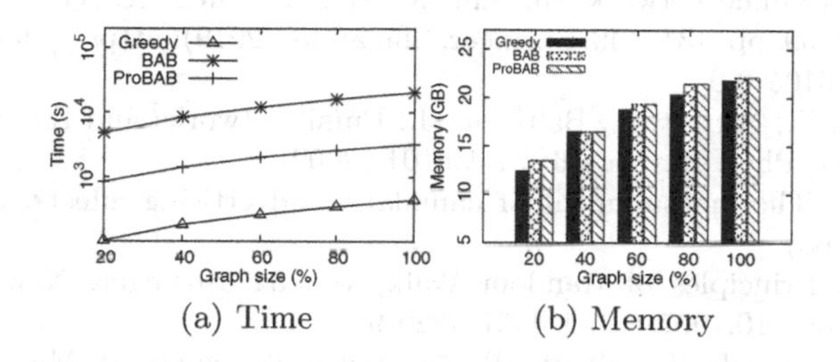

(a) Time (b) Memory

Fig. 7. Scalability test on Gowalla dataset

7 Conclusion

In this paper, we studied the RCIC problem based on a non-submodular influence block model and proved that it is NP-hard to approximate. Then, we proposed a branch-and-bound framework with a sampling-based upper-bound estimation method to solve RCIC problem. To further improve the efficiency, we optimized our framework with a progressive sampling-based greedy method for upper bound estimation. Lastly, we conducted experiments on real-world datasets to verify the efficiency, effectiveness, and scalability of our methods.

Acknowledgements. This work is supported by the Key Project of the National Natural Science Foundation of China (Project Number: U1811263).

References

1. Albert, R., Jeong, H., Barabási, A.L.: Error and attack tolerance of complex networks. Nature **406**(6794), 378 (2000)
2. Bharathi, S., Kempe, D., Salek, M.: Competitive influence maximization in social networks. In: WINE, pp. 306–311 (2007)

3. Borodin, A., Filmus, Y., Oren, J.: Threshold models for competitive influence in social networks. In: WINE, pp. 539–550 (2010)
4. Budak, C., Agrawal, D., El Abbadi, A.: Limiting the spread of misinformation in social networks. In: WWW, pp. 665–674 (2011)
5. Carnes, T., Nagarajan, C., Wild, S.M., van Zuylen, A.: Maximizing influence in a competitive social network: a follower's perspective. In: ACMicec, pp. 351–360 (2007)
6. Feder, G., Just, R.E., Zilberman, D.: Adoption of agricultural innovations in developing countries: a survey. EDCC **33**(2), 255–298 (1985)
7. Habiba, Yu, Y., Berger-Wolf, T.Y., Saia, J.: Finding spread blockers in dynamic networks. In: SNAKDD, pp. 55–76 (2008)
8. Lancaster, T.: The Econometric Analysis of Transition Data, No. 17. Cambridge University Press, Cambridge (1990)
9. Mo, S., Bao, Z., Zhang, P., Peng, Z.: Towards an efficient weighted random walk domination. PVLDB **14**(4), 560–572 (2020)
10. Mo, S., Tian, S., Wang, L., Peng, Z.: Minimizing the spread of rumor within budget constraint in online network. In: Sun, X., He, K., Chen, X. (eds.) NCTCS 2019. CCIS, vol. 1069, pp. 131–149. Springer, Singapore (2019). https://doi.org/10.1007/978-981-15-0105-0_9
11. Newman, M.E., Forrest, S., Balthrop, J.: Email networks and the spread of computer viruses. Phys. Rev. E **66**(3), 035101 (2002)
12. Palda, K.S.: The measurement of cumulative advertising effects. J. Bus. **38**(2), 162–179 (1965)
13. Spitzer, F.: Principles of Random Walk, vol. 34. Springer, New York (2013). https://doi.org/10.1007/978-1-4757-4229-9
14. Taylor, J., Kennedy, R., Sharp, B.: Is once really enough? Making generalizations about advertising's convex sales response function. J. Advert. Res. **49**(2), 198 (2009)
15. Xu, P., Peng, Z., Wang, L.: Proactive rumor control: When impression counts (full version). CoRR abs/2303.10068 (2023)
16. Zhang, P., et al.: Proactive rumor control in online networks. WWW **22**(4), 1799–1818 (2019)
17. Zhang, Y., Li, Y., Bao, Z., Mo, S., Zhang, P.: Optimizing impression counts for outdoor advertising. In: SIGKDD, pp. 1205–1215. ACM (2019)
18. Zhao, Y., Hu, Y., Yuan, P., Jin, H.: Maximizing influence over streaming graphs with query sequence. Data Sci. Eng. **6**(3), 339–357 (2021)

Spatio-Temporal Data

Generative-Contrastive-Attentive Spatial-Temporal Network for Traffic Data Imputation

Wenchuang Peng[1,2], Youfang Lin[1,2], Shengnan Guo[1,2(✉)], Weiwen Tang[1,2], Le Liu[1,2], and Huaiyu Wan[1,2]

[1] School of Computer and Information Technology, Beijing Jiaotong University, Beijing 100044, China
{wchpeng,yflin,guoshn,tangweiwen,liulecs,hywan}@bjtu.edu.cn
[2] Beijing Key Laboratory of Traffic Data Analysis and Mining, Beijing 100044, China

Abstract. Data missing is inevitable in Intelligent Transportation Systems (ITSs). Although many methods have been proposed for traffic data imputation, it is still very challenging because of two reasons. First, the ground truth of missing data is actually inaccessible, which makes most imputation methods hard to be trained. Second, incomplete data would easily mislead the model to learn unreliable spatial-temporal dependencies, which finally hurts the imputation performance. In this paper, we proposes a novel ***G****enerative-****C****ontrastive-****A****ttentive* ***S****patial-****T****emporal* ***N****etwork* (GCASTN) for traffic data imputation. It combines the ideas of generative and contrastive self-supervised learning together to develop a new training paradigm for imputation without relying on the ground truth of missing data. In addition, it introduces *nearest missing interval* to describe missing data and a novel ***M****issing-****A****ware* ***A****ttention* (MAA) mechanism is designed to utilize *nearest missing interval* to guide the model to adaptively learn the reliable spatial-temporal dependencies of incomplete traffic data. Extensive experiments covering three types of missing scenarios on two real-world traffic flow datasets demonstrate that GCASTN outperforms the state-of-the-art baselines.

Keywords: spatial-temporal graph data · traffic data imputation · attention · graph convolution · self-supervised learning

1 Introduction

Traffic data collected in ITSs is usually incomplete due to machine failures, which makes them ineffective in driving the downstream deep learning models and harms the quality of corresponding services. Despite that many imputation approaches have been developed, the traffic data imputation task still has the following two challenges.

© The Author(s), under exclusive license to Springer Nature Switzerland AG 2023
H. Kashima et al. (Eds.): PAKDD 2023, LNAI 13938, pp. 45–56, 2023.
https://doi.org/10.1007/978-3-031-33383-5_4

Firstly, the imputation task faces the dilemma of the absence of ground truth (label) for missing data to guide model training. Some works [18] utilize the supervised learning approach to train imputation models on complete data, which is obviously not practical in real-world applications. Generative Adversarial Network (GAN) based models [11,19] adopt unsupervised adversarial training over the generator and the discriminator to generate complete samples to impute missing data. But they are difficult to train, as the adversarial training process is very unstable. Variational Auto-Encoder (VAE) based models [12,13] learn the generation distributions of data from incomplete observed data by reconstruction, but they usually put some prior assumptions on the distributions (*e.g.*, Gaussian distribution), which is hard to be guaranteed in practice. Recently, self-supervised learning (SSL) is widely used under the advantage that it can capture the patterns in data without requiring additional labels as the supervision information. So it is worthwhile to explore the application of SSL to solve the label-lacking problem of imputation task.

Secondly, it is difficult to learn reliable spatial-temporal dependencies in incomplete traffic data. Actually, learning the spatial-temporal dependencies is the key to modeling traffic data. And many spatial-temporal graph neural networks (STGNNs) [1,7,10] have been proposed to effectively learn the spatial-temporal dependencies of traffic data. However, these technologies are initially designed for complete traffic data. When they are applied for imputation, they cannot be aware of the missing positions in the incomplete traffic data and cannot adjust the learned spatial-temporal dependencies adaptively. That is to say, existing spatial-temporal dependencies modeling technologies trust information from any time slice and any node, so they may capture unreliable and false spatial-temporal dependencies. Therefore, how to let the model be aware of and make full use of the missing information in incomplete traffic data so as to capture reliable spatial-temporal dependencies is challenging.

To address the above challenges, this paper proposes a novel ***G****enerative-****C****ontrastive-****A****ttentive* ***S****patial-****T****emporal* ***N****etwork* (GCASTN) for traffic data imputation. The main contributions of this paper are summarized as follows: 1) A generative-contrastive self-supervised learning is designed, which combines the ideas of generative learning and contrastive learning together, to train the traffic data imputation model without relying on the ground truth of missing data. 2) A novel missing-aware attention mechanism is proposed to capture reliable spatial-temporal dependencies from incomplete traffic data by utilizing the missing information. 3) Experiments on two real-world traffic flow datasets show that our proposed GCASTN model achieves state-of-the-art performance.

2 Related Work

Many imputation methods are proposed to deal with incomplete time series. Traditional statistical methods are the most straightforward which usually use zero values or other statistical values (*e.g.*, mean, mode, and last observation) to fill in the missing positions. Machine learning based methods, including but not limited to K-Nearest Neighbors [8], MissForest [14], Expectation-Maximization [6] algorithm, Multivariate Imputation Chained Equations [16], Low-rank approximation methods [4] are also proposed for time series imputation. However, all the imputation methods mentioned above cannot effectively model the complex spatial-temporal dependencies of traffic data, therefore their imputation capabilities are limited.

The core of accurate spatial-temporal traffic data imputation lies in learning the reliable temporal and spatial dependencies of data. Recently, many deep-learning-based models show advantages in this task. BRITS [2] employs bidirectional recurrent neural networks to model temporal dependence for missing data imputation. IGNNK [17] learn the spatial dependence by graph sampling for recovering the missing data. And STGNNs [1,7,10] that can simultaneously capture the dependencies along temporal and spatial dimensions have also been applied to imputation tasks [18]. Neither of them pays attention to the bias of spatial-temporal dependencies introduced by missing data in the incomplete traffic data. Unlike them, GRIN [5] reduces the introduction of bias with a novel graph neural network (GNN) having special message passing mechanism. In addition, GAN based models [11,19] whose training process is unstable and VAE based generative models [12,13] with unguaranteed predefined distributions are also applied to the imputation task.

3 Preliminaries

Traffic network is expressed as $\mathcal{G} = (\mathcal{V}, \mathcal{E}, \mathbf{A})$, where $\mathcal{V}$ and $\mathcal{E}$ are the sets of nodes (*e.g.* loop detectors or video camera) and edges respectively, $\boldsymbol{A} \in \mathbb{R}^{N \times N}$ is the adjacency matrix representing the proximity between nodes. The signals of traffic network at time slice t is defined as $\boldsymbol{X}_t = (\boldsymbol{x}_{t,1}, \boldsymbol{x}_{t,2}, \dots, \boldsymbol{x}_{t,N})^\top \in \mathbb{R}^{N \times C}$, where C is the number of features. And we use $\boldsymbol{\mathcal{X}} = (\boldsymbol{X}_1, \boldsymbol{X}_2, \dots, \boldsymbol{X}_T) \in \mathbb{R}^{N \times C \times T}$ to denote all signals over T time slices. To indicate the missing positions, we define masking matrix $\boldsymbol{\mathcal{M}} \in \mathbb{R}^{N \times C \times T}$. For each $\boldsymbol{m}_{t,v} \in \mathbb{R}^C$ in $\boldsymbol{\mathcal{M}}$, if $\boldsymbol{x}_{t,v}$ is observable, $\boldsymbol{m}_{t,v}$ is $\mathbf{1}$, otherwise $\boldsymbol{m}_{t,v}$ is $\mathbf{0}$. And then we get corrupted signal matrix: $\tilde{\boldsymbol{\mathcal{X}}} = \boldsymbol{\mathcal{X}} \odot \boldsymbol{\mathcal{M}}$, $\odot$ denotes element-wise dot-product. Further, we introduce the nearest missing interval $\boldsymbol{\Delta} \in \mathbb{R}^{N \times C \times T}$ that indicates the time length between the current missing position and the nearest observable one. To get any $\boldsymbol{\delta}_{t,v} \in \mathbb{R}^C$ in $\boldsymbol{\Delta}$, we first define $\boldsymbol{\delta}^h_{t,v}$ for a node v to denote the time interval between current missing position and its historical closest observable position at time slice t,

$$\boldsymbol{\delta}^h_{t,v} = \begin{cases} \mathbf{1} + \boldsymbol{\delta}^h_{t-1,v}, & t > 1 \text{ and } \boldsymbol{m}_{t,v} = \mathbf{0}, \\ \mathbf{0}, & \boldsymbol{m}_{t,v} = \mathbf{1}, \\ \mathbf{1}, & t = 1 \text{ and } \boldsymbol{m}_{t,v} = \mathbf{0}. \end{cases} \tag{1}$$

Likewise, we calculate the corresponding time interval with regard to the future nearest one $\boldsymbol{\delta}^f_{t,v}$. Finally, we get the nearest missing interval by $\boldsymbol{\delta}_{t,v} = \min(\boldsymbol{\delta}^h_{t,v}, \boldsymbol{\delta}^f_{t,v})$.

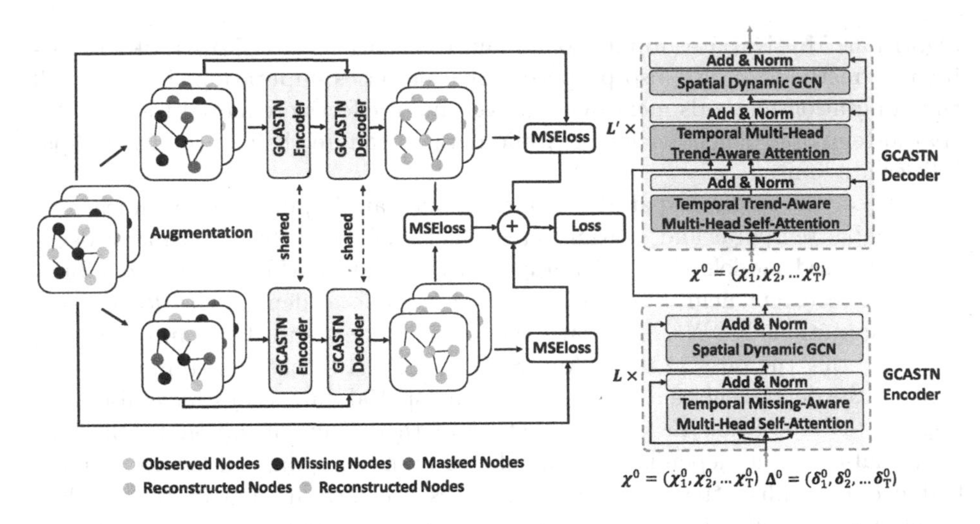

Fig. 1. The architecture of GCASTN.

Problem Statement. The task of traffic data imputation is to learn a function $f(\cdot)$ that uses the corrupted data $\tilde{\mathcal{X}}$, masking matrix $\mathcal{M}$ and the given graph $\mathcal{G}$ to reconstruct complete data $\hat{\mathcal{X}}$. Formally, $\hat{\mathcal{X}} = f(\tilde{\mathcal{X}}, \mathcal{M}, \mathcal{G})$.

4 The GCASTN Model

GCASTN is an Auto-Encoder (AE) designed for traffic data imputation. It follows a novel SSL paradigm that combines generative and contrastive ideas together to train the model effectively without relying on the ground truth of missing data. Similar to classical AE, it consists of an encoder and a decoder. Specifically, the encoder uses the nearest missing interval to help the model learn the reliable spatial-temporal dependencies from incomplete data. Then the decoder reconstructs the original complete data in an autoregressive way. Figure 1 shows the overall architecture of GCASTN which is introduced in detail next.

4.1 Generative-Contrastive Self-Supervised Learning

We propose a novel learning paradigm, *i.e.*, Generative-Contrastive Learning (GCL), to train the traffic data imputation model to solve the first challenge about lacking ground truth. Actually, GCL is a kind of self-supervised learning, which combines generative and contrastive ideas together.

On the one hand, similar to generative self-supervised learning (GL), GCL is based on reconstruction tasks. Specifically, GCL first obtains its input by randomly corrupting the original incomplete data by masking. Then it tries to reconstruct the corresponding complete data via an AE. Under our traffic data imputation scenario, the original data is inherently missing, so less useful information is left after the corrupting process, which leads to the model can not utilize the spatial-temporal dependencies of traffic data well.

Thus, on the other hand, we incorporate the idea of contrastive self-supervised learning (CL) together to make up for the drawback of GL. CL learns data representations by pulling positive sample pairs together and pushing negative sample pairs away, which is helpful for learning the high-level changing trend of the traffic data. To perform contrastive learning, we need to construct positive sample pairs and negative sample pairs. Specifically, with regard to the positive sample pairs, we randomly mask the original incomplete input twice, try to respectively reconstruct the complete data through the same AE, and finally align these two recovered data. Meanwhile, constructing negative pairs is necessary to avoid the mode collapse. However, it is not easy and needs careful design. To relieves the stress of manually designing negative pairs constructing strategies, we treat the above generative self-supervised learning part as an alternative way since the reconstruction learning naturally pushes different samples away. The overall pipeline works as follows.

Firstly, our model takes the original corrupted signal matrix $\tilde{\boldsymbol{\mathcal{X}}}$ and its corresponding masking matrix $\boldsymbol{\mathcal{M}}$ as inputs. Then, we make data augmentation through a two-fold cross random masking strategy (see details in Sect. 4.2), and the two augmented inputs are denoted as $\tilde{\boldsymbol{\mathcal{X}}}'$ and $\tilde{\boldsymbol{\mathcal{X}}}''$. After that, we input both the two augmented inputs into the same AE that consists of an encoder and a decoder to reconstruct the complete traffic data, denoted as $\hat{\boldsymbol{\mathcal{X}}}'$ and $\hat{\boldsymbol{\mathcal{X}}}''$ respectively. Finally, defining α as a hyperparameter, the training loss $\mathcal{L} = \mathcal{L}_{\mathrm{G}} + \alpha\mathcal{L}_{\mathrm{C}}$ consists of two parts. One is the reconstruction loss for GL, *i.e.*, $\mathcal{L}_{\mathrm{G}} = \frac{1}{\|\boldsymbol{\mathcal{M}}\|_0}\sum(\|\hat{\boldsymbol{\mathcal{X}}}' - \tilde{\boldsymbol{\mathcal{X}}}\|_2^2 + \|\hat{\boldsymbol{\mathcal{X}}}'' - \tilde{\boldsymbol{\mathcal{X}}}\|_2^2) \odot \boldsymbol{\mathcal{M}}$. The other one is the contrastive loss for CL, *i.e.*, $\mathcal{L}_{\mathrm{C}} = \frac{1}{\|\boldsymbol{\mathcal{M}}\|_0 + \|1-\boldsymbol{\mathcal{M}}\|_0}\sum(\|\hat{\boldsymbol{\mathcal{X}}}' - \hat{\boldsymbol{\mathcal{X}}}''\|_2^2)$, which aligns positive pairs.

4.2 Data Augmentation via Two-Fold Cross Random Masking

Two-fold cross random masking is a data augmentation strategy specially designed for GCL, which generates a pair of positive samples for an input. The significant advantage of this data augmentation strategy is that all the observable data in the original input are remained to be utilized.

In detail, we first get two all-zero matrices $\tilde{\boldsymbol{\mathcal{X}}}'$ and $\tilde{\boldsymbol{\mathcal{X}}}''$ of the same size as $\tilde{\boldsymbol{\mathcal{X}}}$. Then observed values in $\tilde{\boldsymbol{\mathcal{X}}}$ will replace the zero values at the corresponding identical positions in $\tilde{\boldsymbol{\mathcal{X}}}'$ with the probability at 50%. If an observable value does not exist in $\tilde{\boldsymbol{\mathcal{X}}}'$, it must appear in the same position in $\tilde{\boldsymbol{\mathcal{X}}}''$ to guarantee no useful information loss. Formally,

$$\begin{aligned}\{\boldsymbol{x}_{t,v}|\boldsymbol{x}_{t,v} \in \tilde{\boldsymbol{\mathcal{X}}}'\} \cup \{\boldsymbol{x}_{t,v}|\boldsymbol{x}_{t,v} \in \tilde{\boldsymbol{\mathcal{X}}}''\} &= \{\boldsymbol{x}_{t,v}|\boldsymbol{x}_{t,v} \in \tilde{\boldsymbol{\mathcal{X}}}\} \\ \{\boldsymbol{x}_{t,v}|\boldsymbol{x}_{t,v} \in \tilde{\boldsymbol{\mathcal{X}}}'\} \cap \{\boldsymbol{x}_{t,v}|\boldsymbol{x}_{t,v} \in \tilde{\boldsymbol{\mathcal{X}}}''\} &= \{\boldsymbol{0}\}\end{aligned} \tag{2}$$

Through the above data augmentation, we get two augmented signal matrices $\tilde{\boldsymbol{\mathcal{X}}}'$ and $\tilde{\boldsymbol{\mathcal{X}}}''$ as the positive sample pair.

4.3 GCASTN Encoder

The encoder in our model aims to encode the incomplete input into hidden space by discovering and exploiting the reliable spatial-temporal dependencies in traffic data. Specifically, it contains a stack of L identical spatial-temporal blocks (ST-blocks), and each ST-block includes a temporal Missing-Aware Multi-head Self-Attention (MissMultiSA) block and a spatial Dynamic Graph Convolution Network (DGCN) [7] block with residual connection and layer normalization [15]. Compared to existing technologies, our ST-blocks are able to learn reliable spatial-temporal dependencies from missing data.

Specifically, the input to our encoder includes an augmented sample and its corresponding nearest missing interval. Firstly, we linearly project the augmented sample into a high-dimensional tensor with size d_{model} and then add temporal position embedding [15] and spatial position embedding [7] to obtain $\boldsymbol{\mathcal{X}}^0$. Meanwhile, we obtain $\boldsymbol{\Delta}^0$ by linear mapping the nearest missing interval. Finally, the input to the lth block in our encoder is the output of the $(l-1)^{th}$ block $\boldsymbol{\mathcal{X}}^{l-1} \in \mathbb{R}^{N \times d_{\text{model}} \times T}$ along with $\boldsymbol{\Delta}^0$.

Temporal Missing-Aware Multi-Head Self-Attention. MissMultiSA is a self-attention mechanism specially designed to modelling incomplete data. It makes use of the nearest missing interval information to perceive the missing positions in traffic data, and then adaptively adjusts the learned spatial-temporal dependencies to make the learning process of the model more reliable.

MissMultiSA is the multi-head version of Missing-Aware Attention (MAA). Unlike the widely used Scaled Dot-Product Attention (SDPA) [15], MAA employs two query-key pairs to measure the correlations between the queries and keys. Specifically, MAA in the lth block uses $\boldsymbol{\mathcal{X}}^{l-1}$, *i.e.*, the hidden representations for incomplete traffic data as both the queries $\boldsymbol{Q}$, keys $\boldsymbol{K}$ and values $\boldsymbol{V}$ to perform self-attentions. Considering the computational efficiency, MAA first utilizes Scaled Dot-Product (SDP) to get the weights between queries and keys. SDP is defined as:

$$\text{SDP}(\boldsymbol{Q}, \boldsymbol{K}) = \text{softmax}(\boldsymbol{Q}\boldsymbol{K}^\top / \sqrt{d_{\text{model}}}) \tag{3}$$

However, directly using the above weights to measure the correlations between the elements in data, the model can neither perceive the existence of the missing values nor measure the effect brought by the missing values. This leads to model to capture unreliable dependencies and even to aggregate false information from missing positions. Hence, MAA further utilizes another query-key pairs from the nearest missing interval $\boldsymbol{\Delta}^0$ to get additional weights to revise the former weights. The new weights help the model to accurately perceive the missing information in data, and let the model to consider the corresponding influences. Finally, the output is computed as a weighted sum of the values $\boldsymbol{V}$.

Formally, let $\boldsymbol{Q}_\Delta$ and $\boldsymbol{K}_\Delta$ denote additional queries and keys from the nearest missing interval $\boldsymbol{\Delta}^0$, MAA is expressed as:

$$\text{MAA}(\boldsymbol{Q}_\Delta, \boldsymbol{K}_\Delta, \boldsymbol{Q}, \boldsymbol{K}, \boldsymbol{V}) = (\text{SDP}(\boldsymbol{Q}_\Delta, \boldsymbol{K}_\Delta) \odot \text{SDP}(\boldsymbol{Q}, \boldsymbol{K}))\boldsymbol{V} \tag{4}$$

Then we follows ASTGNN [7] to introduce 1D convolutions in the temporal dimension to learn the local trend information, and extend MAA to multi-head for learning richer information [15], Finally, we get MissMultiSA as follows:

$$\begin{aligned} \text{MissMultiSA}(\boldsymbol{Q}_{\Delta}, \boldsymbol{K}_{\Delta}, \boldsymbol{Q}, \boldsymbol{K}, \boldsymbol{V}) &= \oplus(\text{head}_1, \text{head}_2, \ldots, \text{head}_h)\boldsymbol{W}^O \\ \text{head}_j &= \text{MAA}(\boldsymbol{Q}_{\Delta}\boldsymbol{W}_j^Q, \boldsymbol{K}_{\Delta}\boldsymbol{W}_j^K, \boldsymbol{\Phi}_j^Q \star \boldsymbol{Q}, \boldsymbol{\Phi}_j^K \star \boldsymbol{K}, \boldsymbol{V}\boldsymbol{W}_j^V) \end{aligned} \tag{5}$$

where $\boldsymbol{W}^O$, $\boldsymbol{W}_j^Q$, $\boldsymbol{W}_j^K$ and $\boldsymbol{W}_j^V$ are the learnable parameters, $\boldsymbol{\Phi}_j^Q$ and $\boldsymbol{\Phi}_j^Q$ are the parameters of convolution kernels, while $\star$ is the convolution operation.

Spatial Dynamic Graph Convolution Network. DGCN [7] is a graph neural network for learning spatial dependencies of traffic data. With the aid of the attention mechanism, DGCN can dynamically adjust the correlation strengths among nodes and better aggregate node information, which learn about more reasonable spatial dependencies. Denote the output of MissMultiSA in the lth block of encoder as $\boldsymbol{\mathcal{Z}}^{l-1} = (\boldsymbol{Z}_1^{l-1}, ..., \boldsymbol{Z}_T^{l-1}) \in \mathbb{R}^{N \times d_{model} \times T}$. DGCN calculates the spatial relevant weights $\boldsymbol{S}_t^l$ of the lth block via $\boldsymbol{S}_t^l = \text{SDP}(\boldsymbol{Z}_t^{l-1}, \boldsymbol{Z}_t^{l-1}) \in \mathbb{R}^{N \times N}$. And then it uses the spatial relevant weights to adjust the static normalized adjacency matrix $\tilde{\boldsymbol{A}} = \boldsymbol{D}^{-\frac{1}{2}}\boldsymbol{A}\boldsymbol{D}^{-\frac{1}{2}}$, where $\boldsymbol{D}$ is the degree matrix of the traffic network. Finally DGCN aggregates node information to obtain a new node representation. The whole computation process of DGCN can be defined formally:

$$\boldsymbol{X}_t^l = \sigma((\tilde{\boldsymbol{A}} \odot \boldsymbol{S}_t^l)\boldsymbol{Z}_t^{l-1}\boldsymbol{W}^l) \tag{6}$$

where σ is activation function and $\boldsymbol{W}^l$ is learnable parameters.

4.4 GCASTN Decoder

The decoder aims to decode the learned spatial-temporal dependencies for reconstructing the complete signal matrices. Like the encoder, the decoder stacks $L^{'}$ blocks, and each block contains two temporal attention modules and a DGCN module with residual connection and layer normalization. The first temporal attention module encodes the hidden representation of missing data into an output, and the output is used as the queries of the second temporal attention module. While the corresponding keys and values are obtained from the encoder's output $\boldsymbol{\mathcal{X}}^L$. Different from the operations in the encoder, in our decoder the two temporal attention modules of each block replace MAA with SDPA. After that, the decoder further employs DGCN again to decode the spatial dependencies. Figure 1 shows the details of the encoder and decoder in our model.

5 Experiments

5.1 Datasets and Baselines

We use two real traffic flow datasets including PEMS04 and PEMS08 to evaluate our proposed method. There are 307 nodes in PEMS04 and 170 nodes in PEMS08.

They are collected through California Transportation Agencies Performance Measurement System (PeMS) [3] from 1 January 2018 and 1 July 2016 respectively, which span two months. Initially they are collected every 30 s, and then we aggregate the raw data into 5-min interval. Next we construct corrupted datasets with different missing types and missing rates on the two complete datasets. Here we consider three types of absence: random missing (RM), non-random missing (NM) and block missing (BM) [4]. RM means that data may be missing randomly at any time slice and node. NM is relational in temporal dimension and independent in spatial dimension. It occurs when any single collector does not work within a period (e.g. one day). BM is correlated in all dimensions, indicating failure of collectors in an area for a while. In particular, once the data is missing, all features collected by the collector are missing. For missing rates, we set 20%, 30%, 70% and 90% corresponding to low and high missing rates. And all corrupted datasets are generated by referring to prior works [4].

Table 1. Imputation performances on PEMS04.

model (RM/NM/BM))	20%			30%			70%			90%		
	MAE	RMSE	MAPE	MAE	RMSE	MAPE	MAE	RMSE	MAPE	MAE	RMSE	MAPE
last (RM)	21.42	34.67	14.63	21.81	34.53	14.93	25.34	41.09	17.54*	38.56*	62.67*	29.68*
DCRNN (RM)	21.70	32.38	19.13	24.11	34.69	25.61	47.32	70.60	53.86	78.93	110.03	126.34
GRIN (RM)	17.77	30.89*	12.92	17.94	31.05	12.92	19.49	33.00	14.27	21.78	36.09	16.70
AGCRN (RM)	22.42	34.23	15.64	25.23	38.73	17.08	35.81	63.80	21.62	75.58	135.20	37.16
LATC (RM)	19.63	31.70	15.16	20.08	32.48	15.61	28.76	45.38	21.37	61.60	87.36	67.58
IGNNK (RM)	20.60	32.86	19.18	21.16	33.55	17.85	24.10*	38.15*	22.81	42.02	65.22	43.98
BRITS (RM)	23.13	40.63	20.72	25.14	43.87	23.81	39.46	61.35	45.89	71.39	100.01	97.93
MTAN (RM)	18.36*	29.66	13.32*	19.27*	31.31*	13.25*	24.21	41.99	18.18	58.53	100.48	66.81
GCASTN (RM)	**15.44**	**25.04**	**10.80**	**15.47**	**25.28**	**11.18**	**17.55**	**28.25**	**12.43**	**20.67**	**33.02**	**16.12**
last (NM)	21.73	35.23	14.62	21.81	35.44	14.90	25.57	41.02	17.73*	38.10*	61.11	29.07*
DCRNN (NM)	21.72	32.62	18.07	23.16	33.85	24.04	30.18	48.68	29.71	70.33	101.61	111.97
GRIN (NM)	17.95	31.30	12.77	17.82	31.00	12.88	19.27	32.81	14.10	21.85	36.07	16.72
AGCRN (NM)	19.12*	30.54	14.38	20.00	31.08*	15.12	34.30	63.11	22.11	79.65	137.17	39.89
LATC (NM)	40.34	66.31	21.87	62.58	95.91	32.00	138.64	187.81	67.67	176.56	228.98	82.82
IGNNK (NM)	21.17	33.76	18.60	21.60	33.70	19.63	24.25*	38.03*	21.95	39.85	59.64*	43.04
BRITS (NM)	27.30	45.27	31.06	28.19	46.14	35.84	43.62	64.45	63.10	81.19	105.61	121.96
MTAN (NM)	19.19	30.95*	13.92*	19.17*	31.14	13.38*	29.60	67.85	21.41	94.91	165.32	61.46
GCASTN (NM)	**15.75**	**25.79**	**10.90**	**15.58**	**25.41**	**11.29**	**17.76**	**28.59**	**14.08**	**20.91**	**33.36**	**16.60**
last (BM)	26.65	42.75	18.69*	28.66	46.08	20.34	57.46	92.23	52.57	124.75	170.64	170.36
DCRNN (BM)	20.84*	31.34	17.91	22.02*	32.96	19.05*	28.75*	44.12*	30.32*	69.76	102.35	115.36
GRIN (BM)	20.60	33.96*	19.90	21.04	34.30*	18.82	26.92	41.65	23.64	**60.82**	87.25	76.88
AGCRN (BM)	39.37	71.87	26.64	47.69	85.81	32.07	109.15	174.84	51.30	107.35	132.23	159.26
LATC (BM)	27.07	42.74	20.11	32.84	50.11	23.51	43.46	69.33	35.23	93.16	138.74	**55.16**
IGNNK (BM)	34.41	58.63	42.44	40.52	67.77	57.51	82.56	118.36	128.98	119.38	149.87	209.62
BRITS (BM)	27.27	44.70	33.72	28.56	41.78	29.64	29.33	47.74	34.73	61.68	**83.60**	89.79
MTAN (BM)	38.67	82.23	24.63	43.82	88.24	32.53	90.01	138.92	99.01	120.94	152.39	210.04
GCASTN (BM)	**17.91**	**29.22**	**12.55**	**18.75**	**30.34**	**13.43**	**24.65**	**40.57**	**17.75**	63.77*	94.56*	81.50*

For comparison, we reproduce the following models as baselines: 1) **LAST** simply uses the last observed value to replace the missing value. 2)–3) **DCRNN** [10] and **AGCRN** [1] are STGNNs which can learn spatial-temporal dependencies for imputation. 4) **GRIN** is a STGNNs with novel message passing mechanism. 5) **LATC** [4] is the latest low-rank matrix/tensor completion method. 6) **IGNNK**

[17] learns the spatial dependence by graph sampling for recovering the incomplete data. 7) **BRITS** [2] employs bidirectional recurrent neural networks to impute missing data. 8) **mTAN** [13] is a VAE based model that designs an attention mechanism to map the missing sequences into a fixed-length representation for imputation. Besides, **KNN** [8], **MICE** [16], **missforest** [14] and **GAIN** [19] are also reproduced, but with the space constraints and poor imputation results, this paper does not show them.

Table 2. Imputation performances on PEMS08.

model (RM/NM/BM)	20%			30%			70%			90%		
	MAE	RMSE	MAPE	MAE	RMSE	MAPE	MAE	RMSE	MAPE	MAE	RMSE	MAPE
last (RM)	16.86	26.51	10.46	17.14	27.00	10.63*	20.39	32.26	12.59	31.70	51.60	20.21
DCRNN (RM)	29.42	40.59	22.92	36.67	48.68	30.53	39.04	58.69	41.28	73.82	101.86	123.81
GRIN (RM)	13.49	22.15*	8.82	13.79	22.64	9.28	15.32	24.92	10.24	18.58	30.65	13.00
AGCRN (RM)	17.65	27.13	12.46	18.26	28.08	12.22	28.26	50.14	17.24	63.53	112.38	30.07
LATC (RM)	14.42	23.53	10.66	14.48*	23.70*	10.67	17.04*	27.54*	12.56	21.67*	34.51*	16.35*
IGNNK (RM)	17.72	26.23	20.86	20.05	29.28	33.44	21.50	32.32	28.57	41.99	63.75	65.45
BRITS (RM)	22.91	37.03	30.88	32.57	46.77	51.94	45.25	62.55	72.18	74.04	96.24	133.00
MTAN (RM)	13.99*	22.02	9.39*	18.64	28.99	11.58	18.33	30.24	11.61*	57.02	102.32	56.47
GCASTN (RM)	**11.91**	**19.26**	**7.90**	**12.13**	**19.62**	**8.12**	**13.45**	**21.99**	**9.18**	**17.64**	**28.48**	**12.27**
last (NM)	16.89	26.34	10.64	17.08	26.91	10.90	20.41*	32.27*	12.57	31.12*	50.02*	19.80*
DCRNN (NM)	16.67	24.85	13.19	17.90	26.35	17.94	39.10	58.45	47.26	63.26	92.40	93.71
GRIN (NM)	13.54	22.13	9.10	13.58	22.24	9.28	15.41	25.12	10.21	18.84	31.16	12.90
AGCRN (NM)	14.47*	22.22*	10.67	14.61*	22.69*	10.66	26.55	48.29	15.38	71.87	126.01	33.00
LATC (NM)	15.62	25.45	11.70	15.84	25.99	12.04	26.09	54.63	16.99	112.54	171.73	51.97
IGNNK (NM)	17.88	26.16	25.71	19.28	27.91	26.60	22.20	32.56	27.03	38.48	56.08	55.96
BRITS (NM)	24.86	39.69	38.00	25.50	40.57	40.80	45.58	65.27	77.59	91.78	112.37	167.44
MTAN (NM)	14.51	22.45	9.82*	14.91	23.35	10.01*	22.22	48.57	13.22	119.13	206.23	80.74
GCASTN (NM)	**11.54**	**18.75**	**7.73**	**11.94**	**19.39**	**8.30**	**13.75**	**22.34**	**9.61**	**17.53**	**28.44**	**11.94**
last (BM)	21.67	34.71	13.38	23.36	37.40	14.40*	48.73	78.87	34.43	105.60	147.69	87.95
DCRNN (BM)	75.23	97.26	113.50	58.43	84.37	79.85	88.31	118.51	165.65	114.14	140.72	233.71
GRIN (BM)	16.32	25.71	12.81*	18.80*	30.01*	14.65	24.26	38.05	17.51	**53.55**	**76.73**	**47.65**
AGCRN (BM)	33.42	61.29	20.93	42.18	78.42	22.78	86.62	138.43	37.57	90.27	112.81*	80.75
LATC (BM)	17.04*	27.20*	12.43	17.68	28.14	12.57*	34.78*	55.62*	22.07*	144.90	193.19	64.66*
IGNNK (BM)	29.95	50.20	42.44	36.53	60.47	48.98	75.86	105.87	142.32	112.01	128.26	228.08
BRITS (BM)	52.91	70.99	96.90	58.12	76.79	97.45	70.89	92.36	113.93	89.92*	112.94	169.19
MTAN (BM)	40.23	91.88	20.97	49.93	106.28	26.19	89.64	137.05	97.30	115.15	148.16	182.71
GCASTN (BM)	**14.49**	**23.75**	**9.65**	**15.28**	**25.46**	**10.42**	**22.63**	**38.46**	**15.69**	60.25	88.34	49.43

5.2 Experimental Results

Settings. We divide all datasets into training and validation sets at ratio 8:2 by the time and use Min-Max method to normalize all data, then we construct samples by sliding window on the time axis. And the window length T and sliding step are 12 and 1. During training, the information of observable and masked positions are used as supervision. To prevent overfitting, the observable values in the validation sets are masked off with 10% probability at random. In the evaluation stage, we only evaluate the imputation results on missing positions. And the ground truth of missing positions is only known during the evaluation stage. The evaluation metrics in this paper are mean absolute error (MAE), root mean square error (RMSE) and mean absolute percentage error (MAPE).

We implement our GCASTN model in the PyTorch[1] framework, The relevant code is available in https://github.com/Pumbaa-peng/GCASTN. The number of layers for both decoder and encoder is 4, d_{model} is 64. About MissMultiSA, the number of attention heads is 8 and the convolution kernel size is 3. About training, we use Adam [9] algorithm to train networks, learning rate is 0.001. All experiments are conducted on NVIDIA RTX A4000 GPUs.

Overall Performance. Experimental results are shown in Tables 1 and 2. And in each corrupted dataset, the best results are marked in bold, the second place results are underlined, and the third place results are marked by '*'.

As we can see, our proposed method (GCASTN) achieves the best results for most of missing cases, which indicates that Generative-Contrastive Learning can effectively help models learn the correct spatial-temporal dependencies from incomplete traffic data and apply them to imputation tasks. Meanwhile, GRIN also shows notable interpolation capabilities, and the imputation performance of common STGNNs (DCRNN and AGCRN) are not very satisfactory. The above corresponding experimental results show that effective consideration of the bias introduced by missing locations can improve the ability to model traffic data, which in turn can help improve the model's imputation capabilities. In addition, In the case of BM or higher missing rates, imputation models face greater challenges as learning reasonable spatial-temporal dependencies is more difficult.

Ablation Experiments. In this section we evaluate the two innovations we propose. First we design three variants of **GCASTN** for comparison: i) **-noCL**: It trains the model without the contrastive loss. ii) **-noMAA**: It replaces the MAA with SDPA [15]. iii) **-noM/GCL**: It uses SDPA [15] instead of MAA and does not train models with generative contrast learning.

All settings of the variant models are consistent with GCASTN except the differences mentioned above. Due to space limitations, we only conduct ablation experiments on the PEMS08 dataset and only calculate MAE. The results are shown on Fig. 2.

First, we evaluate the effects of generative-contrastive learning. By comparing -noMAA and -noM/GCL from Fig. 2(a), we can see our proposed GCL has tremendous advantages when applied to imputation tasks. With the help of the GCL, general spatial-temporal graph neural networks can also achieve superior results in imputation tasks. For further exploring the effectiveness of the GCL, we get the experimental results of -noCL and GCASTN and show them on Fig. 2(b). Obviously, GCL incorporating contrastive learning ideas can efficiently improve the imputation accuracy.

Then we assess the Missing-Aware Attention. On Fig. 2(b), GCASTN performs better than -noMAA, which indicates MMA can learn more reliable spatial-temporal dependencies by using nearest missing intervals. To demonstrate the strengths of MAA more visually, the attention scores between locations calculated

[1] https://pytorch.org.

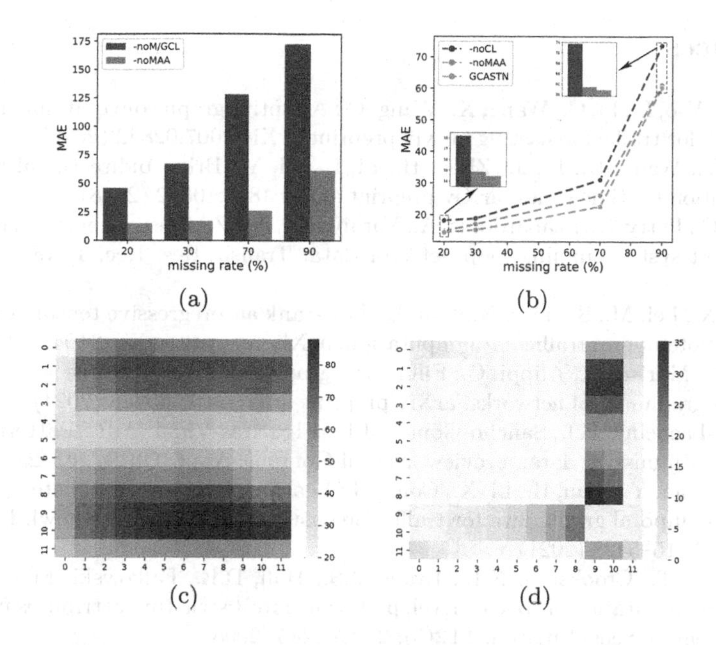

Fig. 2. The results of ablation experiments on PEMS08.

via -noMAA and GCASTN for the same corrupted input sequence are shown in Figs. 2(c) and 2(d) respectively. Comparing Figs. 2(c) and 2(d), it can be seen that GCASTN can clearly perceive missing positions and correctly aggregates reliable information.

6 Conclusion

In this paper, we propose a novel self-supervised method GCASTN for missing traffic data imputation. We combine the generative and contrastive methods in self-supervised learning together to create a novel generative-contrastive learning paradigm, which solves the dilemma of the absence of ground truth (label) for missing data. In addition, we introduce the nearest missing interval and design missing-aware attention to perceive missing positions for capturing reliable spatial-temporal dependencies. Extensive experimental results demonstrate the superiority of GCASTN.

Acknowledgments. This work was supported by the National Natural Science Foundation of China (Grant No. 62202043).

References

1. Bai, L., Yao, L., Li, C., Wang, X., Wang, C.: Adaptive graph convolutional recurrent network for traffic forecasting. arXiv preprint arXiv:2007.02842 (2020)
2. Cao, W., Wang, D., Li, J., Zhou, H., Li, L., Li, Y.: Brits: bidirectional recurrent imputation for time series. arXiv preprint arXiv:1805.10572 (2018)
3. Chen, C., Petty, K., Skabardonis, A., Varaiya, P., Jia, Z.: Freeway performance measurement system: mining loop detector data. Transp. Res. Rec. **1748**(1), 96–102 (2001)
4. Chen, X., Lei, M., Saunier, N., Sun, L.: Low-rank autoregressive tensor completion for spatiotemporal traffic data imputation. arXiv preprint arXiv:2104.14936 (2021)
5. Cini, A., Marisca, I., Alippi, C.: Filling the g_ap_s: multivariate time series imputation by graph neural networks. arXiv preprint arXiv:2108.00298 (2021)
6. García-Laencina, P.J., Sancho-Gómez, J.L., Figueiras-Vidal, A.R.: Pattern classification with missing data: a review. Neural Comput. Appl. **19**(2), 263–282 (2010)
7. Guo, S., Lin, Y., Wan, H., Li, X., Cong, G.: Learning dynamics and heterogeneity of spatial-temporal graph data for traffic forecasting. IEEE Trans. Knowl. Data Eng. **34**(11), 5415–5428 (2021)
8. Hudak, A.T., Crookston, N.L., Evans, J.S., Hall, D.E., Falkowski, M.J.: Nearest neighbor imputation of species-level, plot-scale forest structure attributes from lidar data. Remote Sens. Environ. **112**(5), 2232–2245 (2008)
9. Kingma, D.P., Ba, J.: Adam: a method for stochastic optimization. arXiv preprint arXiv:1412.6980 (2014)
10. Li, Y., Yu, R., Shahabi, C., Liu, Y.: Diffusion convolutional recurrent neural network: data-driven traffic forecasting. arXiv preprint arXiv:1707.01926 (2017)
11. Luo, Y., Zhang, Y., Cai, X., Yuan, X.: E2GAN: end-to-end generative adversarial network for multivariate time series imputation, pp. 3094–3100. AAAI Press (2019)
12. Mattei, P.A., Frellsen, J.: Miwae: deep generative modelling and imputation of incomplete data sets. In: International Conference on Machine Learning, pp. 4413–4423. PMLR (2019)
13. Shukla, S.N., Marlin, B.M.: Multi-time attention networks for irregularly sampled time series. arXiv preprint arXiv:2101.10318 (2021)
14. Stekhoven, D.J., Bühlmann, P.: Missforest-non-parametric missing value imputation for mixed-type data. Bioinformatics **28**(1), 112–118 (2012)
15. Vaswani, A., et al.: Attention is all you need. In: Advances in Neural Information Processing Systems, pp. 5998–6008 (2017)
16. White, I.R., Royston, P., Wood, A.M.: Multiple imputation using chained equations: issues and guidance for practice. Stat. Med. **30**(4), 377–399 (2011)
17. Wu, Y., Zhuang, D., Labbe, A., Sun, L.: Inductive graph neural networks for spatiotemporal kriging. arXiv preprint arXiv:2006.07527 (2020)
18. Yang, B., Kang, Y., Yuan, Y., Huang, X., Li, H.: ST-LBAGAN: spatio-temporal learnable bidirectional attention generative adversarial networks for missing traffic data imputation. Knowl.-Based Syst. **215**, 106705 (2021)
19. Yoon, J., Jordon, J., Schaar, M.: Gain: missing data imputation using generative adversarial nets. In: International Conference on Machine Learning, pp. 5689–5698. PMLR (2018)

Road Network Representation Learning with Vehicle Trajectories

Stefan Schestakov[1(✉)], Paul Heinemeyer[1], and Elena Demidova[2,3]

[1] L3S Research Center, Leibniz Universität Hannover, Hannover, Germany
{schestakov,pheinemeyer}@L3S.de

[2] Data Science and Intelligent Systems Group (DSIS), University of Bonn, Bonn, Germany
elena.demidova@cs.uni-bonn.de

[3] Lamarr Institute for Machine Learning and Artificial Intelligence, Bonn, Germany
https://lamarr-institute.org/

Abstract. Spatio-temporal traffic patterns reflecting the mobility behavior of road users are essential for learning effective general-purpose road representations. Such patterns are largely neglected in state-of-the-art road representation learning, mainly focusing on modeling road topology and static road features. Incorporating traffic patterns into road network representation learning is particularly challenging due to the complex relationship between road network structure and mobility behavior of road users. In this paper, we present TrajRNE – a novel trajectory-based road embedding model incorporating vehicle trajectory information into road network representation learning. Our experiments on two real-world datasets demonstrate that TrajRNE outperforms state-of-the-art road representation learning baselines on various downstream tasks.

1 Introduction

Effective general-purpose representations of road networks are essential for critical machine learning applications in mobility and smart cities, such as traffic inference, travel time estimation, and destination prediction. This demand has recently inspired numerous research works on road network representation learning (e.g., [1,16,17,19]). Whereas existing approaches primarily utilize road network topology and static road features, they often fail to capture complex traffic patterns and mobility behavior of road users. A rich source of complex spatio-temporal traffic patterns, traffic flows, actual-driven speed, and driver road preferences are vehicle trajectories. Thus, integrating vehicle trajectory information into the road network representation can provide valuable information for mobility and smart city applications.

Previous road representation learning approaches (e.g., [1,16,17,19]) have two substantial shortcomings. First, state-of-the-art methods utilize conventional graph representations (e.g., [9,13,15]), which do not consider complex road relationships. For example, for a road leading to an intersection, the importance of the following roads is not equal and depends on user mobility behavior. Second, state-of-the-art road representation models learn static road features, e.g., road

© The Author(s) 2023
H. Kashima et al. (Eds.): PAKDD 2023, LNAI 13938, pp. 57–69, 2023.
https://doi.org/10.1007/978-3-031-33383-5_5

type and speed limit, to infer traffic patterns. However, these features do not directly reflect dynamic traffic conditions. For example, roads with the same speed limit can have vastly different traffic patterns depending on the traffic volume. Recently, few approaches attempted to incorporate trajectories into road network representation learning. Wang et al. [16,17] supplemented random walks with real-world trajectories for learning geo-locality. Wu et al. [19] utilized trajectory data as a supervision signal for graph reconstruction. Further, Chen et al. [1] refined previously learned road embeddings with a route recovery and trajectory discrimination supervision objective using a transformer model. However, existing approaches do not explicitly incorporate trajectory data into their model design and thus fail to incorporate complex traffic and mobility patterns.

We observe two substantial challenges for general-purpose road representation learning. First, conventional graph representation learning methods [9,13,15] are inadequate for road network modeling, as they assume network homophily and do not consider heterogeneous properties of connected roads and complex road relationships. In contrast, connected roads, e.g., a secondary road connected to a primary road, can exhibit highly diverse traffic patterns. Thus, the first challenge is to adapt graph representation methods to road networks with heterogeneous traffic patterns on connected roads. Second, a challenge is to systematically incorporate vehicle trajectories into road representation learning to extract and represent dynamic traffic patterns and complex mobility behavior.

In this paper, we propose a novel Trajectory-based Road Network Embedding model (TrajRNE). TrajRNE includes two modules. First, we propose a novel Spatial Flow Convolution (SFC). SFC aggregates road feature representations based on transition probabilities extracted from vehicle trajectories. Thus, SFC automatically differentiates between relevant and irrelevant road network nodes indicated by the mobility behavior. Moreover, we increase the SFC receptive field by considering the traffic flow of k-hop neighbors. This approach facilitates aggregation of relevant neighbors located at a longer distance without over smoothing with non-relevant neighbors. Second, we propose a novel Structural Road Encoder (SRE) leveraging multitask learning to capture topology, structure, and dynamic traffic. Whereas state-of-the-art road embeddings learn topology using random walks or shortest paths, they do not effectively capture mobility behavior. In contrast, TrajRNE adopts random walks based on the transition probability extracted from real-world trajectories to capture geo-locality and mobility patterns.

In summary, the contributions of our work are as follows:

- We introduce Spatial Flow Convolution and Structural Road Encoder to capture traffic characteristics of road networks from vehicle trajectories.
- We propose TrajRNE[1] – a novel road network representation learning approach, effectively capturing traffic patterns with SFC and SRE methods.
- Our evaluation demonstrates that TrajRNE enables effective general-purpose road network representations. TrajRNE consistently outperforms state-of-the-art baselines on four downstream tasks and two real-world datasets.

[1] Code available at: https://github.com/sonout/TrajRNE.

2 Problem Definition

In this section, we first present the notations and then formally define our task.

Definition 1. *(**Road Network**). We define a road network as a directed graph $G = (\mathcal{V}, \mathcal{A}, \mathcal{F})$. $\mathcal{V}$ is a set of nodes, where each node $v_i \in \mathcal{V}$ represents a road segment. $\mathcal{A}$ is the adjacency matrix, where $\mathcal{A}_{ij} = 1$ implies that a road segment v_j directly follows a road segment v_i, and $\mathcal{A}_{ij} = 0$ otherwise. A road network has a feature set $\mathcal{F} \in \mathbb{R}^{|\mathcal{V}| \times f}$ representing road segment features with dimension f.*

Definition 2. *(**Trajectory**). A trajectory T is a sequence of points representing geographic coordinates from the route driven by a vehicle: $T = [p_1, p_2, \ldots, p_{|T|}]$, where $p_i = (lon_i, lat_i)$ is the i-th point with the longitude lon_i and latitude lat_i and $|T|$ is the trajectory length.*

Given a road network G, we can map a trajectory T to the road network using a map matching algorithm [20], thus obtaining a sequence of road segments.

Definition 3. *(**Road Segment Sequence**). A road segment sequence $R = [v_1, v_2, \ldots, v_N]$ represents the underlying route of a trajectory on a road network, where each $v_i \in \mathcal{V}$ denotes a road segment in the road network $G = (\mathcal{V}, \mathcal{A}, \mathcal{F})$.*

In this work, we target the problem of learning a general-purpose representation of road networks beneficial for various downstream tasks.

Definition 4. *(**Road Network Representation Learning**). Given a road network $G = (\mathcal{V}, \mathcal{A}, \mathcal{F})$ and a set of trajectories $\mathcal{T} = \{T_i\}_{i=1,2,\ldots,|\mathcal{T}|}$, our objective is to learn a representation r_i for each road segment through an unsupervised model F. As a result, we obtain the set of all road representations $S = F(G, \mathcal{T}) \in \mathbb{R}^{|\mathcal{V}| \times d}$ with dimension d.*

3 TrajRNE Approach

In this section, we introduce our proposed Trajectory Road Network Embedding Model (TrajRNE) to learn effective, general-purpose embeddings of road segments in an unsupervised manner. As illustrated in Fig. 1, TrajRNE incorporates two modules, the Spatial Flow Convolution (SFC) and the Structural Road Encoder (SRE). In the following, we present these modules in more detail.

3.1 Spatial Flow Convolution

The Spatial Flow Convolution aggregates roads based on the flow probabilities provided by trajectories. Moreover, we designed the SFC to aggregate over a k-hop neighborhood to leverage distant dependencies. Standard Graph Convolutional Networks (GCNs) commonly assume network homophily, i.e., connected nodes are more similar than distant nodes. However, road networks possess complex dependencies between roads. On the one hand, consecutive roads can indicate different traffic patterns. On the other hand, traffic patterns on distant road

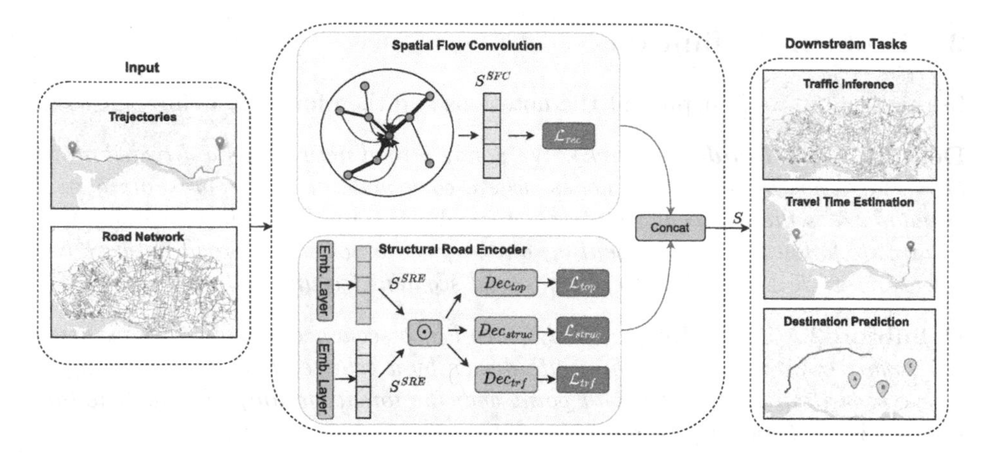

Fig. 1. The proposed TrajRNE architecture incorporating two modules: Spatial Flow Convolution and Structural Road Encoder.

segments can be correlated. Therefore, GCNs are not suitable for learning road network representations.

Inspired by Li et al. [10], who utilized trajectory flows to aggregate spatial traffic information for flow prediction, we design the Spatial Flow Convolution. In contrast to [10], we increase the receptive field by considering the traffic flow of k-hop neighbors and aggregating them within a single layer. This enables us to design an aggregation function, which selectively aggregates local and distant roads based on their importance provided by trajectory flows. We depict our Spatial Flow Convolution in Fig. 2 and compare it to a two-layer GCN. The GCN (left) aggregates all neighbors equally and needs to be stacked, which can lead to over smoothing [12]. The Spatial Flow Convolution (right) aggregates roads based on the vehicle flows (indicated by the thickness of the edges) and thus can weight the aggregation of roads based on their importance. Thus, we consider even distant relationships and tackle the issue of over smoothing by considering only important roads for aggregation.

To obtain the vehicle flow between the roads, we introduce the road transition probability p. Given two road segments v_i and v_j, the road transition probability is the probability of visiting v_j when v_i has been visited. We formally define the road transition probability by:

$$p(v_j|v_i) = \frac{p(v_i \cap v_j)}{p(v_i)}. \tag{1}$$

We estimate p by aggregating the number of transitions in historical trajectories:

$$\hat{p}(v_j|v_i) = \frac{\#transitions(v_i \rightarrow v_j) + \mathcal{A}_{i,j}}{\#total\_visits(v_i) + \sum_{k=0}^{|\mathcal{V}|} \mathcal{A}_{i,k}}, \tag{2}$$

where $\mathcal{A}$ is the adjacency matrix and $\mathcal{A}_{i,j}$ is 1 when the road segment v_j directly follows v_i. That way we keep road segment connections, even in case of sparsity of

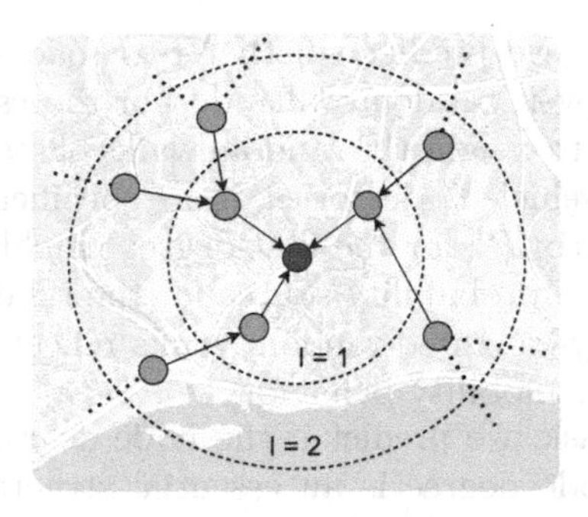

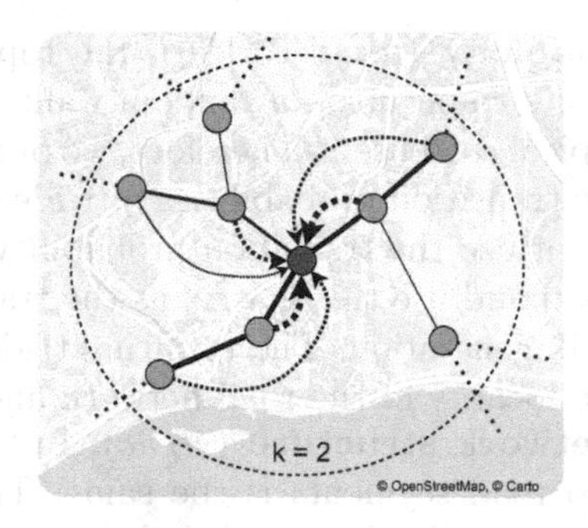

Fig. 2. Left: Graph convolution with two layers, aggregating each node with equal importance. Right: Our proposed approach of aggregating nodes based on traffic flow and hop distance for a maximum hop distance of two.

trajectories. Further, we build a road transition probability matrix $P \in \mathbb{R}^{|\mathcal{V}|\times|\mathcal{V}|}$ containing the transition probability of every road segment pair, i.e., $P_{i,j} = \hat{p}(v_j|v_i)$. The superscript k indicates that for the construction of P we consider all road segment pairs (v_i, v_j) within a k-hop distance, e.g., for P^3 we consider the transition probabilities for roads up to three hops away.

We leverage the transition probability matrix P^k to perform graph convolutions over road networks. We define the spatial flow convolution formally as

$$S^{SFC} = \sigma(P^k \mathcal{F} W), \tag{3}$$

where W is a trainable weight matrix, $\mathcal{F}$ is the set of road features, σ an activation function and S^{SFC} are the obtained road representations.

To train this module in an unsupervised way, we employ the graph reconstruction task. Thus, having obtained the road representations S^{SFC}, we try reconstructing the original adjacency matrix $\mathcal{A}$:

$$\hat{\mathcal{A}} = sigmoid(S^{SFC} \cdot S^{SFC^{\top}}). \tag{4}$$

We employ mean squared error loss to compute the reconstruction loss:

$$\mathcal{L}_{rec} = ||\mathcal{A} - \hat{\mathcal{A}}||^2. \tag{5}$$

The advantage of the reconstruction loss is that it forces the road representations S^{SFC} to learn effective road characteristics and the road network topology.

3.2 Structural Road Encoder

The Structural Road Encoder encodes structural and dynamic traffic properties by training on a multitask prediction objective in a contrastive way. More precisely, we predict whether two road segments are similar regarding three characteristics: topology, network structure, and traffic.

- **Topology** (*top*): To learn the topology, we predict whether two road segments co-occur on a random walk. However, random walks do not represent typical road users. Therefore, we propose to weight the random walks based on the transition probabilities provided by vehicle trajectories. More specifically, we utilize the transition probability matrix P^1 for the first-degree neighborhood and use this matrix as the transition probability source for the random walk generation. The resulting trajectory-weighted random walks reflect the geo-locality of the road network and user mobility behavior.
- **Network structure** (*struc*): In this task, we predict if the node degree of two road segments is the same. The node degree is an essential structural road network feature. It helps to distinguish roads with only one consecutive road segment from, e.g., roads followed by complex intersections.
- **Traffic** (trf): For the third task, we predict whether two road segments have similar traffic. For the traffic label, we utilize a traffic feature extracted from trajectories, i.e., mean traffic speed or volume. As those features are continuous, we divide them into ten equally sized categories and predict whether two road segments fall into the same category. In contrast to previous works learning static road features, we train on features extracted from trajectories, which reflect real-world traffic patterns.

For the training data generation, we sample n trajectory-weighted random walks per road segment, with a walk length of l and a context window of w. For each pair within a window, we set the topology label Y_{top} to 1 and obtain the structure label Y_{struc} and traffic label Y_{trf}. Further, for each positive sample, we create n_{neg} negative samples by randomly selecting road segment pairs, setting $Y_{top} = 0$ and obtaining Y_{struc} and Y_{trf}.

For the SRE training, we input two one-hot-encoded vectors $v_i, v_j \in \mathbb{R}^{|\mathcal{V}|}$ indicating the index of the road segment and encode the input into dense vectors.

$$S_i^{SRE} = Emb(v_i), \tag{6}$$

where S_i^{SRE} is the dense vector representation of the road segment v_i and Emb is the embedding layer, modeled as a fully connected layer. To predict the task labels, we employ the Hadamard product to aggregate the two road embeddings S_i^{SRE} and S_j^{SRE} and input the resulting vector into a task-specific decoder. Then for each $task \in \{top, struc, trf\}$ we obtain a probability output $\mathcal{P}$:

$$\mathcal{P}_{task}(v_i, v_j) = Dec_{task}(S_i^{SRE} \odot S_j^{SRE}), \tag{7}$$

where $\odot$ represents the Hadamard product, and Dec is a task-specific decoder, which we model using a fully connected layer and a sigmoid activation function, i.e., $Dec(x) = sigmoid(FC(x))$. Given the task labels, we can formulate the loss functions for each task as the binary cross-entropy loss:

$$\mathcal{L}_{task}(v_i, v_j) = -[Y_{task} \cdot \log(\mathcal{P}_{task}) + (1 - Y_{task}) \cdot \log(1 - \mathcal{P}_{task})]. \tag{8}$$

The overall loss function of the SRE is defined as the weighted sum of $\mathcal{L}_{top}$, $\mathcal{L}_{struc}$ and $\mathcal{L}_{trf}$ with the corresponding weights $\lambda_{top} + \lambda_{struc} + \lambda_{trf} = 1$:

$$\mathcal{L}_{SRE} = \lambda_{top} \cdot \mathcal{L}_{top} + \lambda_{struc} \cdot \mathcal{L}_{struc} + \lambda_{trf} \cdot \mathcal{L}_{trf}. \tag{9}$$

3.3 TrajRNE Overview

In our proposed TrajRNE model, we train the SFC and SRE modules independently with distinct training objectives. We concatenate the module representations to obtain the final road representation: $S = S^{SFC} \oplus S^{SRE}$, where $\oplus$ is the concatenation operator. As the SFC and SRE representations contain complementary information, they induce more information into the final road representations, making them more effective and general-purpose. Moreover, in contrast to previous work, we incorporate traffic and mobility behavior into the TrajRNE model design, which is essential for various downstream tasks.

4 Experimental Evaluation

The aim of the evaluation is threefold. First, we aim to compare TrajRNE with state-of-the-art unsupervised road embedding models on various road network-related downstream tasks. Second, we aim to evaluate ablation versions of TrajRNE. Third, we aim to assess the impact of the k parameter of SFC, as it influences the receptive field.

4.1 Datasets

We select the trajectory and road network datasets for two cities, namely Porto[2] and San Francisco[3]. The road networks are extracted from OpenStreetMap[4]. We preprocess the trajectory data. In particular, we prune trajectories outside the bounding box of the respective city and remove trajectories containing less than 10 points. Further, we map-match the trajectories [20] to obtain the road segment sequences. Table 1 summarizes the dataset statistics.

4.2 Baselines

We employ state-of-the-art road network representation models and graph representation learning approaches as baselines. For road network representation models, we evaluate **RFN** [7], **IRN2Vec** [16], **HRNR** [19] and **Toast** [1] as baselines. For graph representation learning approaches, we select **GCN** [9], and **GAT** [15] as baselines. We employ the graph reconstruction task proposed in [8] to train GCN and GAT in an unsupervised fashion. We use the parameter values given in the original papers. Note that as we aim to create general-purpose representations enabling a variety of tasks, a comparison with specialized task-specific models is not possible due to task-specific model designs.

---

[2] https://www.kaggle.com/competitions/pkdd-15-taxi-trip-time-prediction-ii.
[3] https://ieee-dataport.org/open-access/crawdad-epflmobility.
[4] https://www.openstreetmap.org/.

Table 1. Statistics of the road network and trajectory datasets.

	Road Network			Trajectory		
	#Intersections	#Road-Segments	Avg. Degree	#GPS-Points	#Trajectories	Coverage
Porto	5,358	11,331	2.4	74,269,739	1,544,234	97.9%
San Francisco	9,739	27,039	3.1	11,219,955	406,456	94.4%

4.3 Downstream Tasks and Evaluation Metrics

We consider four downstream tasks proposed in previous works [1,16,19]. For all downstream tasks, we pre-train the road representation models in an unsupervised manner and use the frozen embeddings to train a simple prediction model for each task. For **Label Classification (LC)** we select the road type as the label. For the road embedding models using the road type feature in the pre-training phase, we leave out that feature to evaluate prediction performance on unseen labels. We adopt a logistic regression classifier as the prediction model and report micro and macro F1 scores, denoted as Mi-F1 and Ma-F1. For **Traffic Inference (TI)**, we predict the average speed on the road segments. We adopt an MLP with a fully connected layer as the prediction model and report Root Mean Squared Error (RMSE) and Mean Absolute Error (MAE). For **Travel Time Estimation (TTE)** given a route, we input the sequence of road embeddings representing the route into a two-layer LSTM and predict the travel time of that route. For evaluation, we adopt RMSE and MAE. Finally, for the **Destination Prediction (DP)** task, we take the first 70% of the trajectory and input the corresponding sequence of road embeddings into a two-layer LSTM to predict the last visited location of the trajectory. We adopt the top-1 and top-5 prediction accuracy, denoted as ACC@1 and ACC@5.

4.4 Experimental Settings

We randomly selected 70% of the trajectory dataset for the representation learning. We used the remaining 30% for the training and evaluation of the trajectory-based downstream tasks **TTE** and **DP**. For those tasks, we further split the remaining trajectory set into 70% for training the prediction models and 30% for evaluation. For the road segment-based tasks **LC** and **TI**, we employed 5-fold cross-validation. We set the embedding dimension to 128 each for the SFC and SRE modules and employed the Adam optimizer with a learning rate of 0.001. For SRE, we used traffic volume for the traffic prediction task and set the weights $\lambda_{top} = \lambda_{struc} = \lambda_{trf} = \frac{1}{3}$. We set $l = 25$, $w = 5$, $n_{neg} = 3$ and performed 1000 walks per node. We trained SRE for ten epochs. For SFC, we set $k = 2$ and trained for 5000 epochs. We discuss parameter selection later in Sect. 4.7.

4.5 Performance Results

Table 2 summarizes the evaluation results for both datasets. As we can observe, our proposed TrajRNE approach consistently outperforms all the baselines on

Table 2. TrajRNE and baselines performance on two datasets and four tasks.

	Task	LC		TI		TTE		DP	
	Metric	Mi-F1	Ma-F1	MAE	RMSE	MAE	RMSE	ACC@1	ACC@5
Porto	GCN	0.660	0.411	14.175	20.361	77.490	110.589	0.246	0.502
	GAT	0.651	0.393	14.238	20.388	77.170	111.562	0.250	0.501
	RFN	0.498	0.087	15.088	20.908	83.125	116.160	0.246	0.492
	IRN2Vec	0.487	0.055	17.366	23.176	77.910	110.606	0.263	0.529
	HRNR	0.540	0.132	13.733	19.798	77.851	113.382	0.254	0.514
	Toast	0.440	0.206	13.793	19.543	78.807	111.270	0.265	0.534
	TrajRNE	**0.682**	**0.496**	**13.228**	**19.215**	**75.495**	**109.067**	**0.270**	**0.546**
San Francisco	GCN	0.663	0.070	10.435	16.294	118.832	273.111	0.027	0.071
	GAT	0.676	0.131	10.113	15.730	109.511	267.021	0.057	0.143
	RFN	0.672	0.125	9.819	15.660	113.185	266.541	0.042	0.112
	IRN2Vec	0.658	0.057	13.336	19.307	104.667	267.479	0.073	0.186
	HRNR	0.692	0.147	10.068	15.265	104.980	263.578	0.078	0.201
	Toast	0.662	0.068	10.122	15.631	109.057	265.514	0.046	0.121
	TrajRNE	**0.759**	**0.475**	**8.437**	**13.334**	**97.776**	**256.812**	**0.097**	**0.238**

both datasets and all tasks, demonstrating that incorporating trajectory information into road representation learning is essential for downstream application. Especially on the **LC** task, where the baselines predict only the most frequent labels, i.e., "residential", with high accuracy, our TrajRNE approach outperforms the baselines by a large margin, in particular on the less frequent classes, as reflected by Ma-F1. It is worth noting that without learning road types explicitly, road embeddings created by TrajRNE enable us to predict the less frequent road types in the dataset with high precision. Comparing both datasets, San Francisco has many more road segments with fewer trajectory data, making the prediction for the most downstream tasks even more challenging. We observe that our approach outperforms the baselines on the San Francisco dataset by a larger margin. This result indicates that our TrajRNE approach can generate more robust road representations even with fewer trajectory data available.

Regarding the baselines, we can observe that road embedding baselines mostly outperform graph representation methods, indicating that generic graph representation methods are unsuitable for road networks. Regarding the road representation baselines, Toast and HRNR outperform IRN2Vec and RFN in many cases, as the former utilize more specific road network-related information, e.g., extracting function zones or traveling semantics.

4.6 Ablation Study

To demonstrate the impact of the TrajRNE modules, we evaluate each module separately, i.e., TrajRNE(SFC) and TrajRNE(SRE). Table 3 presents the ablation study results. As we can observe, the modules indicate different strengths regarding specific tasks. While TrajRNE(SRE) outperforms TrajRNE(SFC) on

Table 3. Ablation study on different tasks on the Porto dataset.

Task	LC		TI		TTE		DP	
Metric	Mi-F1	Ma-F1	MAE	RMSE	MAE	RMSE	ACC@1	ACC@5
TrajRNE	0.682	**0.496**	**13.228**	**19.215**	**75.495**	**109.067**	**0.270**	**0.546**
TrajRNE(SFC)	**0.689**	0.458	13.353	19.719	77.449	113.828	0.255	0.515
TrajRNE(SRE)	0.511	0.254	13.692	19.549	76.609	109.599	0.269	0.544

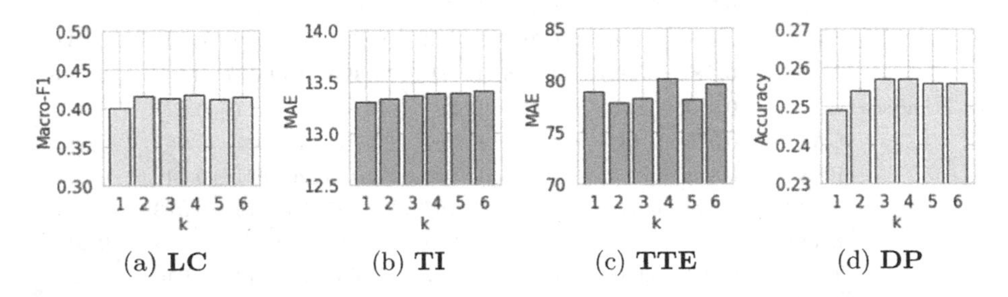

(a) **LC** (b) **TI** (c) **TTE** (d) **DP**

Fig. 3. Impact of the k parameter of the TrajRNE(SFC) module for the Porto dataset on all tasks. Figures with light gray bars indicate the higher values are better, and for the dark gray bars, lower values are better.

the **TTE** and **DP** tasks, TrajRNE(SFC) achieves higher performance on the **LC** task. TrajRNE adopting both modules performs better than the modules isolated, except for the Mi-F1 score on the **LC** task. Regarding **LC**, the slight performance reduction of 1.0% for Mi-F1 is compensated by the increase in Ma-F1 by 8.3%. Overall, these results confirm that the TrajRNE modules provide complementary information and jointly provide the best performance.

4.7 Parameter Study

We examine the k parameter of the SFC module, which influences the receptive field of the method. Thus, with a higher k value, the module can observe a broader neighborhood. We evaluate the SFC module on all selected downstream tasks with varying k. The results are depicted in Fig. 3. We observe that the hyperparameter influence depends on the task. While for the **DP** task higher k value is better, for **TI**, lower k yields better performance. This is because, for **DP**, the distant neighborhood can be more important, as the destination will not typically be located in the local neighborhood. For traffic inference, the direct neighborhood contains the traffic most similar to the target road segment. We select $k = 2$ to balance across the downstream tasks.

5 Related Work

We discuss related work in road network representation and trajectory mining.

Road Network Representation Learning. As road networks are typically modeled as graphs, a natural way to learn representations is to use graph representation learning methods, e.g., GCN [9], and GAT [15]. RFN [7] adapted GCNs to road networks by proposing a relational fusion layer. IRN2Vec [16] used shortest paths to learn the geo-locality and was trained to predict road network tags. HRNR [19] extended graph convolutions by constructing a three-level hierarchical architecture to model road segments, functional, and structural zones. Toast [1] utilized the skip-gram model to learn the graph structure and refined the embeddings using a transformer-based model and an adapted pre-training objective. However, previous works relied on road network topology and static road features to learn road embeddings, which is insufficient for reflecting complex and dynamic traffic patterns and mobility behavior. To overcome these limitations, we extract traffic features and mobility behavior from vehicle trajectories and incorporate this information deeply into our model design. Thus, TrajRNE can learn the complex and dynamic behavior of road users observed in the network. We experimentally demonstrated that TrajRNE outperforms mentioned works on various downstream tasks.

Trajectory Mining. Vehicle trajectories are mined for many road network related tasks [4,18], e.g., functional zones [14,21], travel time on road networks [5] and next location prediction [2,3,11]. Some recent work incorporated trajectories into their model design for different tasks on road networks. Hong et al. [5] created a trajectory-based graph next to a road network graph to learn traffic behavior for travel time estimation jointly. Further, Li et al. [10] integrated flows from historical vehicle trajectories into their Trajectory-based Graph Neural Network model for traffic flow prediction. For short-term traffic speed prediction, Hui et al. [6] replaced the graph convolution networks by sampling trajectories and aggregating features along them. Inspired by Li et al. [10], we designed a graph convolution based on traffic flows and extended the idea by considering the traffic flow of the k-hop neighbors, thus increasing the receptive field.

6 Conclusion

In this paper, we presented TrajRNE – a novel road network representation learning approach incorporating information extracted from trajectories into its model design. TrajRNE comprises static road features, topology, traffic, and user mobility behavior. Specifically, we proposed the Spatial Flow Convolution, aggregating local and distant neighborhoods based on traffic flows. Further, we proposed the Structural Road Encoder, which learns the network topology, structure, and traffic, employing a multitask prediction objective. We incorporated user mobility behavior by weighting random walks with transition probabilities extracted from trajectories. We conducted extensive experiments on real-world datasets and evaluated TrajRNE against state-of-the-art road representation learning and graph representation methods. We demonstrated that TrajRNE consistently outperforms the baselines on four downstream tasks.

Acknowledgements. This work was partially funded by the DFG, German Research Foundation ("WorldKG", 424985896), the Federal Ministry for Economic Affairs and Climate Action (BMWK), Germany ("d-E-mand", 01ME19009B and "ATTENTION!", 01MJ22012D), and DAAD, Germany ("KOALA", 57600865).

References

1. Chen, Y., et al.: Robust road network representation learning: when traffic patterns meet traveling semantics. In: CIKM 2021, pp. 211–220. ACM (2021)
2. Dang, W., et al.: Predicting human mobility via graph convolutional dual-attentive networks. In: WSDM (2022)
3. Feng, J., et al.: Deepmove: predicting human mobility with attentional recurrent networks. In: WWW (2018)
4. Feng, Z., Zhu, Y.: A survey on trajectory data mining: techniques and applications. IEEE Access **4**, 2056–2067 (2016)
5. Hong, H., et al.: Heteta: heterogeneous information network embedding for estimating time of arrival. In: KDD 2020, pp. 2444–2454. ACM (2020)
6. Hui, B., Yan, D., Chen, H., Ku, W.: Trajnet: a trajectory-based deep learning model for traffic prediction. In: KDD 2021, pp. 716–724. ACM (2021)
7. Jepsen, T.S., Jensen, C.S., Nielsen, T.D.: Graph convolutional networks for road networks. In: ACM SIGSPATIAL 2019, pp. 460–463. ACM (2019)
8. Kipf, T.N., Welling, M.: Variational graph auto-encoders. CoRR abs/1611.07308 (2016)
9. Kipf, T.N., Welling, M.: Semi-supervised classification with graph convolutional networks. In: ICLR 2017. OpenReview.net (2017)
10. Li, M., Tong, P., Li, M., Jin, Z., Huang, J., Hua, X.: Traffic flow prediction with vehicle trajectories. In: AAAI 2021, pp. 294–302. AAAI Press (2021)
11. Lin, Y., Wan, H., Guo, S., Lin, Y.: Pre-training context and time aware location embeddings from spatial-temporal trajectories for user next location prediction. In: AAAI 2021, pp. 4241–4248. AAAI Press (2021)
12. Oono, K., Suzuki, T.: Graph neural networks exponentially lose expressive power for node classification. In: ICLR 2020. OpenReview.net (2020)
13. Perozzi, B., Al-Rfou, R., Skiena, S.: Deepwalk: online learning of social representations. In: ACM SIGKDD 2014, pp. 701–710. ACM (2014)
14. Shimizu, T., Yabe, T., Tsubouchi, K.: Enabling finer grained place embeddings using spatial hierarchy from human mobility trajectories. In: SIGSPATIAL (2020)
15. Velickovic, P., Cucurull, G., Casanova, A., Romero, A., Liò, P., Bengio, Y.: Graph attention networks. In: ICLR 2018. OpenReview.net (2018)
16. Wang, M., Lee, W., Fu, T., Yu, G.: Learning embeddings of intersections on road networks. In: SIGSPATIAL, pp. 309–318. ACM (2019)
17. Wang, M., Lee, W., Fu, T., Yu, G.: On representation learning for road networks. ACM Trans. Intell. Syst. Technol. **12**(1), 11:1–11:27 (2021)
18. Wang, S., Bao, Z., Culpepper, J.S., Cong, G.: A survey on trajectory data management, analytics, and learning. ACM Comput. Surv. **54**(2), 39:1–39:36 (2022)
19. Wu, N., Zhao, W.X., Wang, J., Pan, D.: Learning effective road network representation with hierarchical graph neural networks. In: KDD 2020. ACM (2020)
20. Yang, C., Gidófalvi, G.: Fast map matching, an algorithm integrating hidden Markov model with precomputation. Int. J. Geogr. Inf. Sci. **32**(3), 547–570 (2018)
21. Yao, Z., Fu, Y., Liu, B., Hu, W., Xiong, H.: Representing urban functions through zone embedding with human mobility patterns. In: IJCAI 2018 (2018)

Open Access This chapter is licensed under the terms of the Creative Commons Attribution 4.0 International License (http://creativecommons.org/licenses/by/4.0/), which permits use, sharing, adaptation, distribution and reproduction in any medium or format, as long as you give appropriate credit to the original author(s) and the source, provide a link to the Creative Commons license and indicate if changes were made.

The images or other third party material in this chapter are included in the chapter's Creative Commons license, unless indicated otherwise in a credit line to the material. If material is not included in the chapter's Creative Commons license and your intended use is not permitted by statutory regulation or exceeds the permitted use, you will need to obtain permission directly from the copyright holder.

MetaCitta: Deep Meta-Learning for Spatio-Temporal Prediction Across Cities and Tasks

Ashutosh Sao[1(✉)], Simon Gottschalk[1], Nicolas Tempelmeier[2], and Elena Demidova[3,4]

[1] L3S Research Center, Leibniz Universität Hannover, Hanover, Germany
{sao,gottschalk}@L3S.de

[2] Volkswagen Group, Hannover, Germany
nicolas.tempelmeier@volkswagen.de

[3] Data Science & Intelligent Systems Group (DSIS), University of Bonn, Bonn, Germany
elena.demidova@cs.uni-bonn.de

[4] Lamarr Institute for Machine Learning and Artificial Intelligence, Bonn, Germany
https://lamarr-institute.org/

Abstract. Accurate spatio-temporal prediction is essential for capturing city dynamics and planning mobility services. State-of-the-art deep spatio-temporal predictive models depend on rich and representative training data for target regions and tasks. However, the availability of such data is typically limited. Furthermore, existing predictive models fail to utilize cross-correlations across tasks and cities. In this paper, we propose MetaCitta, a novel deep meta-learning approach that addresses the critical challenges of data scarcity and model generalization. MetaCitta adopts the data from different cities and tasks in a generalizable spatio-temporal deep neural network. We propose a novel meta-learning algorithm that minimizes the discrepancy between spatio-temporal representations across tasks and cities. Our experiments with real-world data demonstrate that the proposed MetaCitta approach outperforms state-of-the-art prediction methods for zero-shot learning and pre-training plus fine-tuning. Furthermore, MetaCitta is computationally more efficient than the existing meta-learning approaches.

Keywords: Spatio-Temporal Prediction · Meta-Learning · Pre-training

1 Introduction

Spatio-temporal predictions are critically important for planning and further developing smart cities. For instance, accurate bike and taxi demand prediction can enhance mobility services and city traffic management. Recently, deep learning-based approaches achieved high effectiveness in various spatio-temporal prediction tasks, including traffic forecasting and crowd flow prediction [1,15].

© The Author(s) 2023
H. Kashima et al. (Eds.): PAKDD 2023, LNAI 13938, pp. 70–82, 2023.
https://doi.org/10.1007/978-3-031-33383-5_6

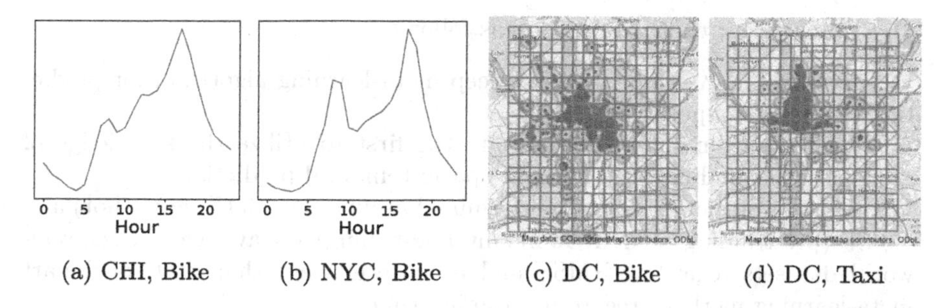

(a) CHI, Bike (b) NYC, Bike (c) DC, Bike (d) DC, Taxi

Fig. 1. (a) and (b) depict bike usage patterns in Chicago (CHI) and New York City (NYC), whereas (c) and (d) illustrate the bike and taxi usage in Washington, D.C. (DC), between 9 and 10 am.

The effectiveness of such approaches depends heavily on the availability of large amounts of training data. However, spatio-temporal data is typically (i) locked in organizational silos and rarely available across different cities and tasks and (ii) not sufficiently exploited for pre-training generalizable models. Consequently, we identify two crucial challenges in the spatio-temporal domain:

- **Lack of data of the target city and prediction task**: Spatio-temporal prediction typically requires data from the specific target city and the task of interest. However, while rich data is available for a few selected cities and tasks, data for the specific city and task of interest is often unavailable.
- **Lack of pre-training strategy in the spatio-temporal domain**: Pre-training requires a large amount of data from different tasks to learn an initialization for the target task, to converge better and faster. However, the adoption of pre-training in the spatio-temporal domain is currently limited.

Meta-learning is typically used to transfer knowledge from multiple cities (e.g., MetaStore [10] and MetaST [13]). However, existing approaches only learn from a specific task (e.g., bike demand prediction) and do not exploit the correlations between the tasks (e.g., taxi and bike demand prediction) and geographic regions. Examples of such correlations are illustrated in Fig. 1a and Fig. 1b, which indicate similar bike usage trends in central business areas in Chicago and New York City. Similarly, Fig. 1c and Fig. 1d indicate similar demand for bikes and taxis across regions. Therefore, we aim to build a predictive model that benefits from incorporating such correlations across tasks and cities.

In this paper, we propose MetaCitta, a novel deep meta-learning approach for spatio-temporal predictions across cities and tasks. In contrast to other approaches [10,13], MetaCitta adopts the knowledge not only from several cities but also different tasks. As mentioned above, different tasks in the same city can exhibit spatial correlations, and a task across cities can exhibit task-specific correlations. Therefore, we learn spatially invariant and task invariant feature representations using the Maximum Mean Discrepancy (MMD) [4]. These representations enable MetaCitta to make accurate predictions when data is unavailable for the target city and task. We also demonstrate how to adopt MetaCitta as a pre-training strategy when some target data is available for fine-tuning.

In summary, our contributions are as follows:

- We propose METACITTA[1], a novel deep meta-learning algorithm for prediction in spatio-temporal networks.
- To the best of our knowledge, we are the first to utilize the knowledge of multiple tasks and cities to improve spatio-temporal prediction.
- METACITTA outperforms best-performing baselines by 9.39% (zero-shot) and 3.86% (pre-training + data-abundant fine-tuning) on average on six real-world datasets regarding RMSE and is more efficient than state-of-the-art meta-learning methods regarding training time.

2 Problem Statement

In the following, we provide a formal definition of the problem of spatio-temporal prediction (based on [12,15]) via meta-learning from multiple cities and tasks.

Definition 1 (City and Region). *We represent a city c as an $m \times n$ grid map [12] with equally sized cells. We refer to each cell as a region.*

For example, we can split the city of Chicago into 20×20 equally sized regions, where each region represents a squared area.

Definition 2 (Spatio-Temporal Image). *A spatio-temporal image (as coined in [12], an image in short) $x_t^{c,\tau}$ of a city c and a task τ is a multi-channel image having the dimension $\mathbb{R}^{q \times m \times n}$ where q is the number of observations relevant for the task τ at a given time point t.*

For example, an image of Chicago for the taxi prediction task contains $q = 2$ observations (taxi pickup and drop-off) for each region at a time point t.

Definition 3 (Spatio-Temporal Prediction). *Given a sequence of spatio-temporal images of length i, $\mathcal{X}_{c,\tau} = \langle x_{t-i-1}^{c,\tau}, x_{t-i}^{c,\tau}, \ldots, x_{t-1}^{c,\tau} \rangle$, where $t-i-1, t-i, \ldots, t-1$ are consecutive time points, spatio-temporal prediction estimates the image at the next time point t, i.e., $x_t^{c,\tau} \in \mathbb{R}^{q \times m \times n}$.*

Given Chicago's spatio-temporal images for the taxi prediction in a given period, we aim to predict the image, i.e., taxi drop-off and pickup, at the next time point.

Definition 4 (Meta-learning from Multiple Cities and Tasks). *Given are a set of cities $\mathcal{C} = \{c_1, c_2, \ldots\}$, a target city $c_{target} \in \mathcal{C}$, as well as a set of tasks $\mathcal{T} = \{\tau_1, \tau_2, \ldots\}$ and a target task $\tau_{target} \in \mathcal{T}$. Training data is available in the form of image sequences for each combination of a city $c \in \mathcal{C}$ and a task $\tau \in \mathcal{T}$ except for the target task τ_{target} of the target city c_{target}: $\mathcal{D} = \{\mathcal{X}_{c,\tau} | c \in \mathcal{C}, \tau \in \mathcal{T}\} \setminus \{\mathcal{X}_{c_{target},\tau_{target}}\}$. The goal is to predict $x_t^{c_{target},\tau_{target}}$ based on $\mathcal{D}$.*

As an example, consider three cities (Chicago (*CHI*), Washington, D.C. (*DC*) and New York City (*NYC*)) and two tasks (bike demand prediction (*bike*) and taxi demand prediction (*taxi*)). With *NYC* as the target city c_{target} and *taxi* as the target task τ_{target}, the following image sequences are available:

[1] https://github.com/ashusao/MetaCitta.

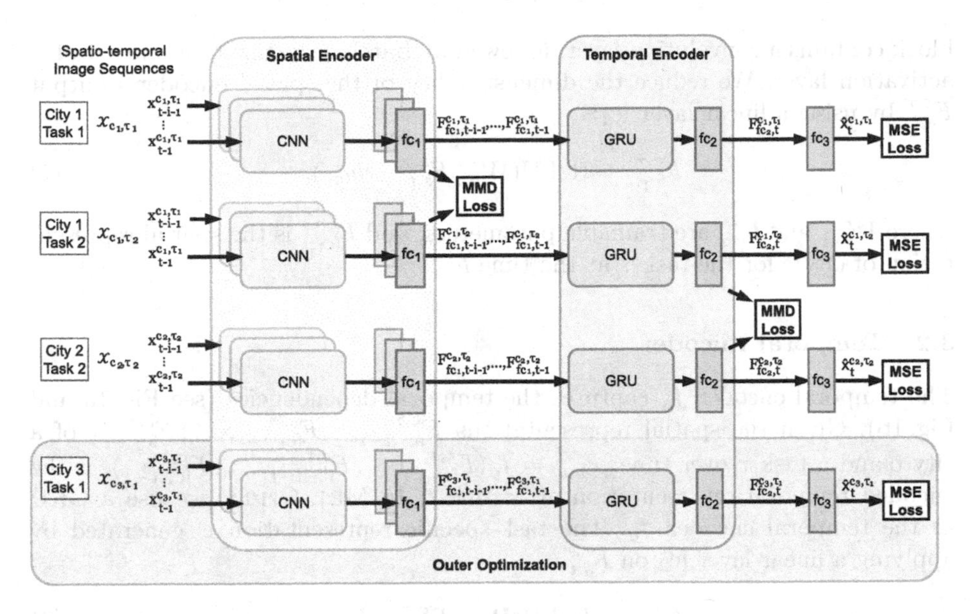

Fig. 2. Training procedure of MetaCitta with three cities (c_1, c_2 and c_3) and two tasks (τ_1 and τ_2). c_3 is the target city and τ_2 is the target task. The fully connected layers fc$_1$, fc$_2$ and fc$_3$ follow Eqs. (1), (2) and (3).

$\mathcal{D} = \{\mathcal{X}_{DC,bike}, \mathcal{X}_{CHI,bike}, \mathcal{X}_{DC,taxi}, \mathcal{X}_{CHI,taxi}, \mathcal{X}_{NYC,bike}\}$. Based on these image sequences, the goal of meta-learning from multiple cities and tasks is to predict the image $x_t^{NYC,taxi}$.

We also consider a variation of Definition 4 where limited data for the target task of the target city is available, specifically $\mathcal{D} = \{\mathcal{X}_{c,\tau} | c \in \mathcal{C}, \tau \in \mathcal{T}\}$ with $\mathcal{X}_{c_{target},\tau_{target}}$ restricted to only a few days.

3 The MetaCitta Approach

MetaCitta extends a typical spatio-temporal network (ST-Net) which has a spatial encoder, a temporal encoder, and a prediction component. Figure 2 illustrates MetaCitta's training with an example of three cities and two tasks. Spatio-temporal image sequences for each city-task pair are encoded by the spatial and temporal encoder. Spatial and task alignment is achieved through alignment losses (MMD loss). The outer optimization step trains on data from the target city (c_3) but not the target task (τ_2).

3.1 Spatial Encoder

The spatial encoder $\boldsymbol{f_s}$ captures the spatial dependencies between regions (see Fig. 1c and Fig. 1d). Given an image $x_t^{c,\tau}$ of a city c and a task τ at the time t, $F_{s,t}^{c,\tau} = \boldsymbol{f_s}(x_t^{c,\tau})$ is the encoded representation of the input image $x_t^{c,\tau}$. In MetaCitta, the spatial encoder $\boldsymbol{f_s}$ is a CNN consisting of three blocks where each block contains a convolution layer, followed by batch normalization and a ReLU activation layer. We reduce the dimensionality of the spatial encoder's output $F_{s,t}^{c,\tau}$ by using a linear layer fc_1:

$$F_{\text{fc}_1,t}^{c,\tau} = \text{ReLU}(\boldsymbol{W}_{\mathbf{fc_1}} F_{s,t}^{c,\tau} + \boldsymbol{b}_{\mathbf{fc_1}}), \tag{1}$$

where $\boldsymbol{W}_{\mathbf{fc_1}}$ and $\boldsymbol{b}_{\mathbf{fc_1}}$ are trainable parameters, and $F_{\text{fc}_1,t}^{c,\tau}$ is the spatial representation of city c for the task τ at the time t.

3.2 Temporal Encoder

The temporal encoder $\boldsymbol{f_g}$ captures the temporal dependencies (see Fig. 1a and Fig. 1b). Given the spatial representations $F_{\text{fc}_1,t-i-1}^{c,\tau}, F_{\text{fc}_1,t-i}^{c,\tau}, \ldots, F_{\text{fc}_1,t-1}^{c,\tau}$ of a city c and a task τ over time, $F_{g,t}^{c,\tau} = \boldsymbol{f_g}(F_{\text{fc}_1,t-i-1}^{c,\tau}, F_{\text{fc}_1,t-i}^{c,\tau}, \ldots, F_{\text{fc}_1,t-1}^{c,\tau})$, is the encoded temporal representation at a time t. In MetaCitta, we use a GRU as the temporal encoder $\boldsymbol{f_g}$. The task-specific representation is generated by applying a linear layer fc_2 on $F_{g,t}^{c,\tau}$:

$$F_{\text{fc}_2,t}^{c,\tau} = \text{ReLU}(\boldsymbol{W}_{\mathbf{fc_2}} F_{g,t}^{c,\tau} + \boldsymbol{b}_{\mathbf{fc_2}}), \tag{2}$$

where $\boldsymbol{W}_{\mathbf{fc_2}}$ and $\boldsymbol{b}_{\mathbf{fc_2}}$ are trainable parameters.

3.3 Prediction

In an ST-Net, a linear operation and activation function transform the output into the desired shape and range. Then, a task-specific loss function is applied to train the network. MetaCitta's ST-Net uses a linear layer fc_3 and the *tanh* activation on the task-specific representation $F_{\text{fc}_2,t}^{c,\tau}$ of a city c and a task τ:

$$\hat{x}_t^{c,\tau} = tanh(\boldsymbol{W}_{\mathbf{fc_3}} F_{\text{fc}_2,t}^{c,\tau} + \boldsymbol{b}_{\mathbf{fc_3}}), \tag{3}$$

where $\boldsymbol{W}_{\mathbf{fc_3}}$ and $\boldsymbol{b}_{\mathbf{fc_3}}$ are trainable parameters, and $\hat{x}_t^{c,\tau} \in \mathbb{R}^{q \times m \times n}$ is the prediction at time t. We utilize the mean squared error as the task-specific loss:

$$L^{c,\tau} = \frac{1}{N} \sum_{i=1}^{N} (\hat{x}_t^{c,\tau} - x_t^{c,\tau})^2, \tag{4}$$

where N is the number of images, $\hat{x}_t^{c,\tau}$ and $x_t^{c,\tau}$ are the predicted and ground truth labels, respectively.

3.4 Training Procedure

Different tasks originate from different distributions. MetaCitta should learn the distribution-invariant properties from the tasks across the cities to perform well on an unseen task. To learn distribution-invariant properties, we use the Maximum Mean Discrepancy (MMD) [4] to minimize the distance between two

Algorithm 1. MetaCitta Training

1: **Input**: Data ($\mathcal{D}$), step size (α), cities ($\mathcal{C}$), tasks ($\mathcal{T}$), target city (c^{target}), target task (τ^{target}), randomly initialized model parameters (θ)
2: **while** *not converged* **do**
3:    **for each** $c_i \in \mathcal{C} \setminus \{c_{target}\}$ **do** ▷ Spatial Alignment
4:       **for each** $\tau_j, \tau_k \in \mathcal{T}$ **do**
5:          Extract pairs $(\mathcal{X}_{c_i,\tau_j}, \mathcal{X}_{c_i,\tau_k})$ from $\mathcal{D}$
6:          Compute $L_{mmd}^{spatial}, L^{c_i,\tau_j}, L^{c_i,\tau_k}$ using Equations (1, 3, 4, 5)
7:          $\theta = \theta - \alpha(\nabla_\theta L^{c_i,\tau_j} + \nabla_\theta L^{c_i,\tau_k} + \nabla_{\theta_s} L_{mmd}^{spatial})$
8:    **for each** $\tau_i \in \mathcal{T}$ **do** ▷ Task Alignment
9:       **for each** $c_j, c_k \in \mathcal{C} \setminus \{c_{target}\}$ **do**
10:          Extract pairs $(\mathcal{X}_{c_j,\tau_i}, \mathcal{X}_{c_k,\tau_i})$ from $\mathcal{D}$
11:          Compute $L^{c_j,\tau_i}, L^{c_k,\tau_i}, L_{mmd}^{task}$ using Equations (2, 3, 4, 6)
12:          $\theta = \theta - \alpha(\nabla_\theta L^{c_i,\tau_j} + \nabla_\theta L^{c_i,\tau_k} + \nabla_{\theta_\tau} L_{mmd}^{task})$
13:    **for each** $\tau_i \in \mathcal{T} \setminus \{\tau_{target}\}$ **do** ▷ Outer Optimization
14:       Extract $\mathcal{X}_{c_{target},\tau_i}$ from $\mathcal{D}$
15:       Compute L^{c_{target},τ_i} using Equation (4)
16:       $\theta = \theta - \alpha \nabla_\theta L^{c_{target},\tau_i}$

distributions. In this way, the network learns the common or invariant properties between the distributions and thus better generalizes to unseen distributions.

Different tasks of a city have the same underlying regions and thus share similar spatial characteristics. Therefore, to extract the spatially invariant representation between the tasks τ_1 and τ_2 in a city c, we apply the MMD constraint to the spatial representations generated by the spatial encoder. As a result, we compute the *spatial alignment loss* $L_{mmd}^{spatial}$:

$$L_{mmd}^{spatial} = MMD(F_{\text{fc}_1,t}^{c,\tau_1}, F_{\text{fc}_1,t}^{c,\tau_2}). \tag{5}$$

Similarly, to extract the task-invariant features from the task τ in two different cities, c_1 and c_2, we apply the MMD constraint on the task-specific representations generated by the temporal encoder. As a result, we compute the *task alignment loss* L_{mmd}^{task}:

$$L_{mmd}^{task} = MMD(F_{\text{fc}_2,t}^{c_1,\tau}, F_{\text{fc}_2,t}^{c_2,\tau}). \tag{6}$$

Finally, to enrich the model with the target city features, we perform knowledge transfer from all available tasks of the target city.

The MetaCitta training procedure is depicted in Fig. 2 and described in Algorithm 1. The algorithm takes the training data $\mathcal{D}$, a set of cities C, a set of tasks $\mathcal{T}$, the target city c^{target} and target task τ^{target} as input. The goal is to learn the model parameters θ, where θ_s are the parameters responsible for generating the spatial representation and θ_τ for the task-specific representation. The algorithm performs the following three steps:

1. The **Spatial Alignment (lines 3–7)** is performed between different tasks of a city on pairs of examples (line 5) and by updating the model parameters using the spatial alignment loss and the task-specific loss (lines 6–7).
2. The **Task Alignment (lines 8–12)** happens by extracting example pairs from the same task but of different cities (line 10), and with an update step using the task-alignment loss and the task-specific loss (lines 10–12).
3. The **Outer Optimization (lines 13–16)** step is done by directly performing an update step on all the available tasks of the target city.

Using the example from Sect. 2 with NYC as c_{target} and $taxi$ as τ_{target}, the spatial alignment considers $(\mathcal{X}_{DC,bike}, \mathcal{X}_{DC,taxi})$ and $(\mathcal{X}_{CHI,bike}, \mathcal{X}_{CHI,taxi})$, while the task alignment involves $(\mathcal{X}_{CHI,bike}, \mathcal{X}_{DC,bike})$, and $(\mathcal{X}_{CHI,taxi}, \mathcal{X}_{DC,taxi})$. The outer alignment updates using $\mathcal{X}_{NYC,bike}$.

Pre-training. Following Definition 4, MetaCitta is tailored towards cases where no data from the target task τ_{target} in the target city c_{target} is available. However, if such data $\mathcal{X}_{c_{target},\tau_{target}}$ is (partially) available, MetaCitta can also be used for pre-training and fine-tuning, i.e., it first learns initialization weights from available data of other cities and tasks $(\{\mathcal{X}_{c,\tau} | c \in \mathcal{C}, \tau \in \mathcal{T}\} \setminus \{\mathcal{X}_{c_{target},\tau_{target}}\})$ and is then fine-tuned on data of the target task $\mathcal{X}_{c_{target},\tau_{target}}$. To use MetaCitta for pre-training, we skip the task alignment and instead perform spatial alignment between all available pairs of source cities and tasks. This method extracts the more general properties of the input data typically captured in the initial layers of a network [8,14], which is required for pre-training.

4 Evaluation Setup

This section describes the datasets, baselines, and experimental settings used to evaluate MetaCitta.

4.1 Datasets

We utilize two tasks (taxi and bike demand) from three cities: NYC, CHI, and DC. Each dataset consists of six months of data, with five months used for training and one month used as a test set. The data for TaxiNYC[2] (36M), TaxiCHI[3] (7M), TaxiDC[4] (3M), and BikeNYC[5] (9M) is from 01/2019 to 06/2019. The data for BikeCHI[6] (2.5M) and BikeDC[7] (1.2M) is from 07/2020 to 12/2020.

[2] https://www1.nyc.gov/site/tlc/about/tlc-trip-record-data.page.
[3] https://data.cityofchicago.org/Transportation/Taxi-Trips-2019/h4cq-z3dy.
[4] https://opendata.dc.gov/documents/taxi-trips-in-2019/explore.
[5] https://ride.citibikenyc.com/system-data.
[6] https://www.divvybikes.com/system-data.
[7] https://www.capitalbikeshare.com/system-data.

4.2 Baselines

We compare MetaCitta to the following seven baselines:

- The Historical Average (**HA**) is calculated by taking the mean value regarding the specific hour for the specific region.
- No pre-training (**NoPretrain**): The network is initialized with random weights and fine-tuned on the target task.
- **Joint**: The model is trained jointly by mixing all the samples.
- **MAML** [3] is a meta-learning method that learns an initialization from multiple tasks. We use the same underlying ST-Net as in MetaCitta.
- **MetaStore** [10] is a MAML-based approach that learns to generate the city-specific parameters based on the city's encoding during training.
- **MetaST** [13] is also based on MAML and learns a pattern-based spatio-temporal memory from source cities.
- **MLDG** [9] extends MAML for domain generalization. It randomly leaves one task out and updates its parameters on the left-out task during training.

4.3 Experimental Settings

In our experiments, each city is divided into 20×20 grid cells of size $1km^2$ each. We use the same underlying ST-Net and parameters for MetaCitta and the baselines (except for HA) to allow for a fair comparison. As spatial encoders, we use CNNs with 32 filters of size 3×3. As temporal encoders, we use GRUs where the input and hidden sizes are set to 256. Their input sequence length is 12 at an interval of 1 hour. The size of the fully connected layers (fc_1, fc_2) is 256. The batch size is 64, and each model is trained for 500 epochs on an NVIDIA GeForce GTX 1080 Ti (11 GB) at a learning rate of $1e^{-5}$ using an Adam optimizer. For the MAML-based approaches, the inner and outer learning rates are set to $1e^{-5}$, and there are 5 update steps.

5 Evaluation

In this section, we evaluate MetaCitta by comparing it to the baselines, by conducting an ablation study, and by training time comparison.

5.1 Comparison with Baselines

MetaCitta was evaluated in two settings: *zero-shot* and *fine-tuning*. Fine-tuning was done in two conditions: *data-limited* (15 days of target data) and *data-abundant* (5 months of target data). Results in Table 1 demonstrate that MetaCitta performs best in both settings and all datasets regarding RMSE.

In the zero-shot setting, HA performs worst, indicating that a simple heuristic approach cannot correctly capture the city's complex spatio-temporal dynamics. In this setting, MLDG is the best-performing baseline in terms of the

Table 1. Comparison of METACITTA to the baselines, where RMSE is the Root Mean Squared Error and MAE is the Mean Absolute Error. IMPROV. (%) is the relative improvement of METACITTA over the best baseline (underlined).

Approach	TaxiNYC		TaxiDC		TaxiCHI		BikeNYC		BikeDC		BikeCHI	
	MAE	RMSE	MAE	RMSE	MAE	RMSE	MAE	RMSE	MAE	RMSE	MAE	RMSE
Zero-shot (no data from the target task of the target city used)												
HA	23.76	104.95	8.20	27.20	10.29	40.30	7.95	24.32	7.11	24.95	7.00	24.78
JOINT	32.11	98.75	**1.72**	10.89	**4.54**	29.34	6.78	24.29	0.34	1.55	0.41	1.56
MAML	31.25	93.99	1.80	10.95	4.86	27.76	6.54	22.95	0.37	1.60	0.42	1.47
METASTORE	30.51	99.91	1.82	9.76	4.81	27.49	8.14	27.90	0.47	1.85	0.41	1.43
METAST	25.64	85.72	1.78	10.62	5.20	25.16	6.84	23.91	0.40	1.63	0.43	1.51
MLDG	21.87	85.08	1.81	10.51	6.04	26.46	6.99	24.27	0.37	1.51	0.48	1.52
METACITTA	**21.26**	**79.77**	1.85	**9.20**	5.34	**22.91**	**6.29**	**21.75**	**0.32**	**1.40**	**0.39**	**1.42**
IMPROV. (%)	2.79	6.24	-2.21	12.46	11.59	13.41	10.01	10.38	13.51	7.28	18.75	6.58
Fine-tuning (data-limited: 15 days of data from target task of target city)												
NOPRETRAIN	170.60	220.13	52.61	72.70	130.39	169.89	55.58	73.18	11.65	4.75	15.53	19.63
JOINT	5.39	22.29	0.89	4.34	**1.79**	9.10	4.28	15.54	0.35	1.05	0.48	1.50
MAML	5.66	23.32	0.89	4.18	1.91	10.26	4.23	15.36	0.57	1.34	0.78	1.55
METASTORE	8.54	38.01	0.99	5.12	2.16	13.74	5.87	20.45	0.45	1.36	0.53	1.68
METAST	6.41	22.58	0.94	4.23	2.02	9.53	4.24	15.17	0.37	1.06	0.66	1.51
MLDG	6.21	24.61	0.94	4.46	2.21	12.87	4.70	17.15	0.57	1.41	0.51	1.56
METACITTA	**5.14**	**20.95**	**0.85**	**4.09**	1.81	**8.96**	**4.14**	**14.76**	**0.31**	**1.03**	**0.45**	**1.35**
IMPROV. (%)	4.64	6.01	4.49	5.76	-1.12	1.54	3.28	5.02	11.42	1.90	6.25	10.0
Fine-tuning (data-abundant: 5 months of data from target task of target city)												
NOPRETRAIN	21.89	40.57	4.60	9.48	7.94	22.05	7.48	19.77	1.10	1.99	2.09	4.01
JOINT	3.43	14.01	0.61	2.78	1.09	6.21	2.09	7.11	0.27	0.87	0.34	0.94
MAML	3.38	14.11	0.70	2.80	1.21	6.19	2.18	7.01	0.31	0.92	0.35	0.96
METASTORE	5.42	24.81	0.77	3.81	1.52	10.19	3.11	11.30	0.31	0.96	0.39	1.17
METAST	3.49	14.40	0.63	2.76	1.19	6.31	2.13	7.02	0.27	0.86	0.34	0.95
MLDG	3.52	14.73	0.66	2.93	1.16	6.42	2.12	7.23	0.29	0.92	0.37	1.05
METACITTA	**3.24**	**13.24**	**0.55**	**2.65**	**1.05**	**5.91**	**1.95**	**6.92**	**0.25**	**0.85**	**0.31**	**0.91**
IMPROV. (%)	5.54	5.49	9.83	4.68	3.70	4.83	6.69	2.67	7.41	2.29	8.82	3.19

RMSE because it is trained to perform particularly well on an unseen task by updating its parameters based on the left-out task.

METACITTA and the pre-trained baselines outperform NOPRETRAIN in fine-tuning due to their ability to extract generic spatio-temporal properties that aid in convergence to the target task. These properties include traffic behavior, such as high demand during peak hours (morning and evening) and low off-peak demand (e.g., after midnight). JOINT, MAML, and METAST perform better than MLDG as they are trained to perform well on all source tasks and adapt quickly to new tasks. METASTORE performs relatively worse in fine-tuning than in the zero-shot setting, as it only fine-tunes the final prediction layers (fc_2 and fc_3), reducing its adaptability to the target task.

On average across datasets, compared to the respective best baselines, METACITTA is 9.39% better than MLDG in the zero-shot setting, 5.04% better than JOINT in the data-limited fine-tuning setting, and 3.86% better than JOINT in

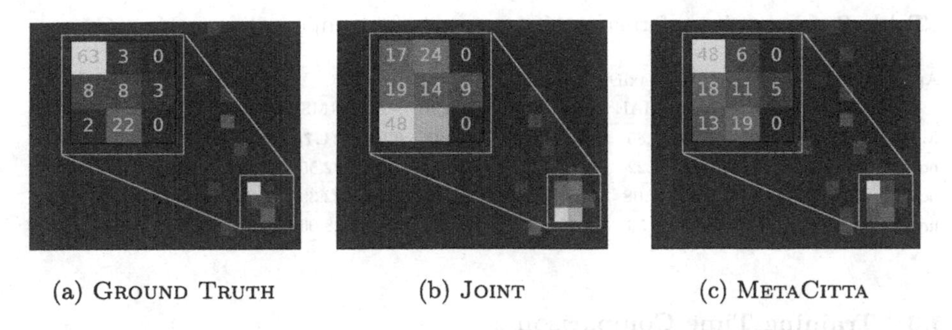

(a) GROUND TRUTH (b) JOINT (c) METACITTA

Fig. 3. Comparison of JOINT and METACITTA on taxi pickup prediction of Chicago, June 6th 2019, between 4 and 5 am. Nine selected regions and their number of taxi pickups in the mentioned hour are shown in detail.

the data-abundant fine-tuning setting regarding the RMSE. According to the MAE, METACITTA exceeds the baselines in all cases except three (e.g., zero-shot on TaxiCHI), where JOINT performs better. Since METACITTA consistently has a lower RMSE, we conclude that METACITTA performs better in predicting extreme values, such as high or low demand values during or after peak periods, while JOINT mainly predicts average values for each region.

To further investigate this behavior, we visualize the predictions of JOINT and METACITTA in the data-limited fine-tuning setting on the TaxiCHI dataset. Figure 3 illustrates the ground truth and the predicted pickup demands in different city regions between 4 and 5 am. The regions to the south and southwest of the zoomed area represent the Chicago Loop area, which is Chicago's central business district. The Northwest region is Chicago's Near North Side, a residential area. While the taxi demand in the Chicago Loop area is high during the day, it is low during off-peak hours. On the other hand, the demand for taxis in the residential area is higher compared to the business area. This is because people use taxis, as public transport does not run at 4 am. While METACITTA correctly captures this shift in demand, as illustrated in Fig. 3c, JOINT continues to forecast average high values in the business district, as observed in Fig. 3b.

5.2 Ablation Study

We analyze the contribution of METACITTA's training components: spatial alignment, task alignment, and outer optimization. By removing one component at a time, we evaluate their effectiveness and present the results in Table 2.

Removing METACITTA's outer optimization step leads to the most significant drop in performance, indicating the importance of having some knowledge of the target city, even if from another task. This knowledge can be obtained directly from other tasks, as observed in Fig. 2a and 2b, where similar regions indicate similar spatio-temporal behavior. Also, the performance decreases when spatial and task alignment is removed, indicating the impact of adding the MMD losses.

Table 2. Change in performance after removal of a component from MetaCitta.

Approach	TaxiNYC		TaxiDC		TaxiCHI		BikeNYC		BikeDC		BikeCHI	
	MAE	RMSE	MAE	RMSE	MAE	RMSE	MAE	RMSE	MAE	RMSE	MAE	RMSE
MetaCitta	**21.26**	**79.77**	1.85	**9.20**	5.34	**22.91**	**6.29**	**21.75**	**0.32**	**1.40**	0.39	**1.42**
no spatial align	23.61	86.75	2.22	9.35	5.02	23.96	6.45	22.55	0.42	1.43	**0.38**	1.43
no task align	46.58	126.72	**1.68**	10.80	**4.65**	27.04	6.30	22.39	**0.32**	1.42	0.39	1.49
no outer optim	31.64	108.38	6.56	18.26	14.52	43.00	9.99	28.06	1.35	3.41	0.77	2.08

5.3 Training Time Comparison

The training times for MetaST, MAML, MLDG, MetaStore, MetaCitta and Joint on the Chicago bike prediction task are 107.21, 106.68, 87.71, 7.99, 5.65 and 3.18 hours, respectively. As MetaCitta applies an extra alignment loss to extract invariant features, it needs slightly more training time than Joint, which has a simplified design. MetaCitta requires less training time than MetaST (94.73%), MAML (94.70%), MLDG (93.56%) and MetaStore (29.28%), as these baselines perform a two-stage optimization requiring time-consuming calculation of higher-order derivatives. Overall, MetaCitta performs most precisely and trains much faster than state-of-the-art meta-learning.

6 Related Work

Deep learning has recently shown great success and is widely adopted for spatio-temporal predictions [1,15]. However, the success of these approaches depends on the availability of large amounts of data, which is typically a bottleneck.

Typically, transfer learning is used to deal with data scarcity, where a network trained for a specific task is fine-tuned on the target task. Recently, meta-learning has gained popularity as a way to learn from multiple tasks. Unlike transfer learning, meta-learning does not specialize in a specific task. Instead, it focuses on finding the parameters from the source tasks, so it can be quickly adapted to the target task (i.e., learn to learn) [6]. Model Agnostic Meta Learning [3] (MAML) and Meta Learning for Domain Generalization [9] (MLDG) are state-of-the-art meta-learning approaches widely used in different domains. MAML performs two optimization steps: task-specific updates on the inner step and global meta-updates on the outer step. MLDG extends MAML by leaving out a task at the inner level and performing meta-updates based on the left-out task.

Motivated by these approaches, several attempts have been made to transfer knowledge in the spatio-temporal domain. [2,5,7,11,12], use transfer learning to transfer knowledge from a data-rich source city to a data-poor target city. MetaStore [10] and MetaST [13] are MAML-based approaches to transfer knowledge from multiple cities to a target city. However, existing approaches transfer the knowledge for a specific task, introduce additional parameters and require small amounts of data from the target city and target task.

In contrast to these methods, MetaCitta learns from different tasks in multiple cities and can be effectively used in zero-shot and fine-tuning settings.

7 Conclusion

In this paper, we presented METACITTA, a novel deep meta-learning approach for spatio-temporal predictions in cases where no or only limited training data of a target city and the task is available. METACITTA leverages knowledge across cities and tasks by learning spatial and task-invariant feature representations. METACITTA outperforms state-of-the-art meta-learning approaches regarding the RMSE in zero-shot and fine-tuning settings and requires 94.70% less training time compared to the state-of-the-art meta-learning approach MAML.

Acknowledgements. This work was partially funded by the DFG, German Research Foundation ("WorldKG", 424985896), the Federal Ministry for Economic Affairs and Climate Action (BMWK), Germany ("d-E-mand", 01ME19009B), and DAAD, Germany ("KOALA", 57600865).

References

1. Bai, L., Yao, L., Li, C., Wang, X., Wang, C.: Adaptive graph convolutional recurrent network for traffic forecasting. In: NIPS (2020)
2. Fang, Z., et al.: When transfer learning meets cross-city urban flow prediction: spatio-temporal adaptation matters. In: IJCAI (2022)
3. Finn, C., Abbeel, P., Levine, S.: Model-agnostic meta-learning for fast adaptation of deep networks. In: International Conference on Machine Learning (ICML) (2017)
4. Gretton, A., Borgwardt, K.M., Rasch, M.J., Schölkopf, B., Smola, A.: A kernel two-sample test. J. Mach. Learn. Res. **13**(1), 723–773 (2012)
5. He, T., Bao, J., Li, R., Ruan, S., Li, Y., Song, L., et al.: What is the human mobility in a new city: transfer mobility knowledge across cities. In: TheWebConf (2020)
6. Huisman, M., van Rijn, J.N., Plaat, A.: A survey of deep meta-learning. Artif. Intell. Rev. **54**(6), 4483–4541 (2021). https://doi.org/10.1007/s10462-021-10004-4
7. Jin, Y., Chen, K., Yang, Q.: Selective cross-city transfer learning for traffic prediction via source city region re-weighting. In: SIGKDD (2022)
8. Lee, H., Grosse, R., Ranganath, R., Ng, A.: Unsupervised learning of hierarchical representations with convolutional deep belief networks. Commun. ACM **54**(10), 95–103 (2011)
9. Li, D., Yang, Y., Song, Y.Z., Hospedales, T.M.: Learning to generalize: meta-learning for domain generalization. In: AAAI (2018)
10. Liu, Y., et al.: Metastore: a task-adaptative meta-learning model for optimal store placement with multi-city knowledge transfer. ACM TIST **12**(3), 1–23 (2021)
11. Wang, L., Geng, X., Ma, X., Liu, F., Yang, Q.: Cross-city transfer learning for deep spatio-temporal prediction. In: IJCAI (2019)
12. Wang, S., Miao, H., Li, J., Cao, J.: Spatio-temporal knowledge transfer for urban crowd flow prediction via deep attentive adaptation networks. IEEE TITS **23**(5), 4695–4705 (2021)
13. Yao, H., Liu, Y., Wei, Y., Tang, X., Li, Z.: Learning from multiple cities: a meta-learning approach for spatial-temporal prediction. In: TheWebConf (2019)

14. Zeiler, M.D., Fergus, R.: Visualizing and understanding convolutional networks. In: Fleet, D., Pajdla, T., Schiele, B., Tuytelaars, T. (eds.) ECCV 2014. LNCS, vol. 8689, pp. 818–833. Springer, Cham (2014). https://doi.org/10.1007/978-3-319-10590-1_53
15. Zhang, J., Zheng, Y., Qi, D.: Deep spatio-temporal residual networks for citywide crowd flows prediction. In: AAAI (2017)

Open Access This chapter is licensed under the terms of the Creative Commons Attribution 4.0 International License (http://creativecommons.org/licenses/by/4.0/), which permits use, sharing, adaptation, distribution and reproduction in any medium or format, as long as you give appropriate credit to the original author(s) and the source, provide a link to the Creative Commons license and indicate if changes were made.

The images or other third party material in this chapter are included in the chapter's Creative Commons license, unless indicated otherwise in a credit line to the material. If material is not included in the chapter's Creative Commons license and your intended use is not permitted by statutory regulation or exceeds the permitted use, you will need to obtain permission directly from the copyright holder.

Deep Graph Stream SVDD: Anomaly Detection in Cyber-Physical Systems

Ehtesamul Azim, Dongjie Wang, and Yanjie Fu(✉)

Department of Computer Science, University of Central Florida, Orlando, FL 32826, USA
{azim.ehtesam,wangdongjie}@knights.ucf.edu, yanjie.fu@ucf.edu

Abstract. Our work focuses on anomaly detection in cyber-physical systems. Prior literature has three limitations: (1) Failing to capture long-delayed patterns in system anomalies; (2) Ignoring dynamic changes in sensor connections; (3) The curse of high-dimensional data samples. These limit the detection performance and usefulness of existing works. To address them, we propose a new approach called deep graph stream support vector data description (SVDD) for anomaly detection. Specifically, we first use a transformer to preserve both short and long temporal patterns of monitoring data in temporal embeddings. Then we cluster these embeddings according to sensor type and utilize them to estimate the change in connectivity between various sensors to construct a new weighted graph. The temporal embeddings are mapped to the new graph as node attributes to form weighted attributed graph. We input the graph into a variational graph auto-encoder model to learn final spatio-temporal representation. Finally, we learn a hypersphere that encompasses normal embeddings and predict the system status by calculating the distances between the hypersphere and data samples. Extensive experiments validate the superiority of our model, which improves F1-score by 35.87%, AUC by 19.32%, while being 32 times faster than the best baseline at training and inference.

1 Introduction

Cyber-physical systems (CPS) have been deployed everywhere and play a significant role in the real world, including smart grids, robotics systems, water treatment networks, etc. Due to their complex dependencies and relationships, these systems are vulnerable to abnormal system events (e.g., cyberattacks, system exceptions), which can cause catastrophic failures and expensive costs. In 2021, hackers infiltrated Florida's water treatment plants and boosted the sodium hydroxide level in the water supply by 100 times of the normal level [3]. This may endanger the physical health of all Floridians. To maintain stable and safe CPS, considerable research effort has been devoted to effectively detect anomalies in such systems using sensor monitoring data [16,19].

Prior literature partially resolve this problem-however, there are three issues restricting their practicality and detection performance. **Issue 1: long-delayed**

© The Author(s), under exclusive license to Springer Nature Switzerland AG 2023
H. Kashima et al. (Eds.): PAKDD 2023, LNAI 13938, pp. 83–95, 2023.
https://doi.org/10.1007/978-3-031-33383-5_7

patterns. The malfunctioning effects of abnormal system events often do not manifest immediately. Kravchik et al. employed LSTM to predict future values based on past values and assessed the system status using prediction errors [5]. But, constrained by the capability of LSTM, it is hard to capture long-delayed patterns, which may lead to suboptimal detection performance. *How can we sufficiently capture such long-delayed patterns?* **Issue 2: dynamic changes in sensor-sensor influence.** Besides long-delayed patterns, the malfunctioning effects may propagate to other sensors. Wang et al. captured such propagation patterns in water treatment networks by integrating the sensor-sensor connectivity graph for cyber-attack detection [17]. However, the sensor-sensor influence may shift as the time series changes due to system failures. Ignoring such dynamics may result in failing to identify propagation patterns and cause poor detection performance. *How can we consider such dynamic sensor-sensor influence?* **Issue 3: high-dimensional data samples.** Considering the labeled data sparsity issue in CPS, existing works focus on unsupervised or semi-supervised setting [10,17]. But traditional models like One-Class SVM are too shallow to fit high-dimensional data samples. They have substantial time costs for feature engineering and model learning. *How can we improve the learning efficiency of anomaly detection in high-dimensional scenarios?*

To address these, we aim to effectively capture spatial-temporal dynamics in high-dimensional sensor monitoring data. In CPS, sensors can be viewed as nodes, and their physical connections resemble a graph. Considering that the monitoring data of each sensor changes over time and that the monitoring data of various sensors influence one another, we model them using a graph stream structure. Based on that, we propose a new framework called Deep Graph Stream Support Vector Data Description (**DGS-SVDD**). Specifically, to capture long-delayed patterns, we first develop a temporal embedding module based on transformer [15]. This module is used to extract these patterns from individual sensor monitoring data and embed them in low-dimensional vectors. Then, to comprehend dynamic changes in sensor-sensor connection, we estimate the influence between sensors using the previously learnt temporal embedding of sensors. The estimated weight matrix is integrated with the sensor-sensor physically connected graph to produce an enhanced graph. We map the temporal embeddings to each node in the enhanced graph as its attributes to form a new attributed graph. After that, we input this graph into the variational graph auto-encoder (VGAE) [4] to preserve all information as final spatial-temporal embeddings. Moreover, to effectively detect anomalies in high-dimensional data, we adopt deep learning to learn the hypersphere that encompasses normal embeddings. The distances between the hypersphere and data samples are calculated to be criteria to predict the system status at each time segment. Finally, we conduct extensive experiments on a real-world dataset to validate the superiority of our work. In particular, compared to the best baseline model, DGS-SVDD improves F1-score by 35.87% and AUC by 19.32%, while accelerating model training and inference by 32 times.

2 Preliminaries

2.1 Definitions

Definition 1. ***Graph Stream.*** A graph object $\mathcal{G}_i$ describes the monitoring data of the CPS at timestamp i. It can be defined as $\mathcal{G}_i = (\mathcal{V}, \mathcal{E}, \mathbf{t}_i)$ where $\mathcal{V}$ is the vertex (i.e., sensor) set of size n; $\mathcal{E}$ is the edge set of size m, and each edge indicates the physical connectivity between two sensors; $\mathbf{t}_i$ is a list that contains the monitoring value of n sensors at the i-th timestamp. A graph stream is a collection of graph objects over the temporal dimension. The graph stream of length L_x at the t-th time segment can thus, be defined as $\mathbf{X}_t = [\mathcal{G}_i, \mathcal{G}_{i+1}, \cdots \mathcal{G}_{i+L_x-1}]$.

Definition 2. ***Weighted Attributed Graph.*** The edge set $\mathcal{E}$ of each graph object in the graph stream $\mathbf{X}_t$ does not change over time, which is a binary set reflecting the physical connectivity between sensors. However, the correlations between different sensors may change as system failures happen. To capture such dynamics, we use $\tilde{\mathcal{G}}_t = (\mathcal{V}, \tilde{\mathcal{E}}_t, \mathbf{U}_t)$ to denote the weighted attributed graph at the t-th time segment. Here, $\mathcal{V}$ is the same the vertex (i.e., sensor) set of size n as defined before; $\tilde{\mathcal{E}}_t$ is the weighted edge set, in which, each item indicates the weighted influence calculated from the temporal information between two sensors; $\mathbf{U}_t$ is the attribute of each vertex, which is also the temporal embedding of each node at the current time segment. Thus, $\tilde{\mathcal{G}}_t$ contains the spatio-temporal information of the system.

2.2 Problem Statement

Our goal is to detect anomalies in cyber-physical systems at each time segment. Formally, assuming that the graph stream data at the t-th segment is $\mathbf{X}_t$, the corresponding system status is y_t. We aim to find an outlier detection function that learns the mapping relation between $\mathbf{X}_t$ and y_t, denoted by $f(\mathbf{X}_t) \rightarrow y_t$. Here, y_t is a binary constant whose value is 1 if the system status is abnormal and 0 otherwise.

3 Methodology

In this section, we give an overview of our framework and then describe each technical part in detail.

3.1 Framework Overview

Figure 1 shows an overview of our framework, named DGS-SVDD. Specifically, we start by feeding the DGS-SVDD model the graph stream data for one time segment. In the model, we first analyze the graph stream data by adopting the transformer-based temporal embedding module to extract temporal dependencies. Then, we use the learnt temporal embedding to estimate the dynamics

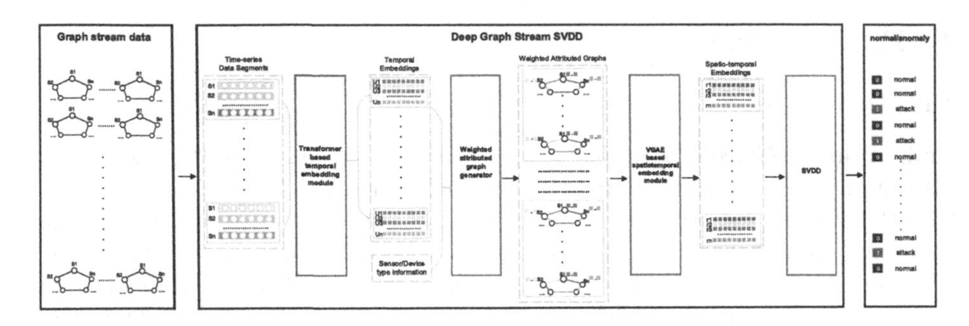

Fig. 1. An overview of our framework. There are four key components: transformer-based temporal embedding module, weighted attributed graph generator, VGAE-based spatiotemporal embedding module, and SVDD-based outlier detector.

of sensor-sensor influence and combine it with information about the topological structure of the graph stream data to generate weighted attributed graphs. We then input the graph into the variational graph autoencoder (VGAE)-based spatial embedding module to get the spatial-temporal embeddings. Finally, we estimate the boundary of the embeddings of normal data using deep learning and support vector data description (SVDD), and predict the system status by measuring how far away the embedding sample is from the boundary.

3.2 Embedding Temporal Patterns of the Graph Stream Data

The temporal patterns of sensors may evolve over time if abnormal system events occur. We create a temporal embedding module that uses a transformer in a predictive manner to capture such patterns for accurate anomaly detection. To illustrate the following calculation process, we use the graph stream data $\mathbf{X}_t$ at the t-th time segment as an example. We ignore the topological structure of the graph stream data at first during the temporal embedding learning process. Thus, we collect the time series data in $\mathbf{X}_t$ to form a temporal matrix $\mathbf{T}_t = [\mathbf{t}_1, \mathbf{t}_2, \cdots, \mathbf{t}_{L_x}]$, such that $\mathbf{T}_t \in \mathbb{R}^{n \times L_x}$, where n is the number of sensors and L_x is the length of the time segment.

The temporal embedding module consists of an encoder and a decoder. For the encoder part, we input $\mathbf{T}_t$ into it for learning enhanced temporal embedding $\mathbf{U}_t$. Specifically, we first use the multi-head attention mechanism to calculate the attention matrices between $\mathbf{T}_t$ and itself for enhancing the temporal patterns among different sensors by information sharing. Considering that the calculation process in each head is the same, we take $head_1$ as an example to illustrate. To obtain the self-attention matrix $\text{Attn}(\mathbf{T}_t, \mathbf{T}_t)$, we input $\mathbf{T}_t$ into $head_1$, which can be formulated as follows,

$$\text{Attn}(\mathbf{T}_t, \mathbf{T}_t) = softmax(\frac{(\mathbf{T}_t \cdot \mathbf{W}_t^Q)(\mathbf{T}_t \cdot \mathbf{W}_t^K)^\top}{\sqrt{L_x}}) \cdot (\mathbf{T}_t \cdot \mathbf{W}_t^V) \quad (1)$$

where $\mathbf{W}_t^K \in \mathbb{R}^{L_x \times d}$, $\mathbf{W}_t^Q \in \mathbb{R}^{L_x \times d}$, and $\mathbf{W}_t^V \in \mathbb{R}^{L_x \times d}$ are the weight matrix for "key", "query" and "value" embeddings; $\sqrt{L_x}$ is the scaling factor. Assuming that we have h heads, we concatenate the learned attention matrix together in order to capture the temporal patterns of monitoring data from different perspectives. The calculation process can be defined as follows:

$$\mathbf{T}_t' = \text{Concat}(\text{Attn}_t^1, \text{Attn}_t^2, \cdots, \text{Attn}_t^h) \cdot \mathbf{W}_t^O \tag{2}$$

where $\mathbf{W}_t^O \in \mathbb{R}^{hd \times d_{\text{model}}}$ is the weight matrix and $\mathbf{T}_t' \in \mathbb{R}^{n \times d_{\text{model}}}$. After that, we input $\mathbf{T}_t'$ into a fully connected feed-forward network constructed by two linear layers to obtain the enhanced embedding $\mathbf{U}_t \in \mathbb{R}^{n \times d_{\text{model}}}$. The calculation process can be defined as follows:

$$\mathbf{U}_t = \mathbf{T}_t' + \text{Relu}(\mathbf{T}_t' \cdot \mathbf{W}_t^1 + \mathbf{b}_t^1) \cdot \mathbf{W}_t^2 + \mathbf{b}_t^2 \tag{3}$$

where $\mathbf{W}_t^1$ and $\mathbf{W}_t^2$ are the weight matrices of shape $\mathbb{R}^{d_{\text{model}} \times d_{\text{model}}}$; $\mathbf{b}_t^1$ and $\mathbf{b}_t^2$ are bias items of shape $\mathbb{R}^{n \times d_{\text{model}}}$.

For the decoder part, we input the learnt embedding $\mathbf{U}_t$ into a prediction layer to predict the monitoring value of the future time segment. The prediction process can be defined as follows:

$$\check{\mathbf{T}}_{t+1} = \mathbf{U}_t \cdot \mathbf{W}_t^p + \mathbf{b}_t^p \tag{4}$$

where $\check{\mathbf{T}}_{t+1} \in \mathbb{R}^{n \times L_x}$ is the prediction value of the next time segment; $\mathbf{W}_t^p \in \mathbb{R}^{d_{\text{model}} \times L_x}$ is the weight matrix and $\mathbf{b}_t^p \in \mathbb{R}^{n \times L_x}$ is the bias item. During the optimization process, we minimize the difference between the prediction $\check{\mathbf{T}}_{t+1}$ and the real monitoring value $\mathbf{T}_{t+1}$. The optimization objective can be defined as follows

$$\min \sum_{t=1}^{L_x} ||\mathbf{T}_{t+1} - \check{\mathbf{T}}_{t+1}||^2 \tag{5}$$

When the model converges, we have preserved temporal patterns of monitoring data in the temporal embedding $\mathbf{U}_t$.

3.3 Generating Dynamic Weighted Attributed Graphs

In CPS, different sensors connect with each other, forming a sensor-sensor graph. As a result, the malfunctioning effects of system abnormality may propagate over time following the graph structure. But, the sensor-sensor influence is not static and may vary as the monitoring data changes are caused by system anomaly events. To capture such dynamics, we want to build up weighted attributed graphs using sensor-type information and learnt temporal embeddings. For simplicity, we take the graph stream data of t-th time segment $\mathbf{X}_t$ as an example to illustrate the following calculation process.

Specifically, the adjacency matrix of $\mathbf{X}_t$ is $\mathbf{A} \in \mathbb{R}^{n \times n}$, which reflects the physical connectivity between different sensors. $\mathbf{A}[i, j] = 1$ when sensor i and j are directly connected and 0 otherwise. From Sect. 3.2, we obtain the temporal embedding $\mathbf{U}_t \in \mathbb{R}^{n \times d_{model}}$, each row of which represents the temporal embedding for each sensor. We assume that the sensors belonging to the same type have similar changing patterns when confronted with system anomaly events. Thus, we want to capture this characteristic by integrating sensor type information into the adjacency matrix. We calculate the sensor type embedding by averaging the temporal embedding of sensors belonging to the type. After that, we construct a type-type similarity matrix $\mathbf{C}_t \in \mathbb{R}^{k \times k}$ by calculating the cosine similarity between each pair of sensor types, k being the number of sensor types. Moreover, we construct the similarity matrix $\check{\mathbf{C}}_t \in \mathbb{R}^{n \times n}$ by mapping $\mathbf{C}_t$ to each element position of $\mathbf{A}$. For instance, if sensor 1 belongs to type 2 and sensor 2 belongs to type 3, we update $\check{\mathbf{C}}_t[1, 2]$ with $\mathbf{C}_t[2, 3]$. We then introduce the dynamic property to the adjacency matrix $\mathbf{A}$ through element-wise multiplication between $\mathbf{A}$ and $\check{\mathbf{C}}_t$. Each temporal embedding of this time segment is mapped to the weighted graph as the node attributes according to sensor information. The obtained weighted attributed graph $\mathcal{G}_t$ contains all spatial-temporal information of CPS for the t-th time segment. The topological influence of this graph may change over time.

3.4 Representation Learning for Weighted Attributed Graph

To make the outlier detection model easily comprehend the information of $\mathcal{G}_t$, we develop a representation learning module based on variational graph autoencoder (VGAE). For simplicity, we use $\mathcal{G}_t$ to illustrate the representation learning process. For $\mathcal{G}_t = (\mathcal{V}, \tilde{\mathcal{E}}_t, \mathbf{U}_t)$, the adjacency matrix is $\tilde{\mathbf{A}}_t$ made up by $\mathcal{V}$ and $\tilde{\mathcal{E}}_t$, and the feature matrix is $\mathbf{U}_t$.

Specifically, this module follows the encoder-decoder paradigm. The encoder includes two Graph Convolutional Network(GCN) layers. The first GCN layer takes $\mathbf{U}_t$ and $\tilde{\mathbf{A}}_t$ as inputs and outputs a lower dimensional feature matrix $\hat{\mathbf{U}}_t$. The calculation process can be represented as follows:

$$\hat{\mathbf{U}}_t = \text{Relu}(\hat{\mathbf{D}}_t^{-1/2} \tilde{\mathbf{A}}_t \hat{\mathbf{D}}_t^{-1/2} \mathbf{U}_t \tilde{\mathbf{W}}_0) \tag{6}$$

where $\hat{\mathbf{D}}_t$ is the diagonal degree matrix of $\mathcal{G}_{\sqcup}$ and $\tilde{\mathbf{W}}_0$ is the weight matrix of the first GCN layer. The second GCN layer estimates the distribution of the graph embeddings. Assuming that such embeddings conform to the normal distribution $\mathcal{N}(\boldsymbol{\mu}_t, \boldsymbol{\delta}_t)$, we need to estimate the mean $\boldsymbol{\mu}_t$ and variance $\boldsymbol{\delta}_t$ of the distribution. Thus, the encoding process of the second GCN layer can be formulated as follows:

$$\boldsymbol{\mu}_t, log(\boldsymbol{\delta}_t^2) = \text{Relu}(\hat{\mathbf{D}}_t^{-1/2} \mathbf{A}_t \hat{\mathbf{D}}_t^{-1/2} \hat{\mathbf{U}}_t \tilde{\mathbf{W}}_1) \tag{7}$$

where $\tilde{\mathbf{W}}_1$ is the weight matrix of the second GCN layer. Then, we use the reparameterization technique to mimic the sample operation to obtain the graph embedding $\mathbf{r}_t$, which can be represented as follows:

$$\mathbf{r}_t = \boldsymbol{\mu}_t + \boldsymbol{\delta}_t \times \boldsymbol{\epsilon}_t \tag{8}$$

where $\boldsymbol{\epsilon}_t$ is the random variable vector, which is sampled from $\mathcal{N}(0, I)$. Here, $\mathcal{N}(0, I)$ represents the high-dimensional standard normal distribution.

The decoder part aims to reconstruct the adjacency matrix of the graph using $\mathbf{r}_t$, which can be defined as follows:

$$\hat{\mathbf{A}}_t = \sigma(\mathbf{r}_t \mathbf{r}_t^\top) \tag{9}$$

where $\hat{\mathbf{A}}_t$ is the reconstructed adjacency matrix and $\mathbf{r}_t \mathbf{r}_t^\top = ||\mathbf{r}_t||\;||\mathbf{r}_t^\top||\cos\theta$.

During the optimization process, we aim to minimize two objectives: 1) the divergence between the prior embedding distribution $\mathcal{N}(0, I)$ and the estimated embedding distribution $\mathcal{N}(\boldsymbol{\mu}_t, \boldsymbol{\delta}_t)$; 2) the difference between the adjacency matrix $\mathbf{A}_t$ and the reconstructed adjacency matrix $\tilde{\mathbf{A}}_t$; Thus, the optimization objective function is as follows:

$$\min \sum_{t=1}^{T} \underbrace{KL[q(\mathbf{r}_t|\mathbf{U}_t, \mathbf{A}_t)||p(\mathbf{r}_t)]}_{\text{KL divergance between } q(.) \text{ and } p(.)} + \overbrace{||\mathbf{A_t} - \tilde{\mathbf{A}}_\mathbf{t}||^2}^{\text{Loss between } \mathbf{A}_t \text{ and } \tilde{\mathbf{A}}_t} \tag{10}$$

where KL refers to the Kullback-Leibler divergence; $q(.|.)$ is the estimated embedding distribution and $p(.)$ is the prior embedding distribution. When the model converges, the graph embedding $\mathbf{r}_t \in \mathbb{R}^{n \times d_{\text{emb}}}$ contains spatio-temporal patterns of the monitoring data for the t-th time segment.

3.5 One-Class Detection with SVDD

Considering the sparsity issue of labeled anomaly data in CPS, anomaly detection demands unsupervised setting. Inspired by deep SVDD [14], we aim to learn a hypersphere that encircles most of the normal data, with data samples located beyond it being anomalous. Due to the complex nonlinear relations among the monitoring data, we use deep neural networks to approximate this hypersphere.

Specifically, through the above procedure, we collecte the spatiotemporal embedding of all time segments, denoted by $[\mathbf{r}_1, \mathbf{r}_2, \cdots, \mathbf{r}_T]$. We input them into multi-layer neural networks to estimate the non-linear hypersphere. Our goal is to minimize the volume of this data-enclosing hypersphere. The optimization objective can be defined as follows:

$$\min_{\mathcal{W}} \quad \underbrace{\frac{1}{n} \sum_{t=1}^{T} ||\phi(\mathbf{r}_t; \mathcal{W}) - c||^2}_{\substack{\text{Average sum of weights, using} \\ \text{squared error, for all normal} \\ \text{training instances (from T segments)}}} + \overbrace{\frac{\lambda}{2} ||\mathcal{W}||_F^2}^{\text{Regularization item}} \tag{11}$$

where $\mathcal{W}$ is the set of weight matrix of each neural network layer; $\phi(\mathbf{r}_t; \mathcal{W})$ maps $\mathbf{r}_t$ to the non-linear hidden representation space; c is the predefined hypersphere center; λ is the weight decay regularizer. The first term of the equation aims to find the most suitable hypersphere that has the closest distance to the center c. The second term is to reduce the complexity of $\mathcal{W}$, which avoids overfitting. As the model converges, we get the network parameter for a trained model, $\mathcal{W}^*$.

During the testing stage, given the embedding of a test sample $\mathbf{r}_o$, we input it into the well-trained neural networks to get the new representation. Then, we calculate the anomaly score of the sample based on the distance between it and the center of the hypersphere. The process can be formulated as follows:

$$s(\mathbf{r}_o) = ||\phi(\mathbf{r}_o; \mathcal{W}^*) - c||^2 \tag{12}$$

After that, we compare the score with our predefined threshold to assess the abnormal status of each time segment in CPS.

4 Experiments

We conduct extensive experiments to validate the efficacy and efficiency of our framework (DGS-SVDD) and the necessity of each technical component.

4.1 Experimental Settings

Data Description. We adopt the SWaT dataset [11], from the Singapore University of Technology and Design in our experiments. This dataset was collected from a water treatment testbed containing 51 sensors and actuators. The collection process lasted 11 days. The system status was normal for the first 7 days and in the final 4 days, it was attacked by a cyber-attack model. The statistical information of the dataset is shown in Table 1. Our goal is to detect attack anomalies as precisely as feasible. The model is trained using only normal data and after the training phase, we validate our model's performance by predicting the status of the testing data containing both normal and anomalous data.

Table 1. Statistics of SWaT Dataset

Data Type	Feature Number	Total Items	Anomaly Number	Normal/Anomaly
Normal	51	496800	0	-
Anomalous	51	449919	53900	7:1

Evaluation Metrics. We evaluate the model performance in terms of precision, recall, area under the receiver operating characteristic curve (ROC/AUC), and F1-score. We adopt the point-adjust way to calculate these metrics. In particular, abnormal observations typically occur in succession to generate anomaly segments and an anomaly alert can be triggered inside any subset of a real window for anomalies. Therefore, if one of the observations in an actual anomaly segment is detected as abnormal, we would consider the time points of the entire segment to have been accurately detected.

Baseline Models. To make the comparison objective, we input the spatial-temporal embedding vector $\mathbf{r}_t$ into baseline models instead of the original data.

There are seven baselines in our work: **KNN** [12]: calculates the anomaly score of each sample according to the anomaly situation of its K nearest neighborhoods. **Isolation-Forest** [8]: estimates the average path length (anomaly score) from the root node to the terminating node for isolating a data sample using a collection of trees.**LODA** [13]: collects a list of weak anomaly detectors to produce a stronger one. LODA can process sequential data flow and is robust to missing data. **LOF** [2]: measures the anomalous status of each sample based on its local density. If the density is low, the sample is abnormal; otherwise, it is normal. **ABOD** [6]: is an angle-based outlier detector. If a data sample is located in the same direction of more than K data samples, it is an outlier; otherwise it is normal data. **OC-SVM** [9]: finds a hyperplane to divide normal and abnormal data through kernel functions. **GANomaly** [1]: utilizes an encoder-decoder-encoder architecture. It evaluates the anomaly status of each sample by calculating the difference between the output embedding of two encoders.

Table 2. Experimental Results on SWaT dataset

Method	Precision (%)	Recall (%)	F1-score (%)	AUC (%)
OC-SVM	34.11	68.23	45.48	75
Isolation-Forest	35.42	81.67	49.42	80
LOF	15.81	93.88	27.06	63
KNN	15.24	96.77	26.37	61
ABOD	14.2	**97.93**	24.81	58
GANomaly	42.12	67.87	51.98	68.64
LODA	75.25	38.13	50.61	67.1
DGS-SVDD	**94.17**	82.33	**87.85**	**87.96**

4.2 Experimental Results

Overall Performance. Table 2 shows experimental results on the SWaT dataset, with the best scores highlighted in **bold**. As can be seen, DGS-SVDD outperforms other baseline models in the majority of evaluation metrics. Compared with the best baseline, DGS-SVDD improves precision by 19%, F1-score by 36% and AUC by 8%. The underlying driver for this success is that DGS-SVDD can capture long-delayed temporal patterns and dynamic sensor-sensor influences in CPS. Another interesting observation is that the detection performance of distance-based or angle-based outlier detectors is poor. A possible reason is that these geometrical measurements are vulnerable to high-dimensional data samples.

Ablation Study. To study the individual contribution of each component of DGS-SVDD, we perform ablation studies, the findings of which are summarized in Table 3 where **bold** indicates the best score. We build four variations of the DGS-SVDD model: 1) We feed unprocessed raw data into SVDD; 2) We only capture temporal patterns; 3) We capture the dynamics of sensor-sensor impact and spatial patterns in CPS; 4) We capture spatio-temporal patterns but discard the dynamics of sensor-sensor influence. We can find that DGS-SVDD outperforms its variants by a significant margin. The observation validates that each technical component of our work is indispensable. Removing the temporal embedding module dramatically degrades the detection performance which renders this module the highest significance. Results from the final experiment show that capturing the dynamics of sensor-sensor influence really boosts model performance.

Table 3. Ablation Study of DGS-SVDD

Method			Precision (%)	Recall (%)	F1-score (%)	AUC (%)
Transformer-based Temporal Embedding Module	Weighted Attributed Graph Generator	VGAE-based Spatiotemporal Embedding Module				
✗	✗	✗	4.61	12.45	6.74	18.55
✓	✗	✗	69.98	64.75	67.26	78.14
✗	✓	✓	12.16	**99.99**	21.68	18.22
✓	✗	✓	87.79	76.68	81.86	82.45
✓	✓	✓	**94.17**	82.33	**87.75**	**87.96**

Robustness Check and Parameter Sensitivity. Figure 2 shows the experimental results for robustness check and parameter sensitivity analysis. To check the model's robustness, we train DGS-SVDD on different percentages of the training data, starting from 10% to 100%. Figure 2(a) shows that DGS-SVDD is stable when confronted with different size of training data. In addition, we vary the dimension of the final spatial-temporal embedding in order to check its impacts. From Fig. 2(b) and 2(c), we can find that DGS-SVDD is barely sensitive to the sliding window length and dimension of the spatio-temporal embeddings. This observation validates that DGS-SVDD is robust to the dimension parameters. A possible reason is that our representation learning module has sufficiently captured spatio-temporal patterns of monitoring data for anomaly detection.

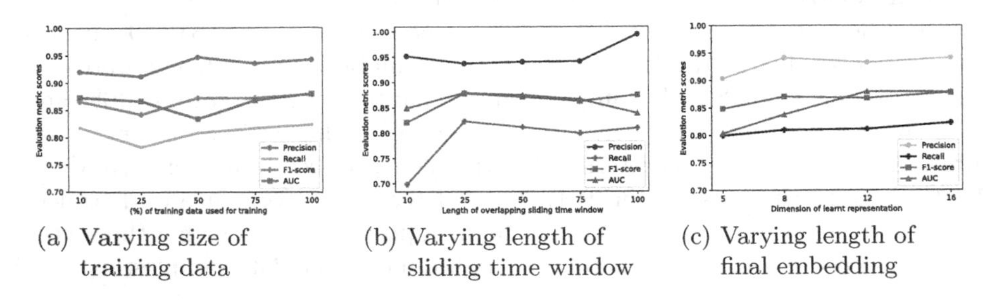

(a) Varying size of training data

(b) Varying length of sliding time window

(c) Varying length of final embedding

Fig. 2. Experimental results for robustness check and parameter sensitivity

Study of Time Cost. We conduct six folds cross-validation to evaluate the time costs of different models. Figure 3 illustrates the comparison results. We can find that DGS-SVDD can be trained at a time competitive with simple models like OC-SVM or LOF while outperforming them by a huge margin as seen from Table 2. This shows that DGS-SVDD effectively learns the representation of each time segment of the graph stream data. Another important observation is that the testing time of DGS-SVDD is consistent with the simpler baselines. A potential reason is that the network parameter $\mathcal{W}^*$, as discussed in Sect. 3.5, completely characterizes our one-class classifier. This allows fast testing by simply evaluating the network ϕ with learnt parameters $\mathcal{W}^*$.

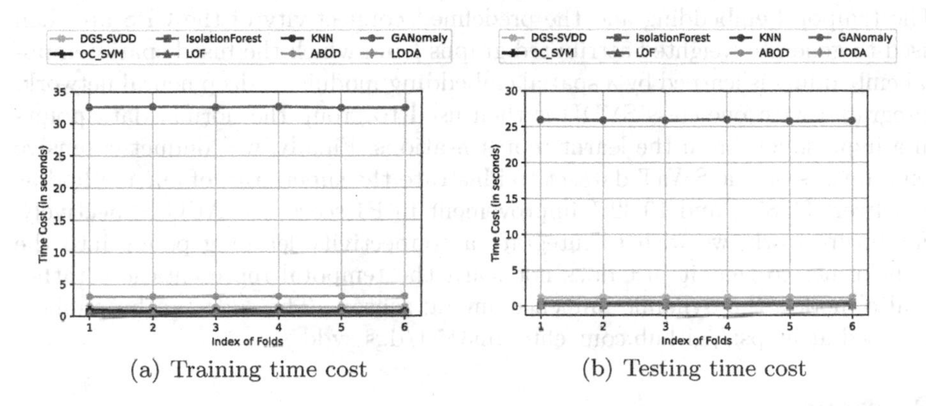

(a) Training time cost (b) Testing time cost

Fig. 3. Comparison of different models in terms of training and testing time cost

5 Related Work

Anomaly Detection in Cyber-Physical Systems. Numerous existing literature have studied the exploitation of temporal and spatial relationships in data streams from CPS to detect anomalous points. For instance, [5] adopts a convolutional layer as the first layer of a Convolutional Neural Network to obtain correlations of multiple sensors in a sliding time window. Further, the extracted features are fed to subsequent layers to generate output scores. [7] proposed a GAN-based framework to capture the spatial-temporal correlation in multidimensional data. Both generator and discriminator are utilized to detect anomalies by reconstruction and discrimination errors.

Outlier detection with Deep SVDD. After being introduced in [14], deep SVDD and its many variants have been used for deep outlier detection. [18] designed *deep structure preservation SVDD* by integrating deep feature extraction with the data structure preservation. [20] proposed a *Deep SVDD-VAE*, where VAE is used to reconstruct the input sequences while a spherical discriminative boundary is learned with the latent representations simultaneously, based on SVDD. Although these models have been successfully applied to detect anomalies in the domain of computer vision, this domain lacks temporal and spatial dependencies prevalent in graph stream data generated from CPS.

6 Conclusion

We propose DGS-SVDD, a structured anomaly detection framework for cyber-physical systems using graph stream data. To this end, we integrate spatiotemporal patterns, modeling dynamic characteristics, deep representation learning, and one-class detection with SVDD. Transformer-based encoder-decoder architecture is used to preserve the temporal dependencies within a time segment. The temporal embedding and the predefined connectivity of the CPS are then used to generate weighted attributed graphs from which the fused spatiotemporal embedding is learned by a spatial embedding module. A deep neural network, integrated with one-class SVDD is then used to group the normal data points in a hypersphere from the learnt representations. Finally, we conduct extensive experiments on the SWaT dataset to illustrate the superiority of our method as it delivers 35.87% and 19.32% improvement in F1-score and AUC respectively. For future work, we wish to integrate a connectivity learning policy into the transformer so that it just does not learn the temporal representation, rather it also models the dynamic influence among sensors. The code can be publicly accessed at https://github.com/ehtesam3154/dgs_svdd.

References

1. Akcay, S., Atapour-Abarghouei, A., Breckon, T.P.: GANomaly: semi-supervised anomaly detection via adversarial training. In: Jawahar, C.V., Li, H., Mori, G., Schindler, K. (eds.) ACCV 2018. LNCS, vol. 11363, pp. 622–637. Springer, Cham (2019). https://doi.org/10.1007/978-3-030-20893-6_39
2. Breunig, M.M., Kriegel, H.P., Ng, R.T., Sander, J.: LOF: identifying density-based local outliers. In: Proceedings of the 2000 ACM SIGMOD International Conference on Management of Data, pp. 93–104 (2000)
3. Bergal, J.: Florida hack exposes danger to water systems (2021). https://www.pewtrusts.org/en/research-and-analysis/blogs/stateline/2021/03/10/florida-hack-exposes-danger-to-water-systems
4. Kipf, T.N., Welling, M.: Variational graph auto-encoders. arXiv preprint arXiv:1611.07308 (2016)
5. Kravchik, M., Shabtai, A.: Detecting cyber attacks in industrial control systems using convolutional neural networks. In: Proceedings of the 2018 Workshop on Cyber-Physical Systems Security and Privacy, pp. 72–83 (2018)
6. Kriegel, H.P., Schubert, M., Zimek, A.: Angle-based outlier detection in high-dimensional data. In: Proceedings of the 14th ACM SIGKDD International Conference on Knowledge Discovery and Data Mining, pp. 444–452 (2008)
7. Li, D., Chen, D., Jin, B., Shi, L., Goh, J., Ng, S.-K.: MAD-GAN: multivariate anomaly detection for time series data with generative adversarial networks. In: Tetko, I.V., Kůrková, V., Karpov, P., Theis, F. (eds.) ICANN 2019. LNCS, vol. 11730, pp. 703–716. Springer, Cham (2019). https://doi.org/10.1007/978-3-030-30490-4_56

8. Liu, F.T., Ting, K.M., Zhou, Z.H.: Isolation forest. In: 2008 Eighth IEEE International Conference on Data Mining, pp. 413–422. IEEE (2008)
9. Manevitz, L.M., Yousef, M.: One-class SVMs for document classification. J. Mach. Learn. Res. **2**(Dec), 139–154 (2001)
10. Martí, L., Sanchez-Pi, N., Molina, J.M., Garcia, A.C.B.: Anomaly detection based on sensor data in petroleum industry applications. Sensors **15**(2), 2774–2797 (2015)
11. Mathur, A.P., Tippenhauer, N.O.: Swat: a water treatment testbed for research and training on ICS security. In: 2016 International Workshop on Cyber-Physical Systems for Smart Water Networks (CySWater), pp. 31–36. IEEE (2016)
12. Peterson, L.E.: K-nearest neighbor. Scholarpedia **4**(2), 1883 (2009)
13. Pevný, T.: LODA: lightweight on-line detector of anomalies. Mach. Learn. **102**(2), 275–304 (2016)
14. Ruff, L., et al.: Deep one-class classification. In: International Conference on Machine Learning, pp. 4393–4402. PMLR (2018)
15. Vaswani, A., et al.: Attention is all you need. CoRR abs/1706.03762 (2017). http://arxiv.org/abs/1706.03762
16. Wang, D., et al.: Hierarchical graph neural networks for causal discovery and root cause localization. arXiv preprint arXiv:2302.01987 (2023)
17. Wang, D., Wang, P., Zhou, J., Sun, L., Du, B., Fu, Y.: Defending water treatment networks: exploiting spatio-temporal effects for cyber attack detection. In: 2020 IEEE International Conference on Data Mining (ICDM), pp. 32–41. IEEE (2020)
18. Zhang, Z., Deng, X.: Anomaly detection using improved deep SVDD model with data structure preservation. Pattern Recogn. Lett. **148**, 1–6 (2021)
19. Zhou, X., Liang, W., Shimizu, S., Ma, J., Jin, Q.: Siamese neural network based few-shot learning for anomaly detection in industrial cyber-physical systems. IEEE Trans. Industr. Inf. **17**(8), 5790–5798 (2020)
20. Zhou, Y., Liang, X., Zhang, W., Zhang, L., Song, X.: VAE-based deep SVDD for anomaly detection. Neurocomputing **453**, 131–140 (2021)

Texts, Web, Social Media

Words Can Be Confusing: Stereotype Bias Removal in Text Classification at the Word Level

Shaofei Shen[1], Mingzhe Zhang[1], Weitong Chen[2](✉), Alina Bialkowski[1], and Miao Xu[1,3]

[1] The University of Queensland, Brisbane, Australia
{shaofei.shen,mingzhe.zhang,alina.bialkowski,miao.xu}@uq.edu.au
[2] University of Adelaide, Adelaide, Australia
t.chen@adelaide.edu.au
[3] RIKEN, Tokyo, Japan

Abstract. Text classification is a widely used task in natural language processing. However, the presence of stereotype bias in text classification can lead to unfair and inaccurate predictions. Stereotype bias is particularly prevalent in words that are unevenly distributed across classes and are associated with specific categories. This bias can be further strengthened in pre-trained models on large natural language datasets. Prior works to remove stereotype bias have mainly focused on specific demographic groups or relied on specific thesauri without measuring the influence of stereotype words on predictions. In this work, we present a causal analysis of how stereotype bias occurs and affects text classification, and propose a framework to mitigate stereotype bias. Our framework detects potential stereotype bias words using SHAP values and alleviates bias in the prediction stage through a counterfactual approach. Unlike existing debiasing methods, our framework does not rely on existing stereotype word sets and can dynamically evaluate the influence of words on stereotype bias. Extensive experiments and ablation studies show that our approach effectively improves classification performance while mitigating stereotype bias.

Keywords: Text Mining · Text Classification · Stereotype Bias · Causal Inference

1 Introduction

Text classification tasks in natural language processing (NLP) can be influenced by **stereotype words**, which are words associated with specific categories or groups based on emotions, politics, or demographic features [1]. Studies have shown that stereotype words such as *pink* and *blue* are often associated with girls and boys, respectively [2]. Additionally, words such as *Varicella* or *Alzheimer* are associated with specific age groups. The distribution of stereotype words across different document categories can result in **stereotype bias** in classification models trained on datasets containing these words [14]. This bias can have significant

© The Author(s) 2023
H. Kashima et al. (Eds.): PAKDD 2023, LNAI 13938, pp. 99–111, 2023.
https://doi.org/10.1007/978-3-031-33383-5_8

implications, particularly in sentiment analysis, where the results can influence decision-making processes.

Stereotype bias in text classification is caused by oversimplified correlations between stereotype words and text categories. However, in text classification, the semantic relationships of the words should be the basis for classification, not the existence of specific words [15]. In this paper, we focus on the problem of detecting and alleviating stereotype bias in text classification. To solve this problem, three challenges need to be addressed. Firstly, identifying the set of stereotype words that can potentially introduce stereotype bias is critical [18]. Such words are usually domain-specific and difficult to identify universally across different domains, given that document classification tasks are often domain-specific, such as in movie reviews or medical research. Secondly, accurately estimating the degree to which a word contributes to stereotype bias in predictions for a document is challenging. This is especially true for complex deep models, and the same words may contribute differently in different documents due to the interdependence between words. Simply removing a particular stereotype word may not eliminate stereotype bias. Thirdly, aside from stereotype bias in classifiers, there may be stereotype bias in widely-used pre-trained word embedding models [24]. Accessing the training data of these models to detect stereotype words is difficult, and it is challenging to measure and reduce stereotype bias without the original training data. Addressing these challenges is essential for developing more inclusive NLP models that are free from stereotype bias.

Previous works have attempted to alleviate various forms of stereotype bias, with a particular focus on removing gender bias in language models [4,24,25]. In [25], the authors reduced gender bias through data augmentation using an occupation word set associated with gender bias. Similarly, [24] proposed reducing gender bias in the word embedding stage. [4] demonstrated bias amplification in language models and used posterior regularization to address it. Other forms of bias, including label bias [17], context-word bias [17], race bias [6], demographic bias [10], and implicit bias [9], have also received attention. However, these works typically rely on fixed word sets considered as stereotype words derived from a different domain, which may not be effective in the current domain. While some works [17] have proposed selecting stereotype words, their selection strategy is based solely on the TextRank score [13], without considering other important metrics, such as word imbalance. Moreover, these works assume that all words contribute equally as stereotype bias creators, without taking into account the fact that different words may contribute differently in different documents.

In this work, a causal graph is constructed to analyze how the stereotype bias from texts and pre-trained models affects classifications, building on previous works. Based on this causal graph, a novel framework is proposed for detecting and alleviating stereotype bias using a counterfactual method to address the three challenges mentioned earlier. Initially, word distribution statistics and word importance (i.e. SHAP value) in prediction are used to determine a dynamic stereotype word set. Subsequently, a fusion model is adopted to learn the relationship between semantics, stereotype words, and text categories. During the prediction stage, the counterfactual approach was used to alleviate the bias from stereotype words. In contrast to previous works, this study utilizes real-time word importance in document-level predictions and domain-level word distributions to identify stereotype words in different document domains, and focuses more on semantically relevant words instead of context words [17]. In summary, the contributions of this work are three-fold.

- We investigate the stereotype bias in text classification from a causal perspective, analyzing how stereotype words from both texts and pre-trained models influence classification results.
- We propose a novel framework to detect and remove stereotype bias, which involves detecting stereotype words based on word importance and word distribution statistics, training a fusion model to learn the relationship between semantics, stereotype words, and text categories, and utilizing a counterfactual approach for unbiased prediction. To the best of our knowledge, this is the first work that systematically addresses the stereotype bias caused by semantic words without relying on a prior thesaurus.
- We conduct extensive experiments to demonstrate the effectiveness of our framework in achieving unbiased classification, and we compare our results with state-of-the-art approaches for unbiased text classification [17].

Related Work. The word-level bias in language models has attracted the growing interest of researchers. Apart from the aforementioned works on gender bias [24, 25], the most recent work handled gender bias via an adaptation perspective and treated gender groups as different domains [3]. Apart from the gender bias, other works also focus on intended bias [22] and the stereotype bias generated from words [3,17,25], word embeddings [24], and pre-trained models [14]. These works inspired us to model the causal relationship among texts, sources of stereotypes bias, and predictions.

As for the debiasing methods, the counterfactual approach is attracting increasing attention. The counterfactual approach utilizes a dummy value as the counterfactual and aims to remove the indirect effect of confounders on the treatment variables [16]. [17] proposed to use a counterfactual method to remove the bias from imbalanced labels and semantically-irrelevant words. [21] removed the bias in fake news classification and [20] mitigated the bias in text understanding and hypothesis inference via counterfactual debiasing. As discussed in the aforementioned challenges, training data is generally unavailable or inaccessible in pre-trained models. In this case, the counterfactual approach is applicable to mitigate the potential stereotype bias in pre-trained models and text classifiers.

2 Methodology

2.1 Problem Formulation

Let $\mathbf{D}$ and $\mathbf{Y}$ denote the text documents and text categories, respectively. Considering a pre-trained word embedding model $\mathbf{h}$ and classification model $\mathbf{g}$, the goal of the text classification is to train the classification model $\mathbf{g}$ to maximize the classification accuracy of $(\mathbf{g} \circ \mathbf{h})(\mathbf{D})$. In the ideal view of the training process, the semantics of documents can be learned through a two-stage model, which first classifies semantics and then performs text classification. Then the semantics will be the main basis of the text classification [14]. However, when training from a pre-trained model, the text classification model will inherit any existing stereotype relationships of the pre-trained model.

To construct the causal relationship among these variables, we analyse the word-level stereotypes first. The words in one document can be divided into three groups: the semantic-irrelevant words which have no contribution to the semantics and the further text category predictions, the normal words which are related to the semantics but will not involve stereotypes in the predictions, and the stereotype words that affect the semantics and introduce the stereotype bias in the predictions meantime. In addition, the pre-trained word embedding model may also involve stereotype bias due to the pre-training dataset. Figure 1(a) demonstrates the causal relationship among these groups of words, semantics, and text category. Ideally, the pre-trained word embedding M should also contribute to the semantics X and be the confounder of causal path $X \rightarrow Y$. However, the causal effect from M on X is hard to estimate and we remove the path $M \rightarrow X$ for easier implementation in the experiments. Considering all the sources of bias, the prediction results can be denoted as:

$$Y_{x(d,s),m,s} = Y(X = x(d,s), M = m, S = s) \tag{1}$$

where Y is a prediction function based on word embedding m, normal words d, stereotype words s, and the specific semantics $x(d, s)$.

The unbiased prediction requires using the semantic as the only direct causal variable to the predictions. Then we need to remove the causal effect from the other two causal variables: stereotype bias from pre-trained model $\mathbf{M}$ and stereotype words $\mathbf{S}$ as shown in Fig. 1(b). Then the debiasing goal can be denoted as:

$$Y_{x(d,s)} = Y(X = x(d,s)) \tag{2}$$

where Y is only decided on the semantic $x(d, s)$ of the texts.

Based on the analysis and causal relationship in Fig. 1, our framework contains three stages: stereotype word set construction, fusion model training to learn the causal effect from the sources of bias, and unbiased prediction.

2.2 Stereotype Words Detection

To mitigate the stereotype bias from the training documents, the first stage of our framework is to select the potential words that may lead to stereotypes. As Fig. 1 shows, we assume the stereotype words have a direct causal effect on the semantics and predictions at the same time. Therefore, we can focus on the words that contribute to predictions and utilize the **word importance on predictions** to detect stereotype words. [17] proposes to use the TextRank-based method to calculate the word importance in the document. However, TextRank can only select the keywords and does not consider whether these words affect

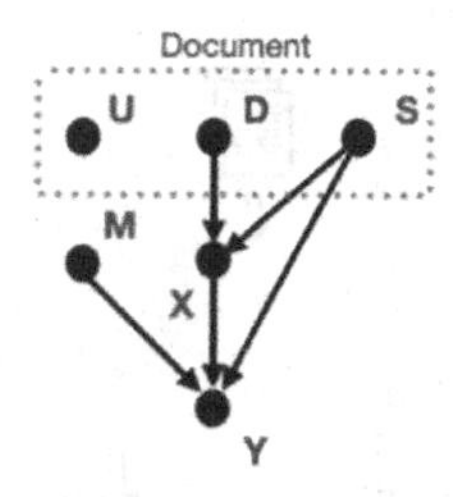

(a) Causal graph after involving stereotypes

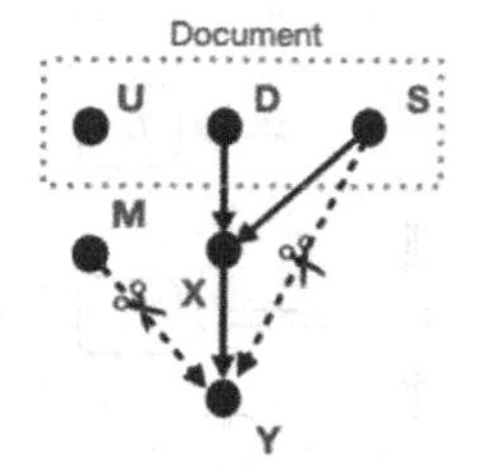

(b) Cut off stereotypes to semantics in training

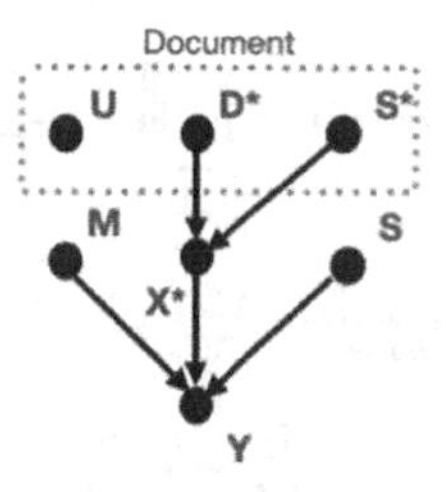

(c) Cut off stereotypes to predictions in test

Fig. 1. Conventional text classification and debiased text classification. The words in the document are composed of three parts: **D** donates the normal words, **S** donates the potential stereotype words from the dataset, and **U** donates the semantic-irrelevant words. **M** denotes the stereotypes in existing pre-trained models, and **X** is the semantic embedding of the document and **Y** is the prediction results of the documents. (a) shows a causal graph after introducing stereotype bias from texts and pre-trained models, (b) illustrates the goal of mitigating stereotype bias: removing the direct causal effect from **S** and **M** to **Y**, and (c) is our proposed method of unbiased predictions via a counterfactual approach. In language models, the causal effects from words **D** and **S** to predictions **Y** needs the mediator variable **X**. Therefore, we focus on the causal effect in path $S \rightarrow Y \leftarrow X$ instead of the path $S \rightarrow Y \leftarrow D$. Moreover, in this causal graph, we remove the path $M \rightarrow X$ for easier implementation.

downstream tasks while we are aiming to get the word importance on the downstream predictions in this work. Therefore, we adopt the post-training SHAP value, which can provide the feature importance (i.e. word importance in this work) for the predictions [7,12] and provide the contribution of each word to the predictions based on the same prediction model. Moreover, another characteristic to select the stereotype words is the word distribution in different classes.

Specifically, as shown in stage 1 of Fig. 2, after the initial training stage, we calculate the SHAP values as the word importance for the words in training data and select the set of words $\mathbf{D} + \mathbf{S}$ which contributes to the predictions. Then to select the potential stereotype words, we calculate the word distributions in each document class and rank them via information entropy:

$$H(w) = -\sum_{c \in C} p(c|w) \log p(c|w) \tag{3}$$

where w is all semantic-relevant words and C is the text category set. Lower $H(w)$ means a more imbalanced distribution in different text classes and a higher potential to involve stereotypes in the predictions. Then we set the proportion of stereotype words as a parameter and select the percentage of data from the ranking of $H(w)$.

2.3 Fusion Model Training

After selecting the potential stereotype words, we build a fusion model to estimate the direct causal effect from the pre-trained model and stereotype words,

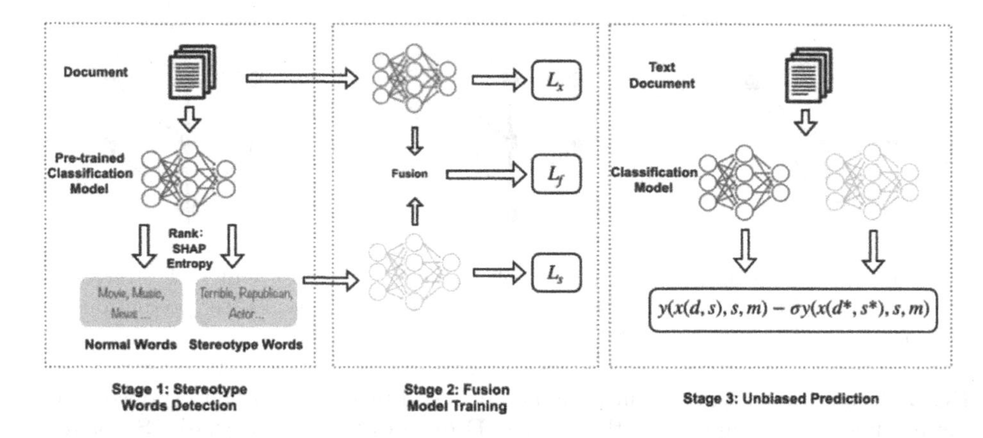

Fig. 2. The proposed framework to mitigate stereotype bias in text classification. This framework contains three stages: stereotype word detection after the first training; fusion model training to learn the causal relationships of stereotype bias and prediction results; unbiased prediction to alleviate stereotype bias. Moreover, the blue and yellow models in the framework indicate the pre-trained and de-biased classification model respectively (Color figure online)

which is shown as $S \rightarrow Y$ and $M \rightarrow Y$ in Fig. 1. Inspired by [20,21], we build two models $\hat{y}_{d,s}$ and $\hat{y}_s$. We use the original texts as input of the first model and predict the corresponding text categories to capture the causal relationships in the paths $S \rightarrow X \leftarrow D$ and $S \rightarrow Y$. This model is trained to learn the causal effect from semantic-relevant words to the semantic and then to the prediction results. To train this model, we use the cross-entropy loss function as shown in

$$L_x = \sum_c y_c \log(\hat{y}_{d,s,c}) \tag{4}$$

where y_c and $\hat{y}_{d,s,c}$ means the ground-truth and prediction probability on the text category c.

Another model $\hat{y}_s$ is used to estimate the causal effect in $S \rightarrow Y$. In this model, we preserve the stereotype words $\mathbf{S}$ and semantic-irrelevant words $\mathbf{U}$ and mask the normal words $\mathbf{D}$ as the model input. The output of the model is still the corresponding text category predictions. Then we optimize the second model by the following loss function:

$$L_s = \sum_c y_c \log(\hat{y}_{s,c}) \tag{5}$$

where $\hat{y}_{s,c}$ means the prediction probability of the word-based classification model on the text category c. Moreover considering that the stereotype influence from the pre-trained model is intrinsic and does not affect the training process, we assume that the overall prediction is a linear combination of stereotype influence $\hat{y}_m$ and the fusion of $\hat{y}_{x(d,s)}$ and $\hat{y}_s$:

$$\begin{aligned} P(Y=y|x(d,s),m,s) &= y(x(d,s),m,s) \\ &= f(\hat{y}_{x(d,s)},\hat{y}_s,\hat{y}_m) \\ &= f(\hat{y}_{x(d,s)},\hat{y}_s)+\hat{y}_m \end{aligned} \tag{6}$$

As for the fusion model of $f(\hat{y}_{x(d,s)},\hat{y}_s)$, we adopt the following fusion strategy to combine the model predictions: $\hat{y}_{x(d,s)} + \alpha\tanh(\hat{y}_s)$, where α is a hyperparameter, σ is sigmoid function, and tanh is the tanh activation function. And the corresponding loss function of this fusion model is set to be:

$$L_f = \sum_c y_c \log(f(\hat{y}_{x(d,s)},\hat{y}_s)) \tag{7}$$

2.4 Unbiased Prediction

The third stage of our framework is to mitigate the stereotype bias from the text category predictions. As we have mentioned in Sect. 3.1, we aim to remove the direct causal effect from the set of stereotype words **S** and pre-trained word embedding model **M**. The total effect (**TE**) stands for all the direct and indirect causal effects of the causal variable on the outcome, which can be denoted as:

$$\text{TE} = P(Y=y|x(d,s),m,s) - P(Y=y|x(d^*,s^*),m^*,s^*) \tag{8}$$

Then the direct causal effect of m and s can be represented by the natural direct effect (**NDE**):

$$\text{NDE} = P(Y=y|x(d^*,s^*),m,s) - P(Y=y|x_{c^*,s^*},m^*,s^*) \tag{9}$$

where d^*,s^*, and m^* represent the counterfactual value of **D**, **S**, and m respectively. Specifically, the counterfactual values d^* and s^* can be obtained by the masked values based on the training dataset and the value of m^* can be set as any value and will not influence the indirect effects shown in (10).

Finally, as shown in Fig. 1 (b), we aim to cut all the direct causal effects from **M** and **S** to the text categories **Y**. Therefore, we use the total indirect effects (**TIE**) to remove the stereotype bias from the pre-trained model and stereotype words:

$$\begin{aligned} \text{TIE} &= \text{TE} - \text{NDE} \\ &= P(Y=y|x(d,s),m,s) - P(Y=y|x(d^*,s^*),m,s) \\ &= y(x(d,s),m,s) - \sigma y(x(d^*,s^*),m,s) \\ &\approx f(\hat{y}_{x(d,s)},\hat{y}_s) - \sigma f(\hat{y}_{x(d^*,s^*)},\hat{y}_s) \end{aligned} \tag{10}$$

where σ is a hyperparameter to control the influence of the stereotype bias on the prediction results.

3 Experiments

3.1 Settings

To validate the effectiveness of our debiasing framework, we conduct experiments on multiple text classification datasets using different classifiers and mitigate the stereotype bias via our proposed framework. Then we compare our results with two state-of-the-art works on bias mitigation [17,22]. In experiments, we concentrate on the effectiveness of our proposed methods on classification results and word-level fairness. The code of the experiments is available[1].

Baseline. We choose three representative text classifiers as the baselines of our framework. The first one is **TextCNN** [5] which is based on the convolutional neural network (CNN) to extract the textual features and the TextCNN requires word embedding as the input, which can utilize pre-trained word embeddings. Another model is **TextRCNN** [8], which uses the bi-directional recurrent networks to capture the contextual information and utilize CNN for future feature extraction. The last one is **RoBERTa** [11], which uses dynamic masking and a larger pre-training set than BERT. RoBERTa can reach better generalization and robustness in text classification tasks. The three models all require pre-trained word embedding as inputs to involve the stereotype bias from the pre-trained model in downstream text classification tasks. To compare with the SOTA debiasing works, we choose two methods: **IPS-Weight** [22] and **CORSAIR** [17] for comparison. IPS-Weight uses the inverse propensity score as the instance weights to reduce the intended bias while CORSAIR removes the bias from the context words and imbalanced classes by removing counterfactual predictions.

Dataset. We conduct experiments on nine text classification datasets. Among these datasets, six datasets are the same as used in [17]: HyperPartisan, Twitter, ARC, SCIERC, Economy, and Parties. In addition, we also use an Amazon product review datasets [23]. We adopt the same pre-processing procedures on these datasets as [17].

Evaluation. We evaluate the framework from two perspectives: classification performance and word-level fairness. Considering the class imbalance of our dataset, we use Macro-F1 to measure the text classification performance. As for the word fairness, we adopt the evaluation framework shown in [19]. For each word in the dataset, we compare the prediction distribution of the data that contains this word with the even distribution and calculate the Jensen-Shannon divergence (JS). We use the average JS of all the words as the fairness metric.

Parameters. Then we use the grid search method to decide the specific value of parameters in the experiments. We set the batch size as 32 for the training and test data and then we use the Adam optimizer with a learning rate of 5e-4 for all three classification models during both the initial training and fusion training stages which have 20 epochs respectively. Then we set the proportion of

[1] https://github.com/DATA-Transpose/StereotypeWords.

the stereotype words as 5%, and set the α in the fusion training stage as 0.1. As for the parameter σ in the unbiased predictions, we search for the best σ from 0 to 2 with a stride of 0.05 based on the validation results. The experiments are conducted on three servers with NVIDIA RTX A5000 GPUs and the results are the average results of three rounds of experiments using different seeds.

3.2 Classification Performance

Table 1 shows our proposed methods' classification performance (Macro F1 score) and the comparison with two SOTA methods. The higher results mean a better classification performance. The rows of **BASELINE** stand for the results without any debiasing methods. The rows of **KEYWORD** and **IMBWORD** represent the results of ablation studies where we regard all the words that have positive contributions to predictions as stereotype words in the experiments of KEYWORD and we mark all the words that have large entropy as stereotype words in the rows of KEYWORD.

From the results, our proposed methods have average improvements of 4.23%, 4.88%, and 4.82% using TextCNN, TextRCNN, RoBERTa from the baselines across the nine datasets. As for the two comparison methods, the improvements of our proposed methods are much more significant and stable across different datasets. Then compared with the results of two ablation studies, the proposed method considering both the word importance and entropy can reach better classification results in most of the datasets and the results are slightly lower than the ablation methods in SCIERC and Parties datasets. Moreover, the results that

Table 1. Classification performances compared with the State-of-the-art methods (%)

Model	Method	HYP	TWI	ARC	SCI	ECO	PRT	AMA
TextCNN	BASELINE	59.63	80.41	38.80	44.25	56.16	57.70	72.71
	IPS	45.48	63.76	13.87	9.79	44.19	57.75	65.60
	CORSAIR	51.20	69.26	17.57	22.06	58.23	55.16	67.33
	KEYWORD	66.12	74.62	48.72	43.13	57.97	57.88	72.84
	IMBWORD	65.95	75.03	41.65	**46.32**	60.08	**58.90**	72.50
	PROPOSED	**66.88**	**82.24**	**52.96**	45.14	**60.67**	57.94	**73.46**
TextRCNN	BASELINE	60.33	69.84	52.08	62.29	56.42	**60.68**	72.45
	IPS	36.89	60.58	11.23	9.79	44.19	52.60	68.57
	CORSAIR	48.88	74.72	22.21	23.03	56.04	57.84	64.19
	KEYWORD	**74.59**	80.08	55.96	62.01	61.91	57.78	**73.50**
	IMBWORD	73.61	77.31	54.24	**63.59**	61.78	59.15	73.28
	PROPOSED	70.55	**82.11**	**58.31**	62.00	**63.51**	58.97	72.83
RoBERTa	BASELINE	65.60	71.18	24.92	26.54	60.36	**64.33**	89.45
	IPS	50.66	70.77	15.88	9.79	44.19	54.12	74.15
	CORSAIR	59.62	68.17	17.17	20.30	57.57	53.64	71.67
	KEYWORD	72.23	74.80	27.52	31.34	64.91	64.25	89.83
	IMBWORD	73.23	76.64	23.35	30.46	**65.46**	63.91	89.67
	PROPOSED	**75.80**	**79.17**	**28.71**	**33.04**	65.45	63.61	**90.33**

Table 2. Word-level fairness of proposed methods

Model	Method	HYP	TWI	ARC	SCI	ECO	PRT	AMA
TextCNN	BASELINE	17.87	19.46	42.35	39.91	18.20	13.80	16.27
	IPS	19.63	20.41	45.18	47.78	21.54	13.14	16.39
	CORSAIR	**16.28**	19.45	40.76	35.67	16.58	**13.00**	**15.82**
	KEYWORD	17.12	18.64	41.73	37.12	19.85	17.41	16.34
	IMBWORD	17.18	**18.27**	41.83	39.37	17.60	16.14	16.16
	PROPOSED	17.21	18.60	**37.00**	**37.11**	**16.31**	13.70	16.44
TextRCNN	BASELINE	19.08	19.11	41.48	37.34	20.00	13.54	16.30
	IPS	**16.05**	19.22	43.32	**35.53**	**15.33**	13.17	16.17
	CORSAIR	16.94	19.14	43.49	37.79	16.97	**13.12**	**16.15**
	KEYWORD	17.50	18.21	41.60	37.23	20.74	20.12	16.22
	IMBWORD	17.48	**16.52**	**41.50**	36.43	20.11	19.88	16.32
	PROPOSED	17.87	18.03	41.89	36.26	20.16	14.45	16.20
RoBERTa	BASELINE	17.60	18.78	43.44	42.75	19.40	14.57	16.41
	IPS	17.43	18.42	42.21	40.36	17.09	14.36	16.48
	CORSAIR	18.58	18.69	46.27	**39.89**	17.36	**13.97**	16.52
	KEYWORD	17.86	18.23	40.81	41.51	**16.93**	14.16	16.34
	IMBWORD	17.73	18.63	**40.70**	41.36	16.93	14.69	16.37
	PROPOSED	**17.17**	**17.90**	41.11	41.47	19.76	14.10	**16.34**

only consider the word importance(KEYWORD) result in a higher Macro F1 score than the results of IMBWORD in all three baseline models, which implies that semantic words are one source of bias in text classification.

3.3 Stereotype Word Fairness

Table 2 shows the word-level fairness of our proposed methods, ablation studies, and two comparison techniques. Considering that the stereotype words are not fixed among different classification models, we calculate the fairness of all the words instead of the stereotype words in the texts in Table 2. Lower results mean better fairness in the prediction results.

Compared with the baselines, our proposed methods can reach average improvements of 1.64, 0.28, and 0.73 respectively across all the datasets. From the results, we can find our proposed methods can reach lower fairness metrics in TextCNN and RoBERTa models in most datasets. The improvements in fairness are not as significant as the improvements in F1 scores because we use the fairness on the whole word set in the documents instead of the stereotype word set. The larger stereotype word set easily leads to a smaller word fairness. Compared with three methods: CORSAIR, KEYWORD, and IMBWORD, our proposed methods select a smaller and more accurate stereotype word set for debiasing and can reach competitive results.

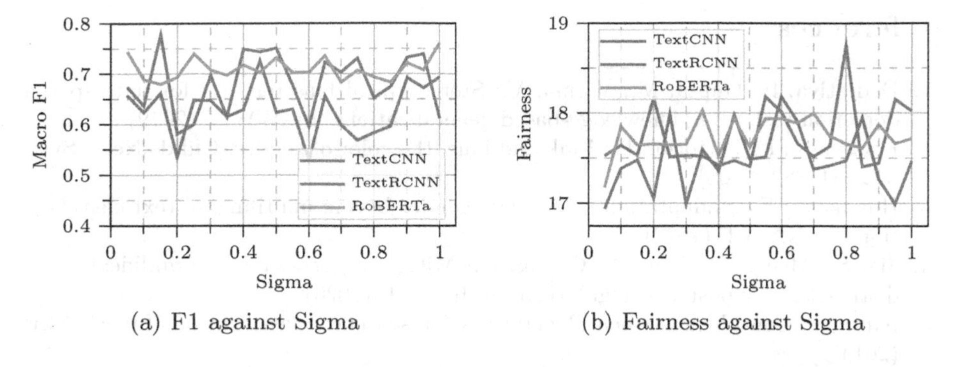

(a) F1 against Sigma (b) Fairness against Sigma

Fig. 3. F1 and Fairness results under different proportion of stereotype words

3.4 Proportion of Stereotype Words

In this section, we implement further experiments on the influences of stereotype word proportion in the document set. We choose HyperPartisa and try 20 different proportions from 5%, 10%, to 100%. The Macro F1 scores and fairness are recorded in Fig. 3. For the TextRCNN model, the proportion of 15% can reach the highest F1 scores of more than 0.78 and rather low fairness around 17.00 while the best stereotype word proportion for the TextCNN model is around 50%, where the classification F1 is the highest: 0.73 and the word fairness is about 17.50. In addition, for the RoBERTa model, the F1 score can reach 0.75 when we set the proportion as 5%. In the meantime, the fairness is around 17.17 under this proportion. The different proportions show the effects of pre-trained embedding models and classification models on the stereotype bias. Similar to HyperPartisa, in other datasets, we can also select the best stereotype word proportions that can retain higher F1 scores and lower word fairness.

4 Conclusion

In this work, we follow previous works and focus on the potential word-level stereotype bias in text classification. We analyse the generation of bias in a causal view and propose a novel framework for bias mitigation. Our framework includes stereotype detection, fusion model training and unbiased prediction. Different from previous works, our framework can detect words that have a direct contribution to the predictions and does not rely on an external thesaurus. The experiments show better and more stable performances on multiple datasets and the ablation studies prove the effectiveness of the two parts in stereotype word detection. Moreover, we also explore the influences of the proportions of selected stereotype words. In future work, we will refine and weaken our assumptions on the proposed causal graph. We will include the causal path from the pre-trained model to the semantic variables and model the corresponding causal effects.

References

1. Badjatiya, P., Gupta, M., Varma, V.: Stereotypical bias removal for hate speech detection task using knowledge-based generalizations. In: WWW (2019)
2. Frassanito, P., Pettorini, B.: Pink and blue: the color of gender. Childs Nerv. Syst. **24**, 881–882 (2008)
3. Huang, X.: Easy adaptation to mitigate gender bias in multilingual text classification. In: NAACL (2022)
4. Jia, S., Meng, T., Zhao, J., Chang, K.: Mitigating gender bias amplification in distribution by posterior regularization. In: ACL (2020)
5. Kim, Y.: Convolutional neural networks for sentence classification. In: EMNLP (2014)
6. Kiritchenko, S., Mohammad, S.M.: Examining gender and race bias in two hundred sentiment analysis systems. In: NAACL-HLT (2018)
7. Kokalj, E., Skrlj, B., Lavrac, N., Pollak, S., Robnik-Sikonja, M.: BERT meets shapley: extending SHAP explanations to transformer-based classifiers. In: EACL (2021)
8. Lai, S., Xu, L., Liu, K., Zhao, J.: Recurrent convolutional neural networks for text classification. In: AAAI (2015)
9. Liu, H., Jin, W., Karimi, H., Liu, Z., Tang, J.: The authors matter: Understanding and mitigating implicit bias in deep text classification. arXiv preprint arXiv:2105.02778 (2021)
10. Liu, J., et al.: Fair representation learning: An alternative to mutual information. In: Zhang, A., Rangwala, H. (eds.) KDD (2022)
11. Liu, Y., et al.: Roberta: a robustly optimized BERT pretraining approach. arXiv preprint arXiv:1907.11692 (2019)
12. Lundberg, S.M., Lee, S.: A unified approach to interpreting model predictions. In: NIPS (2017)
13. Mihalcea, R., Tarau, P.: Textrank: bringing order into text. In: EMNLP (2004)
14. Nadeem, M., Bethke, A., Reddy, S.: Stereoset: measuring stereotypical bias in pretrained language models. In: ACL/IJCNLP (2021)
15. Nasukawa, T., Yi, J.: Sentiment analysis: capturing favorability using natural language processing. In: K-CAP (2003)
16. Pearl, J.: Direct and indirect effects. In: Probabilistic and Causal Inference: The Works of Judea Pearl, pp. 373–392 (2022)
17. Qian, C., Feng, F., Wen, L., Ma, C., Xie, P.: Counterfactual inference for text classification debiasing. In: ACL/IJCNLP (2021)
18. Sun, P., Wu, B., Li, X., Li, W., Duan, L., Gan, C.: Counterfactual debiasing inference for compositional action recognition. In: MM (2021)
19. Sweeney, C., Najafian, M.: A transparent framework for evaluating unintended demographic bias in word embeddings. In: ACL (2019)
20. Tian, B., Cao, Y., Zhang, Y., Xing, C.: Debiasing NLU models via causal intervention and counterfactual reasoning. In: AAAI (2022)
21. Wu, J., Liu, Q., Xu, W., Wu, S.: Bias mitigation for evidence-aware fake news detection by causal intervention. In: Amigó, E., Castells, P., Gonzalo, J., Carterette, B., Culpepper, J.S., Kazai, G. (eds.) SIGIR (2022)
22. Zhang, G., Bai, B., Zhang, J., Bai, K., Zhu, C., Zhao, T.: Demographics should not be the reason of toxicity: mitigating discrimination in text classifications with instance weighting. In: Jurafsky, D., Chai, J., Schluter, N., Tetreault, J.R. (eds.) ACL (2020)

23. Zhang, X., Zhao, J.J., LeCun, Y.: Character-level convolutional networks for text classification. In: NIPS (2015)
24. Zhao, J., Wang, T., Yatskar, M., Cotterell, R., Ordonez, V., Chang, K.: Gender bias in contextualized word embeddings. In: NAACL-HLT (2019)
25. Zhao, J., Wang, T., Yatskar, M., Ordonez, V., Chang, K.: Gender bias in coreference resolution: Evaluation and debiasing methods. In: NAACL-HLT (2018)

Open Access This chapter is licensed under the terms of the Creative Commons Attribution 4.0 International License (http://creativecommons.org/licenses/by/4.0/), which permits use, sharing, adaptation, distribution and reproduction in any medium or format, as long as you give appropriate credit to the original author(s) and the source, provide a link to the Creative Commons license and indicate if changes were made.

The images or other third party material in this chapter are included in the chapter's Creative Commons license, unless indicated otherwise in a credit line to the material. If material is not included in the chapter's Creative Commons license and your intended use is not permitted by statutory regulation or exceeds the permitted use, you will need to obtain permission directly from the copyright holder.

Knowledge-Enhanced Hierarchical Transformers for Emotion-Cause Pair Extraction

Yuwei Wang[1], Yuling Li[1], Kui Yu[1(✉)], and Yimin Hu[2]

[1] School of Computer Science and Information Engineering, Hefei University of Technology, Hefei 230601, China
{wyw,lyl95}@mail.hfut.edu.cn, yukui@hfut.edu.cn
[2] Institute of Intelligent Machines, Hefei Institutes of Physical Science, Chinese Academy of Sciences, Hefei, China
ymhu@iim.ac.cn

Abstract. Emotion-cause pair extraction (ECPE) aims to extract all potential pairs of emotions and corresponding cause(s) from a given document. Current methods have focused on extracting possible emotion-cause pairs by directly analyzing the given documents on the basis of a large training set. However, there are many hard-matching emotion-cause pairs that require commonsense knowledge to understand. Exploiting only the given documents is insufficient to capture the latent semantics behind these hard-matching emotion-cause pairs, which may downgrade the performance of existing ECPE methods. To fill this gap, we propose a Knowledge-Enhanced Hierarchical Transformers framework for the ECPE task. Specifically, we first inject commonsense knowledge into the given documents to construct the knowledge-enhanced clauses. To incorporate the injected knowledge into the clause representations, we then develop a hierarchical Transformers module that leverages two different types of transformer blocks to encode knowledge-enriched clause representations at both global and local stages. Experimental results show that our method achieves state-of-the-art performance.

Keywords: Emotion-cause pair extraction · Commonsense knowledge · Hierarchical Transformers

1 Introduction

Emotion-cause pair extraction (ECPE) plays an important role in natural language processing and has been widely used in various realistic applications including customer reviews [11] and conversation analysis [14]. ECPE aims to simultaneously identify both emotion clauses and their corresponding cause clauses in documents. As shown in Fig. 1, given a document consisting of four clauses (c_1, c_2, c_3, c_4), the goal of ECPE is to extract a set of emotion-cause pairs at the clause level: $(c_3, c_2), (c_4, c_4)$.

To achieve this goal, a series of ECPE models have been proposed in recent years. Xia et al. [17] proposed a two-step framework, which utilizes two subtasks to extract all possible emotion and cause clauses individually and then

© The Author(s), under exclusive license to Springer Nature Switzerland AG 2023
H. Kashima et al. (Eds.): PAKDD 2023, LNAI 13938, pp. 112–123, 2023.
https://doi.org/10.1007/978-3-031-33383-5_9

performs emotion-cause pairing and filtering to eliminate invalid emotion-cause pairs. Since the two-step method suffers from the problems of error propagation and high computational costs, recent studies are more inclined to solve the ECPE task by designing end-to-end models, which identify emotion-cause pairs in one step [1,3,7,15]. Existing one-step methods can be divided into two categories: (1) Classification models. While these models have different model structures, they all share a similar underlying idea, i.e., reframing the ECPE task as a classification problem of candidate emotion-cause pairs [1,3,4,7]; (2) Unified sequence labeling models. Instead of viewing the ECPE task as a clause pair classification problem, Yuan et al. [20] and Cheng et al. [2] reframe the task as a sequence labeling problem, which allows extracting emotion-cause pairs through one pass of sequence labeling.

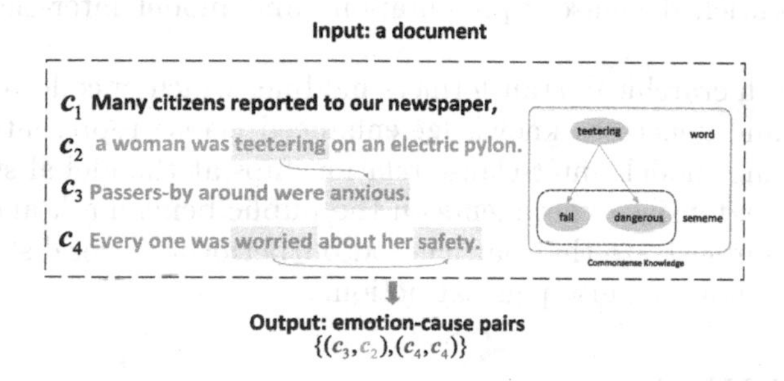

Fig. 1. An example of the ECPE task. (c_3, c_2) means clause c_2 is the cause for the emotion clause c_3, and (c_4, c_4) means c_4 is both an emotion and a cause clause.

Although these methods have achieved compelling results, their efforts have been focused on extracting possible emotion-cause pairs by directly analyzing the documents on the basis of a large training set. In this way, existing ECPE methods can identify the easy-matching emotion-cause pairs that require only a relatively shallow understanding. As shown in Fig. 1, it is possible to identify the easy-matching emotion-cause pair (c_4, c_4) based on the shallow semantic relevance between `worried` and `safety`. In practice, however, there are many hard-matching emotion-cause pairs that require commonsense or basic factual knowledge to understand. Exploiting only the given documents is insufficient to capture the latent semantics behind these hard-matching emotion-cause pairs, which may downgrade the performance of existing ECPE models. For example, to extract the hard-matching emotion-cause pair (c_3, c_2) in Fig. 1, one needs to grasp the factual knowledge: the concept `teetering` is associated with two latent semantics, i.e., `fall` and `dangerous`. The two semantics have shallow relevance with `anxious`, which further reveals the potential association between `teetering` and `anxious`.

Motivated by this observation, we propose a novel **K**nowledge-**E**nhanced **H**ierarchical **T**ransformer (KEHT) framework, which utilizes external commonsense knowledge to capture relations between emotion and cause clauses. Specifically, we first inject commonsense or factual knowledge into the given documents and obtain the knowledge-enhanced clauses. To incorporate the injected knowledge into the clause representations, we then develop a hierarchical Transformers module that leverages two different types of transformer blocks to encode knowledge-enriched clause representations at both global and local stages. At the local stage, a word-level transformer is developed to encode commonsense knowledge and generate the knowledge-enriched clause representations by aggregating the latent semantics of words. At the global stage, a clause-level transformer is designed to model the inter-clause relationships by exploiting the latent associations of clauses.

In summary, our main contributions are as follows:

- We propose a knowledge-enhanced hierarchical transformer method, which encodes commonsense knowledge at local and global stages to learn knowledge-enriched clause representations and model inter-clause relationships.
- We design a hierarchical Transformers module, which encodes commonsense knowledge and generates knowledge-enhanced clause representations at the local stage and models inter-clause relationships at the global stage.
- We conduct extensive experiments on the public benchmark and experimental results demonstrate that our method outperforms several state-of-the-art methods of emotion-cause pair extraction.

2 Related Work

The emotion-cause pair extraction task stems from the emotion cause extraction (ECE) task [6,12,18], which aims to extract causes behind emotions. Since the emotions in the ECE task must be annotated in advance, Xia and Ding [17] presented the ECPE task, which needs to identify both emotion and cause clauses simultaneously in a given document.

To solve the ECPE task, some two-step methods have been proposed [17] but it has two shortcomings: the errors in the first step may affect the performance of the second step, and it is hard to jointly optimize the performance of the two steps. Therefore, recent research focuses on developing one-step approaches to handle the ECPE task [1,3,4,9,15,16] and these approaches can be broadly divided into two categories: classification models and unified sequence labeling models. **Classification models** [1,3,4,9,15,19]. Ding et al. [4] transform the ECPE task into the emotion-pivot cause extraction problem in the sliding window. Wei et al. [15] tackled the ECPE task from a rank perspective. Bao et al. [1] performed clause representation learning by incorporating both fine-grained and coarse-grained semantic features. **Unified sequence labeling models** [2,16,20]. Yuan et al. [20] framed the ECPE task as a sequence labeling problem and presented a tagging scheme.

However, these methods mentioned above all put their efforts into extracting possible emotion-cause pairs by directly analyzing the documents on the basis of a large training set, while there are many hard-matching emotion-cause pairs that require commonsense or basic factual knowledge to understand. Exploiting only the given documents is insufficient to capture the latent semantics behind these hard-matching emotion-cause clauses.

3 Proposed Method

3.1 Overall Architecture

In the emotion-cause pair extraction task, our goal is to extract a set of emotion-cause pairs $P = \{(c_1^{emo}, c_1^{cau}), (c_2^{emo}, c_2^{cau}), \cdots\}$ from a given document D, where (c_i^{emo}, c_i^{cau}) is the i-th emotion-cause pair, and c_i^{emo}, c_i^{cau} are the i-th emotion clause and the i-th cause clause, respectively. Note that an emotion clause may be associated with multiple causes, and a cause clause may also correspond to multiple emotion clauses. The input document $D = [c_1, c_2, ..., c_{|D|}]$ contains $|D|$ clauses, and each clause consists of multiple words, i.e., $c_i = [w_i^1, w_i^2, \cdots, w_i^{|c_i|}]$, where w_i^j denotes the j-th word in the i-th clause.

To handle the ECPE task, we propose a knowledge-enhanced hierarchical transformers (KEHT) framework, which is illustrated in Fig. 2. There are three parts in the architecture: (1) **Commonsense Knowledge Injection** is to extract commonsense knowledge related to the given document from external Knowledge graphs and construct knowledge-enhanced clauses; (2) **Knowledge-Enhanced Clause Encoding** is to encode knowledge-enhanced clauses at both global and local stages to generate knowledge-enrich clause representations; (3) **Emotion-Cause Pair Extraction** is to utilize the learned clause representations to extract the final emotion-cause pairs.

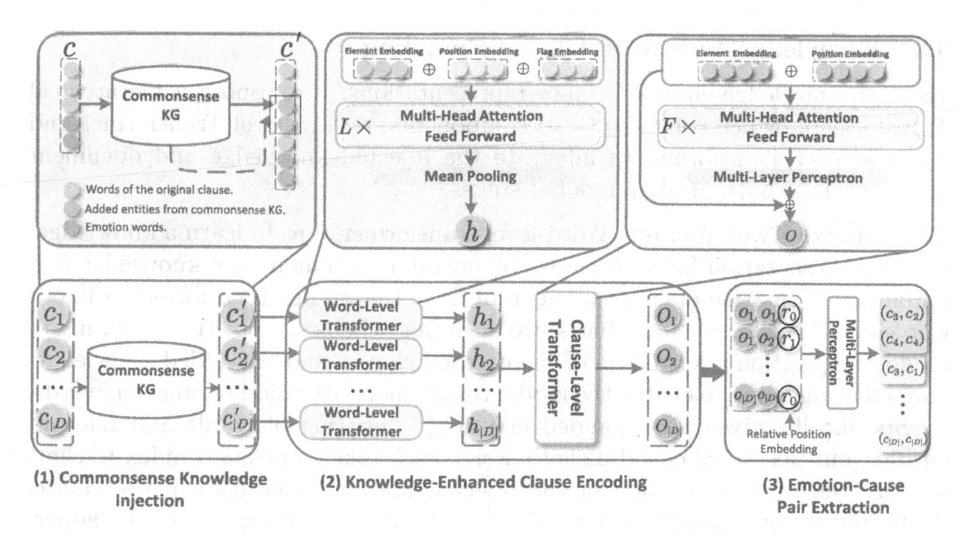

Fig. 2. An overview of our framework KEHT. ⊕ represents element-wise sum.

3.2 Commonsense Knowledge Injection

As discussed above, understanding complex clauses requires ECPE models to be equipped with external commonsense knowledge. To this end, we adopt the well-known sememe knowledge graph HowNet [5] to enrich the semantic information of the input documents. Due to its significance in understanding the nature of semantics in human languages, HowNet has been widely used in various natural language processing tasks such as sentiment analysis and word sense disambiguation. In HowNet, each Chinese word is annotated with its minimum semantic units called sememes, and each word corresponds to a sememe set. For example, `apple` is associated with two meanings (i.e., brand and fruit), and it has a list of sememes including {`computer`, `able`, `bring`, $\cdots$}.

Formally, given an input clause $c_i = \{w_i^1, w_i^2, \cdots, w_i^{|c_i|}\}$, we extract the sememes of each word (if exist) in HowNet. These extracted sememes are considered as external knowledge that helps to understand the semantics of input documents. We then inject these knowledge into the original clause c_i and obtain the expanded clause $c_i' = [w_i^1, w_i^2, \cdots, w_i^j \cup \{e_i^{j,k}\}_{k=1}^K, \cdots, w_i^{|c_i|}]$, where $\{e_i^{j,k}\}_{k=1}^K$ is the sememe set of word w_i^j in the KG. Since too many additional entities may cause noisy information, we restrict the maximum number of added entities to K, and $K = 2$ works well. Figure 3 shows an example of the expanded clause.

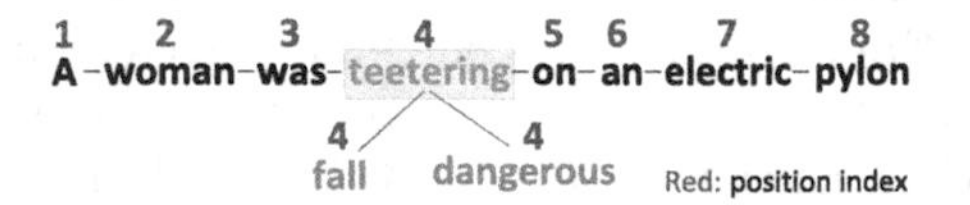

Fig. 3. The process of commonsense knowledge injection.

3.3 Knowledge-Enhanced Clause Encoding

To learn knowledge-enriched clause representations, we propose a hierarchical Transformers, which consists of two components: a word-level Transformer and a clause-level Transformer to integrate the injected knowledge and document information at both local and global stages.

Word-Level Transformer. Word-level transformer aims to learn a knowledge-enrich representation for each clause by encoding commonsense knowledge and semantic information of clauses. Different from the vanilla Transformer [13], our word-level Transformer introduces two encoding schemes: a position encoding to model the position information of the original clauses and the added knowledge, and a flag encoding to enable the model to pay more attention to emotion words.

Specifically, given an expanded clause c_i' consisting of words and relevant entities, our positional encoding is to assign each token a position index to characterize the relative position of the injected knowledge entities in the clause. If the token corresponds to a word $w_i^j \in c_i$, its position index is the superscript j of the word. If the word w_i^j can find corresponding knowledge entities $\{e_i^{j,k}\}_{k=1}^K$, their position indexes are the superscript of the start character in its associated word w_i^j. Formally, the position indices p_i of clause c_i'

are $p_i = [1, 2, \cdots, j \cup [j]_{\times K}, \cdots, |c_i|]$, where $[j]_{\times K}$ denotes a sequence containing K identical indices. Figure 3 illustrates an example of our proposed positional encoding. The clause is expanded by two knowledge entities: `fall` and `dangerous`, which are extracted from commonsense KG based on the word `teetering`. The entities are assigned the same index as the word `teetering`, and therefore its semantic information can be aggregated equally by the word-level Transformer.

Emotion words in clauses play an important role in modeling the semantic information of clauses. Based on this idea, we introduce a flag encoding to highlight the importance of emotion words (all emotion words have already been annotated in the dataset). The flag encoding is to assign 0 or 1 for each word to indicate whether the token is an emotion word. This is formalized as:

$$f_i^j = \begin{cases} 1, & x_i^j \in E \\ 0, & x_i^j \notin E \end{cases}, \quad x_i^j \in c_i', \tag{1}$$

where E is the set of emotion words.

For each token in c_i', we construct its input embedding z_j by summing its element embedding, position embedding, and flag embedding as $z_j = x_j^{ele} + x_j^{pos} + x_j^{flg}$, where x_j^{ele} is the element embedding of the j-th token in the clause c_i', x_j^{pos} is the position embedding corresponding to its position index and x_j^{flg} is the flag embedding corresponding to its flag index. So the clause c_i' can be represented as $Z_{c_i'}^0 = [z_1, \cdots, z_i, \cdots]$, which will be fed into the word-level Transformer.

Similar to the original Transformer, the word-level Transformer consists of L stacked transformer blocks. The l-th transformer block takes z_i^l as the input and outputs the hidden representation z_i^{l+1} of the l-th layer about the word w_j^i, which is the input sequence of the $l+1$ transformer block. The calculation process is as follows:

$$\begin{aligned} \hat{Z}_{c_i'}^l &= \mathrm{LN}(Z_{c_i'}^l + \mathrm{MHA}(Z_{c_i'}^l)), \\ Z_{c_i'}^{l+1} &= \mathrm{LN}(\hat{Z}_{c_i'}^l + \mathrm{MLP}(\hat{Z}_{c_i'}^l)), \end{aligned} \tag{2}$$

where LN is the Layer-Norm unit, and MLP is the Multi-layer Perceptron. We leverage Multi-Head Attention (MHA) to encode the input representations of words, which can be formalized as:

$$\mathrm{MHA}(Z_{c_i'}^l) = W^O \mathop{\|}_{k=1}^{H} \mathrm{ATT}_i(Z_{c_i'}^l W_k^Q, Z_{c_i'}^l W_k^K, Z_{c_i'}^l W_k^V), \tag{3}$$

where $W_k^Q, W_k^K, W_k^V, W^O \in \mathbb{R}^{d \times d}$ are learnable weight matrices; d is the embedding dimension; k denotes the number of attention heads; $\|$ is the concatenation operation. For i-th attention head, the scale dot-product is as follows:

$$\begin{aligned} &\mathrm{ATT}_i(Z_{c_i'}^l W_k^Q, Z_{c_i'}^l W_k^K, Z_{c_i'}^l W_k^V) \\ &= \mathrm{softmax}\left(\frac{Z_{c_i'}^l W_k^Q (Z_{c_i'}^l W_k^K)^T}{\sqrt{d}}\right) Z_{c_i'}^l W_k^V. \end{aligned} \tag{4}$$

Finally, we can obtain the clause representation h_i of the clause c'_i by averaging the hidden state of the final layer word representations:

$$h_i = \frac{1}{|c'_i|}\sum_{j=1}^{|c'_i|} z_j^L, \tag{5}$$

where $|c'_i|$ is the number of tokens in the expanded clause c'_i.

Clause-level Transformer. In this part, we aim to model inter-clause relationships and learn interactive representations of clauses. Specifically, we formalize the clause-level representation of the document D as $H_D = \{h_1, \cdots, h_i, \cdots\}$, where h_i is the knowledge-enriched representation of the i-th clause in D. We use the absolute position to formalize its position indices as: $[1, 2, \cdots, |D|]$. For each clause in D, we construct its input embedding s_i by summing its element embedding and position embedding as $s_i = h_i^{ele} + h_i^{pos}$, where h_i^{ele} is h_i and h_i^{pos} is the position embedding corresponding to its position index. The document D can be represented as $S_D = \{s_1, \cdots, s_i, \cdots, s_{|D|}\}$, which will be fed into the clause-level Transformer. The clause-level Transformer consists of F stacked transformer blocks, which can be formalized as:

$$\begin{aligned}
\text{MHA}(S_D^f) &= W^P \mathop{\Vert}_{m=1}^{H} \text{ATT}_i(S_D^f W_m^Q, S_D^f W_m^K, S_D^f W_m^V), \\
\hat{S}_D^f &= \text{LN}(S_D^f + \text{MHA}(S_D^f)), \\
S_D^{f+1} &= \text{LN}(\hat{S}_D^f + \text{MLP}(\hat{S}_D^f)),
\end{aligned} \tag{6}$$

where $W_m^Q, W_m^K, W_m^V, W^P \in \mathbb{R}^{d\times d}$ are trainable weight matrices, and $\Vert$ denotes the concatenation operation. For i-th attention head, the scale dot-product is performed as:

$$\begin{aligned}
&\text{ATT}_i(S_D^f W_m^Q, S_D^f W_m^K, S_D^f W_m^V) \\
&= \text{softmax}\left(\frac{S_D^f W_m^Q (S_D^f W_m^K)^T}{\sqrt{d}}\right) S_D^f W_m^V.
\end{aligned} \tag{7}$$

Finally, we couple the original clause representations and the weighted clause representations. The final clause representations can be formulated as:

$$\boldsymbol{O}_D = S_D + S_D^F W_D, \tag{8}$$

where $W_D \in \mathbb{R}^{d\times d}$ is a learnable weight matrix; $\boldsymbol{O}_D = \{\boldsymbol{o}_1, \cdots, \boldsymbol{o}_i, \cdots\}$, and $\boldsymbol{o}_i$ is the final clause representation of the clause c_i, which integrates both the semantic information and the inter-clause relationships in the document.

After obtaining the final clause representations, we use two MLP layers to predict whether each clause is an emotion/cause clause or none of both:

$$\hat{y}_i^{emo} = \sigma(\boldsymbol{o}_i W_{emo} + b_{emo}), \tag{9}$$

$$\hat{y}_i^{cau} = \sigma(\boldsymbol{o}_i W_{cau} + b_{cau}), \tag{10}$$

where W_{emo} , $W_{cau} \in \mathbb{R}^{d\times 1}$ and b_{emo}, $b_{cau} \in \mathbb{R}$ are weight matrixes and bias vectors for the emotion prediction MLP layer and cause prediction MLP layer respectively, and $\sigma(\cdot)$ is the logistic function.

3.4 Emotion-Cause Pair Extraction

For all clauses from document $D = \{c_1, c_2, \cdots, c_{|D|}\}$, we consider the combination of all clauses. As a result, there are $|D| * |D|$ possible cases of clause pairs, which can be formalized as $\mathcal{P} = \{(c_1, c_1), \cdots, (c_i, c_j), \cdots, (c_{|D|}, c_{|D|})\}$, where $\mathcal{P}$ is the set of all candidate pairs, and (c_i, c_j) means that cause clause c_j and emotion clause c_i can formulate a candidate pair. For each candidate clause pair (c_i, c_j), we can get their clause pair representation $\boldsymbol{p}_{ij}$, as follows:

$$\boldsymbol{p}_{ij} = ReLU([\boldsymbol{o}_i; \boldsymbol{o}_j; r_{j-i}]W_p + b_p), \tag{11}$$

where W_p and b_p are learnable weight matrix and bias vector, $\boldsymbol{o}_i$ and $\boldsymbol{o}_j$ represents the clause c_i, c_j embedding respectively, r_{j-i} represents the relative position embedding of clause pair (c_i, c_j). For each relative position $\beta \in \{-k, \cdots, +k\}$, its embeddings r_m can be obtained by:

$$r_\beta = \sum_{j=-k}^{+k} exp(-(j-\beta)^2) \cdot r'_j, \tag{12}$$

where $j \in \{-k, \cdots, +k\}$ denotes one of all possible relative position values, and r'_j is randomly intialized via sampling from a uniform distribution.

Then we use a MLP layer to predict the score of each candidate pair $\boldsymbol{p}_{ij}$:

$$\hat{y}_{ij} = tanh\left(\boldsymbol{p}_{ij} W_s + b_s\right), \tag{13}$$

where $W_s \in \mathbb{R}^{d\times 1}$ and $b_s \in \mathbb{R}$ are weight matrix and bias vector respectively.

Our method is trained in an end-to-end manner. We use a cross-entropy loss function $\mathcal{L}_{pair}$ for the emotion-cause pair extraction task, and two cross-entropy loss functions $\mathcal{L}_{emo}$ and $\mathcal{L}_{cau}$ for two sub-tasks: emotion clause extraction task and cause clause extraction task, respectively. Finally, the objective function can be defined as the sum of the above loss functions, as follows:

$$\mathcal{L} = \mathcal{L}_{pair} + \lambda(\mathcal{L}_{emo} + \mathcal{L}_{cau}). \tag{14}$$

where hyper-parameter λ controls the tradeoff between emotion-cause pair extraction and sub-tasks: emotion clause extraction and cause clause extraction.

4 Experiments

4.1 Datasets and Metrics

We conduct our experiments on the benchmark dataset which is released by Xia and Ding [17]. The dataset is constructed based on an emotion cause extraction corpus [8] which consists of 1,945 Chinese documents. We summarize the detailed statistics in Table 1. We repeat the experiments 15 times, and report the average results of precision (P), recall (R), and F1-score (F1) on the main task: emotion-cause pair extraction (ECPE) and two sub-tasks: emotion clause extraction (EE) and cause clause extraction (CE).

Table 1. Statistics of the dataset. "Doc." is the abbreviation for "Document".

# Doc. with one emotion-cause pair	1746
# Doc. with two emotion-cause pairs	177
# Doc. with three or more emotion-cause pairs	22
Avg. of clauses per document	14.77
Max. of clauses per document	77

4.2 Baselines

To confirm the effectiveness of our method, we compare our method with following baseline methods. **Indep** [17] is a two-step method that extracts emotion clauses and cause clauses respectively, then pair them and select the final emotion-cause pairs. **Inter-CE** [17] is an variant of Indep which uses emotion extraction to improve cause extraction. **Inter-EC** [17] is an variant of Indep which uses cause extraction to improve emotion extraction. **RankCP** [15] is a unified framework to tackle ECPE task from a ranking perspective. **ECPE-2D** [3] represents the emotion-cause pairs by a 2D representation scheme, and integrates the emotion-cause pair representation, interaction and prediction into a joint framework. **PTN** [16] is a neural network which tackles the complete emotion-cause pair extraction in one unified tagging task. **MGSAG** [1] aims to alleviate the position bias problem by incorporating fine-grained and coarse-grained semantic features jointly. **MaCa** [19] extracts emotion-cause pairs as a procedure of sequence modeling.

4.3 Implementation Details

In this paper, we adopt $\mathrm{BERT}_{Chinese}$ as the basis in this work. For both word-level and clause-level Transformer, we set the layer number of each transformer block to 1, attention heads to 1, the embedding size to 768, the hidden size to 3072, and we add dropout with the rate of 0.1 to reduce overfitting.. The dimension of relative position embedding is set to 50 and the maximum relative position value k to 12. Moreover, we use the Adam optimizer [10] as the optimizer with an initial learning rate of $3e-5$. We implement our method and further conduct it on a server with 2 NVIDIA GeForce RTX 3090 GPUs.

4.4 Comparison with ECPE Methods

Table 2 reports the comparative results on the ECPE task and two sub-tasks, i.e., EE and CE. To be fair, here we adopt the experimental results of all methods using the BERT model (if the model uses the BERT), and all experimental results are obtained from the references. From the table, we can observe that:

- Compared to other ECPE baselines, KEHT achieves the best performance. In particular, for the ECPE task, KEHT improves over the best baseline by 2.58% for F_1 score, and 4.32% for R score. For the EE task, KEHT improves

Table 2. Comparison of our method with other baseline methods.

	ECPE			EE			CE		
	F_1	P	R	F_1	P	R	F_1	P	R
Indep	0.5818	0.6832	0.5082	0.8210	0.8375	0.8071	0.6205	0.6902	0.5673
Inter-CE	0.5901	0.6902	0.5135	0.8300	0.8494	0.8122	0.6151	0.6809	0.5634
Inter-EC	0.6128	0.6721	0.5705	0.8230	0.8364	0.8107	0.6507	0.7041	0.6083
RankCP	0.7360	0.7119	0.7630	0.9057	0.9123	0.8999	0.7615	0.7461	0.7788
ECPE-2D	0.6889	0.7292	0.6544	0.8910	0.8627	0.9221	0.7123	0.7336	0.6934
PTN	0.6650	0.7600	0.5918	0.8360	0.8447	0.8278	0.6799	0.7175	0.6470
MGSAG	0.7521	0.7743	0.7321	0.8717	0.9208	0.9211	0.7712	**0.7979**	0.7568
MaCa	0.7387	**0.8047**	0.7215	0.8704	0.8819	0.8955	0.7435	0.7841	0.7260
Ours	**0.7779**	0.7526	**0.8062**	**0.9455**	**0.9481**	**0.9431**	**0.7858**	0.7672	**0.8068**

Table 3. Comparison of different supervised signals for KEHT.

Loss Function	F_1	P	R
$\mathcal{L}_{pair}$	0.7545	0.7291	0.7831
$\mathcal{L}_{pair} + \lambda(\mathcal{L}_{emo} + \mathcal{L}_{cau})$	**0.7779**	**0.7526**	**0.8062**

over the best baseline by 3.98% for F_1 score, 2.73% for P score, and 2.1% for R score. For the CE task, KEHT still outperforms other baselines by 1.46% for F_1 score, and 2.8% for the R score. The experimental results demonstrate the effectiveness of utilizing external commonsense knowledge to build knowledge-enhanced texts for extracting emotion-cause pair.

- KEHT outperforms other ECPE methods on the ECPE task and two sub-tasks. The main reason would be that other methods only extract emotion-cause pairs by directly analyzing the documents on the basis of a large training set and this is insufficient to model the complex relations between clauses. On the contrary, our KEHT utilizes external commonsense knowledge to build knowledge-enhanced texts and encode texts at both word level and clause level. We can also observe that MGSAG outperforms us on the P score of the CE task and MaCa outperforms us on the P score of the ECPE task. The main reason would be that MGSAG and MaCa may abandon some true emotion-cause pairs.

Effect of Two-Level Supervision. We use two-level supervised signals to train our method: a low-level signal $\mathcal{L}_{emo} + \mathcal{L}_{cau}$ at the output of the clause-level Transformer (see Eq. 9), and a high-level signal $\mathcal{L}_{pair}$ at the classification stage (see Eq. 13). Table 3 shows the effect of two-level supervision. The results show that training with two-level supervision improves extraction performance. This indicates that two-level supervision is helpful for learning clause representations and facilitates the emotion-cause pair extraction process.

Effect of Commonsense Knowledge Injection. We remove the knowledge graph which is used for injecting external commonsense knowledge to construct the knowledge-enhanced texts. Figure 4 reports that method without external knowledge results in a drop. The F_1 decreases 1.49%, and P, R drop 1.36% and

1.22%, respectively. It shows that utilizing external commonsense knowledge can enrich semantic information and significantly improve extraction performance.

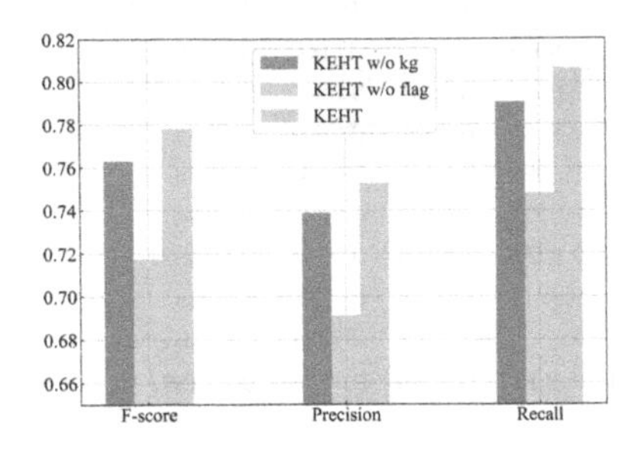

Fig. 4. Comparative results of KEHT which removes different components

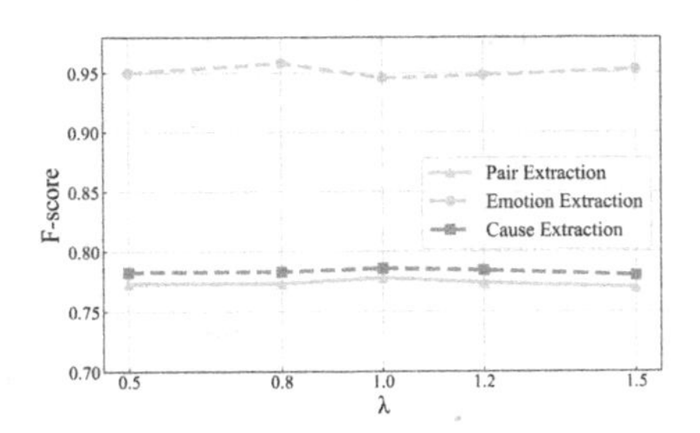

Fig. 5. Results with various values of λ.

Effect of Flag Embedding. We further investigate the importance of flag embedding. Figure 4 shows that by removing the flag embedding part in KEHT, the performance results in a significant drop. The F_1 decreases 6.05%, and P, R drop 6.15% and 5.83%, respectively. We can conclude that flag embedding is useful for our method to focus on more important information.

Effect of Hyperparameter for Loss Function. To explore the influence of two sub-tasks, we conduct a deeper analysis of the hyperparameter λ which is used in Eq. 14. We vary the values of λ from 0.5 to 1.5, and the results are shown in Fig. 5. We can observe that with the increase of λ, the performance of our method increases and then decreases slightly, and reaches a peak value at $\lambda = 1$ on the emotion-cause pair extraction and the cause clause extraction and at $\lambda = 0.8$ on the emotion clause extraction. This suggests that the emotion-cause pair extraction task and two sub-tasks play an equally important role.

5 Conclusion and Future Work

In this paper, we propose a novel KEHT method to tackle the emotion-cause pair extraction task. Our method effectively incorporates external commonsense knowledge into the clause representations via a hierarchical Transformers module, which leverages two different types of transformer blocks to encode knowledge-enriched clause representations at both global and local stages. In future work, we would like to tackle the problem of the documents with more than one emotion-cause pair by exploring the more fine-grained roles of the clauses.

Acknowledgements. This work is supported by the National Key Research and Development Program of China (under grant 2020AAA0106100).

References

1. Bao, Y., Ma, Q., Wei, L., Zhou, W., Hu, S.: Multi-granularity semantic aware graph model for reducing position bias in emotion cause pair extraction. In: Findings of ACL, pp. 1203–1213 (2022)
2. Cheng, Z., Jiang, Z., Yin, Y., Li, N., Gu, Q.: A unified target-oriented sequence-to-sequence model for emotion-cause pair extraction. IEEE/ACM Trans. Audio Speech Lang. Process. **29**, 2779–2791 (2021)
3. Ding, Z., Xia, R., Yu, J.: ECPE-2D: emotion-cause pair extraction based on joint two-dimensional representation, interaction and prediction. In: ACL, pp. 3161–3170 (2020)
4. Ding, Z., Xia, R., Yu, J.: End-to-end emotion-cause pair extraction based on sliding window multi-label learning. In: EMNLP, pp. 3574–3583 (2020)
5. DongZ, D., HAO, C.: Hownet and the computation of meaning (2006)
6. Fan, C., et al.: A knowledge regularized hierarchical approach for emotion cause analysis. In: EMNLP-IJCNLP, pp. 5614–5624 (2019)
7. Fan, C., Yuan, C., Du, J., Gui, L., Yang, M., Xu, R.: Transition-based directed graph construction for emotion-cause pair extraction. In: ACL, pp. 3707–3717 (2020)
8. Gui, L., Xu, R., Wu, D., Lu, Q., Zhou, Y.: Event-driven emotion cause extraction with corpus construction. In: EMNLP, pp. 145–160. World Scientific (2018)
9. Huang, W., Yang, Y., Peng, Z., Xiong, L., Huang, X.: Deep neural networks based on span association prediction for emotion-cause pair extraction. Sensors **22**(10), 3637 (2022)
10. Kingma, D.P., Ba, J.: Adam: a method for stochastic optimization. arXiv preprint arXiv:1412.6980 (2014)
11. Mittal, A., Vaishnav, J.T., Kaliki, A., Johns, N., Pease, W.: Emotion-cause pair extraction in customer reviews. arXiv preprint arXiv:2112.03984 (2021)
12. Turcan, E., Wang, S., Anubhai, R., Bhattacharjee, K., Al-Onaizan, Y., Muresan, S.: Multi-task learning and adapted knowledge models for emotion-cause extraction. arXiv preprint arXiv:2106.09790 (2021)
13. Vaswani, A., et al.: Attention is all you need. In: Advances in Neural Information Processing Systems, vol. 30 (2017)
14. Wang, F., Ding, Z., Xia, R., Li, Z., Yu, J.: Multimodal emotion-cause pair extraction in conversations. arXiv preprint arXiv:2110.08020 (2021)
15. Wei, P., Zhao, J., Mao, W.: Effective inter-clause modeling for end-to-end emotion-cause pair extraction. In: ACL, pp. 3171–3181 (2020)
16. Wu, Z., Dai, X., Xia, R.: Pairwise tagging framework for end-to-end emotion-cause pair extraction. Front. Comp. Sci. **17**(2), 1–10 (2023)
17. Xia, R., Ding, Z.: Emotion-cause pair extraction: a new task to emotion analysis in texts. In: ACL, pp. 1003–1012 (2019)
18. Yan, H., Gui, L., Pergola, G., He, Y.: Position bias mitigation: a knowledge-aware graph model for emotion cause extraction. arXiv preprint arXiv:2106.03518 (2021)
19. Yang, C., Zhang, Z., Ding, J., Zheng, W., Jing, Z., Li, Y.: A multi-granularity network for emotion-cause pair extraction via matrix capsule. In: CIKM, pp. 4625–4629 (2022)
20. Yuan, C., Fan, C., Bao, J., Xu, R.: Emotion-cause pair extraction as sequence labeling based on a novel tagging scheme. In: EMNLP, pp. 3568–3573 (2020)

PICKD: In-Situ Prompt Tuning for Knowledge-Grounded Dialogue Generation

Rajdeep Sarkar[1(✉)], Koustava Goswami[2], Mihael Arcan[1], and John McCrae[1]

[1] University of Galway, Galway, Ireland
{r.sarkar1,mihael.arcan,john.mccrae}@universityofgalway.ie
[2] Adobe Research Bangalore, Bangalore, India
koustavag@adobe.com

Abstract. Generating informative, coherent and fluent responses to user queries is challenging yet critical for a rich user experience and the eventual success of dialogue systems. Knowledge-grounded dialogue systems leverage external knowledge to induce relevant facts in a dialogue. These systems need to understand the semantic relatedness between the dialogue context and the available knowledge, thereby utilising this information for response generation. Although various innovative models have been proposed, they neither utilise the semantic entailment between the dialogue history and the knowledge nor effectively process knowledge from both structured and unstructured sources. In this work, we propose PICKD, a two-stage framework for knowledgeable dialogue. The first stage involves the *Knowledge Selector* choosing knowledge pertinent to the dialogue context from both structured and unstructured knowledge sources. PICKD leverages novel *In-Situ* prompt tuning for knowledge selection, wherein prompt tokens are injected into the dialogue-knowledge text tokens during knowledge retrieval. The second stage employs the *Response Generator* for generating fluent and factual responses by utilising the retrieved knowledge and the dialogue context. Extensive experiments on three domain-specific datasets exhibit the effectiveness of PICKD over other baseline methodologies for knowledge-grounded dialogue. The source is available at https://github.com/rajbsk/pickd.

Keywords: Language Model Prompting · Knowledge grounded Dialogue Systems · Knowledge Graphs

1 Introduction

With the proliferation of personal assistants (Siri, Alexa, etc.), research on dialogue systems has gained a lot of traction. Inducing relevant information in responses leads to a fluent, engaging and coherent conversation with a dialogue system. While language models help develop fluent dialogue systems [16], such systems lack the necessary tools for generating accurate responses. Researchers have utilised external knowledge from either unstructured knowledge sources such as Wikipedia articles [1,13], domain-grounded documents [14] or structured sources like Knowledge Graphs (KGs) [19,22] for generating informative responses.

© The Author(s), under exclusive license to Springer Nature Switzerland AG 2023
H. Kashima et al. (Eds.): PAKDD 2023, LNAI 13938, pp. 124–136, 2023.
https://doi.org/10.1007/978-3-031-33383-5_10

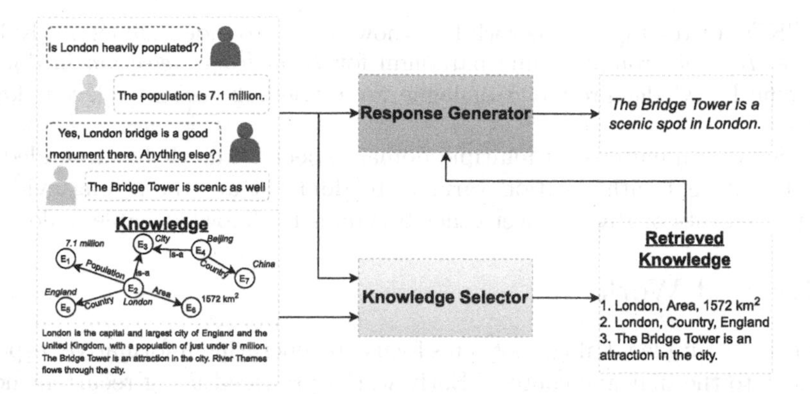

Fig. 1. Modular overview of the PICKD framework. The *Knowledge Selector* retrieves dialogue context-relevant knowledge from the structured and unstructured knowledge available. The retrieved knowledge and along with the dialogue context is then sent to the *Response Generator* for producing informative and fluent responses.

Regardless of the recent improvements in generic dialogue systems, the systems must interpret the semantic relatedness between the dialogue context and the external knowledge. For example, in Fig. 1, while responding to utterance 1, the model needs to understand the context encompasses "London" and then utilise the knowledge triple "London, has population, 7.1 million" when generating the response text. Similarly, when responding to utterance 3, the agent needs to register the contextual overlap between the dialogue history and the knowledge sentence "The London Bridge and the Bridge Tower are major attractions in London.". It should then leverage this information while generating utterance 4.

While prior works have posited novel frameworks to address this task, they tend to cater exclusively to unstructured knowledge sources like paragraphs/documents or solely to structured sources like KGs. Zhou et al. [21] proposed the KdConv dataset, which spans three domains and includes both structured and unstructured knowledge. They also proposed adapting Seq2Seq and the Hierarchical Recurrent Encoder-Decoder Framework (HRED) with knowledge memory for response generation. Wang et al. [11] built upon this work with their RT-KGD framework, which uses a Heterogeneous Graph Transformer Network to capture information flow and BART [4] for response generation. However, these models do not represent dialogue context and knowledge elements in the same semantic space, leading to irrelevant facts in responses.

We present a novel framework called PICKD for generating knowledgeable dialogues with context-relevant knowledge selection using *In-Situ* prompt tuning. The framework includes a *Knowledge Selector* which is initially trained using the *In-Situ* prompt tuning approach for retrieving relevant knowledge from structured and unstructured sources that align with the dialogue context. The *Knowledge Selector* is based on the RoBERTa [5] architecture and only the prompt tokens and classification heads are learned during training. The retrieved knowledge, along with the dialogue context, is then passed to the *Response Generator* which utilises the BART architecture for producing knowledgeable and coherent responses. In summary, our contributions are as follows:

- PICKD, a two-stage framework for Knowledge Grounded Dialogue (KGD).
- Novel *In-Situ* prompt tuning paradigm for *Knowledge Selection* in domain-specific knowledge-grounded dialogue for retrieving context-relevant knowledge.
- Extensive evaluations on multiple domain-specific knowledge-grounded dialogue datasets with ablation variants to demonstrate the strong evidential improvements of PICKD over other baselines for knowledgeable dialogue.

2 Related Work

Knowledge-grounded dialogue systems focus on generating informative responses pertinent to the dialogue context. Early works proposed using recurrent neural network-based sequence-to-sequence models for dialogue generation [9,10]. With the success of the transformer architecture, researchers explored transformer-based models for dialogue understanding and response generation [8,16]. Zhang et al. [15] proposed Dialo-GPT, a GPT-2 [7] based model trained on large-scale dialogue corpus for dialogue response generation.

However, models trained on open-domain data suffer from hallucinations and produce factually inconsistent responses. This necessitated the need for knowledge-grounded dialogue systems. Dinan et al. [1] proposed the Wizard of Wikipedia dataset wherein dialogues are grounded on Wikipedia articles. Additionally, they proposed a novel Transformer Memory Network for knowledgeable response generation. Following this, researchers explored the utilisation of reinforcement learning [17], personalisation memory [2] and knowledge internalisation [12] based methodologies for knowledgeable dialogue using unstructured knowledge sources. On the other hand, Zhu et al. [22] suggested grounding dialogue systems on structured sources such as KGs. Following this, novel frameworks [19,20] leveraging structured knowledge were proposed for response generation. Another line of work has been exploring prompting frameworks [6,18] for response generation. In these prompt-based models, the dialogue context and learnable prompt tokens are sent to language models for knowledgeable dialogue.

A major drawback of such models is the inability to utilise structured and unstructured knowledge in unison. Zhou et al. [21] proposed using a memory-based Hierarchical Recurrent Neural Network for grounding dialogues on external knowledge. Following this, Wang et al. [11] suggested RT-KGD, a novel framework utilising dialogue transition for knowledgeable dialogue generation. However, RT-KGD does not represent knowledge elements and the dialogue history in the same semantic vector space leading to an information gap between the dialogue context and the knowledge elements. In this work, we propose PICKD a novel knowledge-grounded dialogue generation framework. PICKD employs a *Knowledge Selector* for retrieving appropriate knowledge aligned with the dialogue context for response generation. The *Knowledge Selector* is trained using a novel *In-Situ* prompt tuning framework, wherein the prompt tokens are injected into the knowledge and dialogue input. The *Response Generator* module of PICKD then exploits this retrieved knowledge and the dialogue context for response generation.

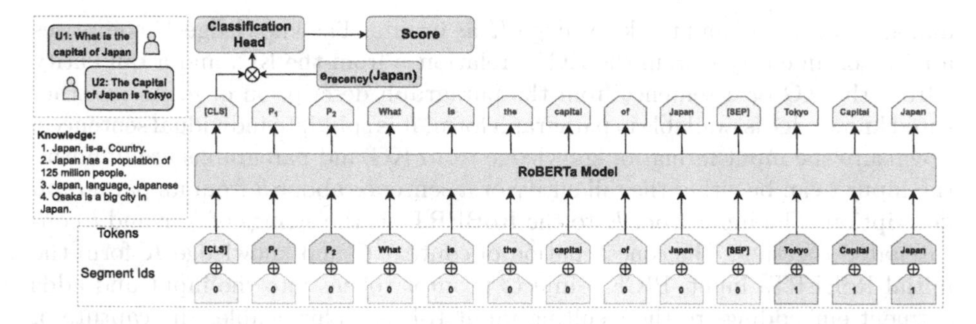

Fig. 2. Overview of the *In-Situ* prompting tuning framework which uses the RoBERTa architecture as its base model. The input token embeddings are augmented with segment token embeddings, and the output of the $[CLS]$ token embedding is then sent to a classification layer along with the recency embedding of the head entity of the knowledge element. This classification layer learns to assign higher scores to relevant knowledge. Here, $\bigoplus$ denotes addition, while $\bigotimes$ denotes the concatenation operation.

3 Methodology

We begin with formally defining the problem statement. Thereafter, we introduce the *Knowledge Selector* module, our novel *In-Situ* prompt tuning framework for knowledge selection. Finally, we describe the workings of the *Response Generator* for knowledgeable dialogue generation.

3.1 Formal Problem Definition

We are given the dialogue context $C = \{u_1, u_2, u_3, ..., u_{n-1}\}$, where u_i is an utterance from the conversation history, triples $\{(h_1, r_1, t_1), (h_2, r_2, t_2), ..., (h_{|K_k|}, r_{|K_k|}, t_{|K_k|})\}$ from a KG, where K_k is the number of triples in the KG and descriptive text set $S = \{S_1, S_2, ..., S_{S_k}\}$, where S_k is the number of articles in the set and each article comprises of multiple text sentences. The objective is to generate a coherent response u_n that is not only grammatically correct but also informative, leveraging the context and the knowledge sources available.

3.2 Contextual Prompting for Knowledge Selection

The *Knowledge Selector* in PICKD retrieves knowledge from structured and unstructured sources to be leveraged by the *Response Generator*. The *Knowledge Selector* module is trained to understand the semantic congruence between the dialogue context and knowledge elements using the novel *In-Situ* prompt tuning paradigm. The knowledge facts from structured and unstructured sources are then ranked according to their relevance by the module, and appropriate knowledge is then sent to the *Response Generator* (Fig. 2).

Prompting Architecture. The *Knowledge Selector* module of PICKD employs a pre-trained RoBERTa model as its backbone. The RoBERTa[1] model takes the

[1] https://huggingface.co/hfl/chinese-roberta-wwm-ext.

dialogue context C and the knowledge K as inputs. The knowledge K comprises a triple of an entity h from the KG, a relation r from the KG, and a tail entity t from the KG or a sentence from the paragraph description of h. As unstructured knowledge is available in paragraph form, it is split into individual sentences. To ensure the input format of knowledge from KG and paragraphs are uniform, the input t can be either the tail entity or a sentence about h from its paragraph description. The input from K to the RoBERTa is the strings of h, r and t concatenated together. The concatenation of context C and knowledge K form the initial RoBERTa input. PICKD injects prompt tokens into the input and adds segment embeddings to the resulting input tokens. This enables the capture of semantic dependencies between the context and knowledge elements. The token embeddings then pass through a RoBERTa layer, and the final representation of the $[CLS]$ token along with the recency embedding of h is sent to the classification layer for ranking. Recency embeddings are learnable vectors that capture the relevance of h in the dialogue history, similar to positional embeddings. This ensures that head entities mentioned recently have more influence during response generation. Formally, we define the setup as follows:

$$P_{prompt} = P_1, P_2, ..., P_{k_{prompt}} \tag{1}$$
$$D_{dial} = D_1 D_2 ... D_n \tag{2}$$
$$K_{know} = K_1 K_2 ... K_k \tag{3}$$
$$T_{input} = [CLS] P_{prompt} D_{dial} [SEP] K_{know} \tag{4}$$
$$h_{CLS}, h_{p_1}, ..., h_{K_k} = \text{RoBERTa}(T_{input}) \tag{5}$$
$$v = MLP(h_{CLS}; \mathbf{e}_{recency}) \tag{6}$$
$$\text{score} = Softmax(v) \tag{7}$$

where P_i, D_j and K_k denote the ith prompt token, j^{th} dialogue history token and the kth knowledge tokens. The input to the RoBERTa model is composed of the prompt tokens, dialogue tokens and knowledge tokens concatenated together. A $[CLS]$ token is added to the beginning of the sequence denoting start of the sequence and a $[SEP]$ token is used to separate the prompt and dialogue tokens from the knowledge tokens as shown in Eq. 4. This input is sent through a RoBERTa layer as shown in Eq. 5 wherein segment embeddings are added to the token embeddings before realising the contextual embeddings. PICKD then concatenates the RoBERTa representation of $[CLS]$ token with the head entity recency embedding and is then sent through the classification MLP layer as detailed in Eq. 6 and 7. The model is trained by minimising the cross-entropy loss. During inference, the knowledge elements are ranked following Eq. 7. The top-k elements are then sent to the *Response Generator* for response generation.

3.3 BART Fine-Tuning for Response Generation

The *Knowledge Selector* retrieves context-relevant knowledge to be utilised by the *Response Generator* for response generation. The *Response Generator* is trained to utilise the semantic information from the dialogue context and the retrieved knowledge to generate the knowledgeable response. Similar to Wang et al. [11], PICKD employs the BART architecture[2] [4] for response generation.

More formally, the input of the BART model consists of the dialogue context and the knowledge retrieved. At the Nth turn, input from the dialogue context takes the form "*[CLS]* u_1 *[SEP]* u_2 ... *[SEP]* u_{N-1} *[SEP]*", where *[CLS]* and *[SEP]* tokens are special tokens denoting the start of the sequence and sentence boundary token respectively. Knowledge input is constructed as "*[CLS]* K_1 *[SEP]* K_2 *[SEP]*... $K_{k_{know}}$ *[SEP]*", where k_{know} is the cardinality of the retrieved knowledge set and K_i is the knowledge fact defined as the concatenation of the head entity, relation and tail entity. The context concatenated with the knowledge is sent as input to the BART encoder layer as detailed in Eq. 8.

$$\begin{aligned} h_1^C, h_2^C, h_3^C...h_n^C = \text{BART}_{enc}([CLS]u_1[SEP]u_2...u_{N-1}[SEP] \\ [CLS]K_1[SEP]K_2...K_{k_{know}}[SEP]) \end{aligned} \tag{8}$$

$$G = \text{BART}_{dec}(h_1^C; h_2^C; h_3^C; ...; h_n^C) \tag{9}$$

$$\mathcal{L}_{decoder} = -\frac{1}{|Y|}\sum_{t=1}^{|Y|} log(Y_t = G_t) \tag{10}$$

where h_i^C is the contextual representation of the ith token in the input sequence and n is the length of the concatenated input. The encoded representations are then sent to a BART decoder layer for knowledgeable response generation as defined in Eq. 9, where G is the response representation. The autoregressive BART_{dec} model is then trained by minimising the cross-entropy loss over the ground-truth tokens as described in Eq. 10, where G_t is the ground-truth token and Y_t is the predicted token at timestep t.

4 Experimental Setup

This section details the setup and the evaluation of PICKD against the baseline methodologies on the KGD task.

4.1 Datasets

We conduct experiments on the KdConv dataset [21] to evaluate the effectiveness of PICKD. KdConv is a knowledge-grounded dialogue dataset spanning film, music and travel domains. The dialogue utterances are annotated with knowledge

Table 1. KdConv dataset characteristics

	Film	Music	Travel
# Train dialogues	1,200	1,200	1,200
# Dev dialogues	150	150	150
# Test dialogues	150	150	150
# KG entities	7,477	4,441	1,154
# KG rels	4,939	4,169	7
# KG triples	89,618	56,438	10,973

[2] https://huggingface.co/fnlp/bart-base-chinese.

from either a KG or from paragraphs. The are 1,200, 150 and 150 dialogues in the training, development and test set in each domain. The dialogues in the dataset have an average of 19.0 utterances and 10.1 annotated knowledge triples. The choice of this dataset enables robust evaluation of PICKD on multiple domains in the presence of both structured and unstructured knowledge sources. The dataset statistics are detailed in Table 1.

4.2 Baseline Methods

Following Wang et al. [11] and Zhou et al. [21], we consider the following methodologies as the baselines for the task:

- Seq2Seq [21]: An attention-based encoder-decoder framework trained to generate the response text conditioned on the dialogue history.
- HRED [21]: It employs a contextual Recurrent Neural Network (RNN) to inject the historical utterances into the context state. The model then generates the response conditioned on the dialogue context.
- Seq2Seq + Know [21]: An encoder-decoder framework trained to generate responses conditioned on the dialogue history and the external knowledge.
- HRED + Know [21]: Similar to HRED, however, the architecture generates a response conditioned on the dialogue context and the external knowledge.
- BART [11]: An encoder-decoder framework employing the BART architecture for generating dialogue responses.
- RT-KGD [11]: A relation-transition aware knowledge grounded dialogue generation framework. RT-KGD employs Heterogeneous Graph Transformers to inject KG information into a BART model for response generation.

4.3 Evaluation Metrics

Automatic Metrics: Following previous works [11,21], we adopt perplexity, bleu-N and distinct-N metrics for automatic model evaluation. Perplexity evaluates the capacity of the model to generate the ground-truth response. Lower perplexity denotes better confidence in generating ground-truth responses. Blue-N measures the N-gram overlap between the generated response and the ground-truth response, while distinct-N assess the diversity of N-gram tokens in the response generated.

Manual Metrics: Following Zhou et al. [21] and Wang et al. [11], we use fluency and coherence as the metrics for human evaluation. Fluency measures whether the generated responses are grammatically correct, fluent and human-like. On the other hand, coherence measures if the generated responses are coherent to the dialogue context and consistent with the available knowledge.

4.4 Implementation Details

We utilise the PyTorch and the Huggingface libraries for developing PICKD. The *Knowledge Selector* is trained with a learning rate of 1e-4 and batch size 8

Table 2. Performance of different methodologies on domain-grounded datasets for knowledgeable dialogue generation. We report the performance of PICKD using the average of 5 different runs of the entire framework.

Model	PPL	BLEU-1/2/3/4				Dist-1/2/3/4			
Film									
Seq2Seq	23.88	26.97	14.31	8.53	5.30	2.32	6.13	10.88	16.14
HRED	24.74	27.03	14.07	8.30	5.07	2.55	7.35	14.12	21.86
Seq2Seq+Know	25.56	27.45	14.51	8.66	5.32	2.85	7.98	15.09	23.17
HRED+Know	26.27	27.94	14.69	8.73	5.40	2.86	8.08	15.81	24.93
BART	**2.66**	28.54	13.28	14.21	11.00	2.46	14.12	25.72	36.12
RT-KGD	2.86	32.11	22.21	16.68	13.18	**3.05**	16.34	31.36	44.68
PICKD$_{ablated}$	5.36	38.31	30.66	26.26	22.84	2.79	17.35	31.56	43.61
PICKD	4.89	**40.69**	**33.11**	**28.58**	**25.46**	2.96	**19.08**	**35.23**	**48.97**
Music									
Seq2Seq	16.17	28.89	16.56	10.63	7.17	2.52	7.02	12.69	18.78
HRED	16.82	29.92	17.31	11.17	7.52	2.71	7.71	14.07	20.97
Seq2Seq+Know	17.12	29.60	17.26	11.36	7.84	3.93	12.35	23.01	34.23
HRED+Know	17.69	29.73	17.51	11.59	8.04	3.80	11.7	22.00	33.37
BART	**2.46**	31.65	23.04	18.22	15.05	2.80	13.69	24.73	34.59
RT-KGD	2.47	40.75	31.26	25.56	21.64	**4.18**	**17.38**	30.05	41.05
PICKD$_{ablated}$	5.94	38.01	30.86	26.57	23.59	3.07	15.12	29.19	40.65
PICKD	4.55	**41.56**	**33.81**	**28.50**	**25.97**	2.84	16.17	**31.10**	**43.77**
Travel									
Seq2Seq	10.44	29.61	20.04	14.91	11.74	3.75	11.15	19.01	27.16
HRED	10.90	30.92	20.97	15.61	12.30	4.15	12.01	20.52	28.74
Seq2Seq+Know	10.62	37.04	27.28	22.16	18.94	4.25	13.64	24.18	34.35
HRED+Know	11.15	36.87	26.68	21.31	17.96	3.98	13.31	24.06	34.35
BART	1.83	34.77	29.11	25.69	23.33	2.70	13.39	21.92	29.53
RT-KGD	**1.61**	47.56	41.46	37.40	34.31	**3.58**	15.50	26.10	35.72
PICKD$_{ablated}$	3.40	46.68	41.23	37.90	35.53	2.46	16.55	26.88	34.74
PICKD	2.41	**52.40**	**47.37**	**44.19**	**41.88**	2.92	**17.21**	**27.67**	**36.31**

for 5 epochs using the Adam optimiser [3]. k_{prompt} is set to 100 for all domains. The *Response Generator* is trained with a learning rate of 1e-4 and batch size of 8 using Adam optimiser with warmup for 1,000 steps. Our choice of these hyperparameters remains constant across all the domains. k_{know} is set to 7, 6 and 3 for the film, music and travel domain. We utilise beam search decoding with beam width 5 for response generation. It is essential to note that we conduct a two-stage training of PICKD. In the first stage, the *Knowledge Selector* is trained to retrieve appropriate facts. Once this learning phase is complete, the *Response Generator* is trained for knowledgeable response generation.

5 Empirical Results

This section reports the performance of PICKD on the knowledgeable dialogue generation task. We also conduct ablation studies by changing different parameters of PICKD that can potentially impact the performance. For completeness,

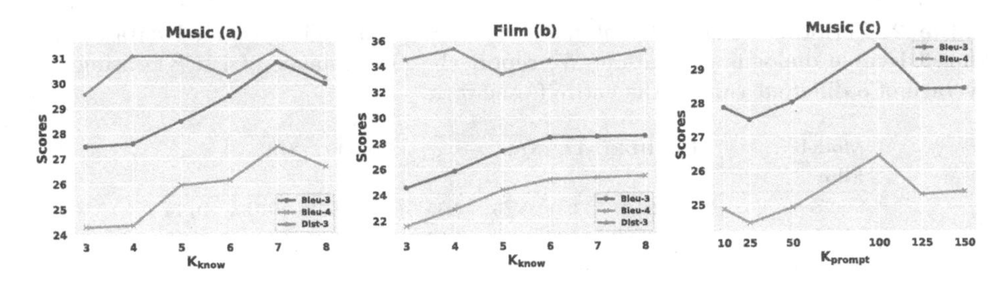

Fig. 3. Performance impact of PICKD with varying context size (k_{know}) for the *Knowledge Selector* on (a) Music and (b) Film domain. (c) showcases the performance of PICKD with changing the number of prompt tokens.

we explore the performance of different methodologies on human evaluation metrics and conduct error analysis on the generation results.

5.1 Automatic Evaluation

Table 2 showcases the performance of different baselines against PICKD . PICKD outperforms all the competing methodologies on the targeted datasets in the Bleu-N metric. The results indicate the generated responses have a high textual overlap with the reference ground-truth responses as higher Bleu-N scores indicate good and fluent responses. PICKD outperforms other methodologies on the Dist-2,3,4 metrics on the film and travel domains while on the Dist-3,4 metrics on the music domain. This indicates the responses generated by PICKD are not bland and are more engaging. Nevertheless, it is interesting to notice that RT-KGD has better perplexity than PICKD over all three domains which is due to the fact that during the training of the *Response Generator*, generation is more constrained by the necessity to make factually correct answers over fluent, inaccurate responses. Furthermore, we evaluate the performance of PICKD when the prompt tokens and recency embeddings are removed. During the training of the *Knowledge Selector*, only the classification heads are fine-tuned, resulting in a variant of the model referred to as $\text{PICKD}_{\text{ablated}}$. Our results indicate that the performance of $\text{PICKD}_{\text{ablated}}$ is worse than the original model, highlighting the importance of the prompt tokens and recency embeddings in the PICKD framework. Overall, we find that PICKD generated responses have a higher overlap with the ground-truth responses.

5.2 Impact of Prompt Length

The *Knowledge Selector* uses the novel *In-Situ* prompt tuning paradigm for retrieving context-relevant knowledge. The module injects prompt tokens within the context-knowledge text, and the prompt tokens along with the classification heads are then trained for the ranking task. A lower number of prompt tokens leads to fewer trainable parameters, thereby limiting the performance of PICKD.

In comparison, a higher number of prompt tokens would lead to the truncation of knowledge text due to the limited length processing capacity of the RoBERTa model. Hence, empirically setting the value of K_{prompt} is essential for good performance. As seen in Fig. 3c, the performance of PICKD follows a bell-shaped curve with increasing K_{prompt}, illustrating the validity of our hypothesis (as discussed above). Hence, it is essential to choose the value of K_{prompt} which maximises the performance of PICKD in different domains.

5.3 Impact of Knowledge Length

The performance of *Response Generator* is dependent on the knowledge triples retrieved by the *Knowledge Selector* module. To investigate this, we conduct an ablation study by varying the knowledge memory size(k_{know}) retrieved by the module in film (Fig. 3a) and music (Fig. 3b) domains. The results show that in the music domain, the performance initially improves with the increase in k_{know}, as more relevant knowledge is retrieved. However, performance deteriorates as k_{know} is increased beyond a certain threshold due to the introduction of irrelevant knowledge. In contrast, in the film domain, the model's performance stabilizes on higher values of k_{know} due to the limited length processing capacity of BART. As the knowledge and dialogue context are concatenated for input to the BART model, increasing the knowledge memory size beyond a certain threshold leads to knowledge truncation, resulting in performance saturation.

5.4 Manual Evaluation

We conduct qualitative evaluations on 50 examples from each of the three domain-specific datasets using two human annotators. Following Zhou et al. [21], the annotators are asked to rate the generated responses on fluency and coherence using a three-point scale (0, 1, 2). The results, shown in Table 3, demonstrate that PICKD outperforms other methods on all datasets, particularly in terms of coherence, indicating its ability to generate responses with relevant knowledge and appropriate context. The high Fleiss' Kappa (κ) indicates the robustness of evaluation due to high inter-annotator agreement.

Table 3. Human evaluation results.

Model	Fluency	Coherence
Film/κ	0.78	0.73
HRED+Know	1.74	0.22
RT-KGD	1.89	0.35
PICKD	**1.94**	**1.25**
Music/κ	0.91	0.93
HRED+Know	1.49	0.38
RT-KGD	1.84	0.46
PICKD	**1.91**	**1.15**
Travel/κ	0.95	0.94
HRED+Know	1.35	0.44
RT-KGD	1.64	1.06
PICKD	**1.70**	**1.44**

5.5 Error Analysis

This section sheds light on interesting scenarios underlying the internal working of PICKD. Table 4 showcases two examples wherein the results produced by PICKD and other baseline models differ from the ground-truth results.

Table 4. Response generation analysis of PICKD and other baselines.

Conversation 1	Conversation 2
U1: Have you heard the song "Everyone Will"?	... U1: Yes, and we know her because she played the second female lead in Blue Bridge of Souls
U2: I've heard it, it's a very good R&B style song	U2: Oh, and this movie is a classic, and it was released in America
U1: How long is it?	U1: Do you know what year it was released?
True Knowledge: Head="Everyone Will" Relation="Song Duration", Tail="2:58"	True Knowledge: Head="Blue Bridge of Souls" Relation="release time", Tail="May 17, 1940"
Human: 2 min and 58 s	Human: Yes, 1940
HRED+Know: The duration is 4 min and 46 s	HRED+Know: Yes, it was on September 14, 2011
RT-KGD: 3 min and 14 s	RT-KGD: You know, it was released on March 25, 1940
PICKD: 2:58, who sang this song?	PICKD: Yes, it was September 9, 1939

The response generated by PICKD in the first example demonstrates effective utilisation of the knowledge triple, resulting in a grammatically correct and semantically consistent response. However, such responses may adversely impact automatic evaluation scores, leading to underestimated model performance. In contrast, baseline methodologies generate factually incorrect responses with ineffective utilisation of knowledge. The second example illustrates a scenario where the responses generated by the baseline models are fluent but factually inconsistent, possibly due to incorrect knowledge retrieval by the *Knowledge Retriever*. This highlights the challenge of effectively retrieving relevant knowledge and generating factually correct responses. Notably, even though the RT-KGD model produces inaccurate responses, it may score higher on automatic metrics due to the presence of lexical overlap with the human response. Such cases underscore the need to consider multiple evaluation metrics to avoid overestimating the performance of models.

6 Conclusion

In this work, we propose PICKD, a two-stage framework for knowledge-grounded dialogue. PICKD employs the novel *In-Situ* prompt tuning mechanism enabling the selection of appropriate knowledge suited to the dialogue context. The second stage of PICKD engages the BART model as the encoder-decoder framework for knowledgeable response generation. We conduct extensive analysis and depict the performance on three domain-specific knowledge-grounded dialogue datasets and exhibit the improved performance in knowledgeable response generation. In future, we plan on exploring prompting mechanisms in the *Response Generator* and extend PICKD to prompt-tuning based end-to-end models for knowledge-grounded dialogue.

Acknowledgements. This work is supported by a grant from The Government of Ireland Postgraduate Fellowship, Irish Research Council under project ID GOIPG/2019/3480. The work is also co-supported by Science Foundation Ireland under grant number SFI/12/RC/2289 2 (Insight).

References

1. Dinan, E., Roller, S., Shuster, K., Fan, A., Auli, M., Weston, J.: Wizard of Wikipedia: knowledge-powered conversational agents. In: ICLR (2019)
2. Fu, T., Zhao, X., Tao, C., Wen, J., Yan, R.: There are a thousand hamlets in a thousand people's eyes: enhancing knowledge-grounded dialogue with personal memory. In: ACL (2022)
3. Kingma, D.P., Ba, J.: Adam: a method for stochastic optimization. In: ICLR (2015)
4. Lewis, M., et al.: BART: denoising sequence-to-sequence pre-training for natural language generation, translation, and comprehension. In: ACL (2020)
5. Liu, Y., et al.: RoBERTa: a robustly optimized BERT pretraining approach. CoRR abs/1907.11692 (2019)
6. Liu, Z., et al.: Multi-stage prompting for knowledgeable dialogue generation. In: Findings of the ACL (2022)
7. Radford, A., Wu, J., Child, R., Luan, D., Amodei, D., Sutskever, I., et al.: Language models are unsupervised multitask learners. OpenAI blog **1**(8), 9 (2019)
8. Roller, S., et al.: Recipes for building an open-domain chatbot. In: EACL (2021)
9. Serban, I.V., Sordoni, A., Bengio, Y., Courville, A.C., Pineau, J.: Building end-to-end dialogue systems using generative hierarchical neural network models. In: AAAI (2016)
10. Sordoni, A., et al.: A neural network approach to context-sensitive generation of conversational responses. In: NAACL (2015)
11. Wang, K., et al.: RT-KGD: relation transition aware knowledge-grounded dialogue generation. In: ISWC (2022)
12. Wu, Z., Bi, W., Li, X., Kong, L., Kao, B.: Lexical knowledge internalization for neural dialog generation. In: Proceedings of the ACL (2022)
13. Yang, C., et al.: TAKE: topic-shift aware knowledge selection for dialogue generation. In: COLING (2022)
14. Zhang, S., Du, Y., Liu, G., Yan, Z., Cao, Y.: G4: grounding-guided goal-oriented dialogues generation with multiple documents. In: Proceedings of the Second DialDoc@ACL 2022 (2022)
15. Zhang, Y., et al.: DIALOGPT : Large-scale generative pre-training for conversational response generation. In: ACL (2020)
16. Zhao, X., Wang, L., He, R., Yang, T., Chang, J., Wang, R.: Multiple knowledge syncretic transformer for natural dialogue generation. In: WWW (2020)
17. Zhao, X., Wu, W., Xu, C., Tao, C., Zhao, D., Yan, R.: Knowledge-grounded dialogue generation with pre-trained language models. In: Webber, B., Cohn, T., He, Y., Liu, Y. (eds.) EMNLP (2020)
18. Zheng, C., Huang, M.: Exploring prompt-based few-shot learning for grounded dialog generation. CoRR abs/2109.06513 (2021)
19. Zhou, H., Huang, M., Liu, Y., Chen, W., Zhu, X.: EARL: informative knowledge-grounded conversation generation with entity-agnostic representation learning. In: EMNLP (2021)
20. Zhou, H., Young, T., Huang, M., Zhao, H., Xu, J., Zhu, X.: Commonsense knowledge aware conversation generation with graph attention. In: IJCAI (2018)

21. Zhou, H., Zheng, C., Huang, K., Huang, M., Zhu, X.: KdConv: a Chinese multi-domain dialogue dataset towards multi-turn knowledge-driven conversation. In: ACL (2020)
22. Zhu, W., Mo, K., Zhang, Y., Zhu, Z., Peng, X., Yang, Q.: Flexible end-to-end dialogue system for knowledge grounded conversation. CoRR abs/1709.04264 (2017)

Fake News Detection Through Temporally Evolving User Interactions

Shuzhi Gong[1(✉)], Richard O. Sinnott[1], Jianzhong Qi[1], and Cecile Paris[2]

[1] The University of Melbourne, Melbourne, VIC 3010, Australia
{shuzhi,rsinnott,jianzhong.qi}@unimelb.edu.au
[2] Data61 CSIRO, Sydney, NSW 1710, Australia
Cecile.Paris@data61.csiro.au

Abstract. Detecting fake news on social media is an increasingly important problem, because of the rapid dissemination and detrimental impact of fake news. Graph-based methods that encode news propagation paths into tree structures have been shown to be effective. Existing studies based on such methods represent the propagation of news through static graphs or coarse-grained graph snapshots. They do not capture the full dynamics of graph evolution and hence the temporal news propagation patterns. To address this issue and model dynamic news propagation at a finer-grained level, we propose a temporal graph-based model. We join this model with a neural Hawkes process model to exploit the distinctive self-exciting patterns of true news and fake news on social media. This creates a highly effective fake news detection model that we named SEAGEN. Experimental results on real datasets show that SEAGEN achieves an accuracy of fake news detection of over 93% with an advantage of over 2.5% compared to other state-of-the-art models.

Keywords: Fake News Detection · Dynamic Graph Embedding

1 Introduction

Fake news created with malicious intent can lead to a substantially negative impact on society, especially during major events such as the U.S. presidential election and the COVID-19 pandemic. To combat the negative impact, various methods have been proposed including exploiting the news content [4], the characteristics of the users involved [17] and the message propagation patterns [1]. In this paper, we focus on detecting fake news propagated on social media platforms such as Twitter through its dissemination and user interactions patterns.

News propagation on social media can be represented by graph-based models where the social media posts (or users) are represented as nodes, while replies, retweets, or other dissemination actions are represented as edges. Many existing graph-based fake news detection models [1,19] use a static graph that shows the complete spatial propagation network after a (fake) news item has been spread. However, these spatial structure-based approaches have largely oversimplified the temporal structure associated with the message propagation, i.e., the

© The Author(s), under exclusive license to Springer Nature Switzerland AG 2023
H. Kashima et al. (Eds.): PAKDD 2023, LNAI 13938, pp. 137–148, 2023.
https://doi.org/10.1007/978-3-031-33383-5_11

sequence and interval of the messages propagated along the timeline. For example, in Fig. 1, three news propagation graphs (each node represents a post) share the same tree structure, but have different temporal patterns. The propagation graphs in Figs. 1a and 1b differ in the time order when the nodes v_2, v_3, and v_4 are added to the graphs. For the propagation graphs in Figs. 1a and 1c, whilst their time orders are the same, nodes v_1 to v_4 in Fig. 1c are much closer in time.

Further, it has been observed that the fake news propagation process exhibits a viral nature and has different stages in terms of people's attention and reactions, resulting in a unique life cycle [16]. In particular, fake news tend to exhibit a sudden increase in the propagation process, while true news have a much smoother process. A sudden increase can be related to the *self-exciting phenomenon* [13] caused by social bot promotions or people's rapid actions to question or correct false information [16].

Such observations motivate us to model the temporal evolving nature of user interactions (i.e., news propagation process) as the basis for detecting fake news.

Sequence-based methods (e.g., [8]) flatten the propagation graph into a chronological sequence of events. Models such as Recurrent Neural Networks (RNN) and BERT [3] can then be applied to learn temporal patterns. A limitation of such methods is that they largely overlook the graph structure of the news propagation patterns.

Other studies [2] use propagation graph snapshots to model both the spatial and temporal propagation patterns. This method only captures the graphs at selected time points, hence they may miss the exact time when a drastic change in the propagation graph occurs. Besides, these studies ignore the self-exciting phenomenon associated with fake news.

To better model the temporal propagation patterns for fake news detection, we propose the **S**elf-**E**xciting-**A**ware **G**raph **E**volution **N**etwork (SEAGEN) model, based on the temporal interactions associated with news propagation processes. We represent news propagation on social media using temporal graphs, where social media posts are nodes and user interactions are the edges. Different from existing graph-based methods, we encode the graph evolution process by integrating local sub-graph modeling and global evolution modeling. Our model

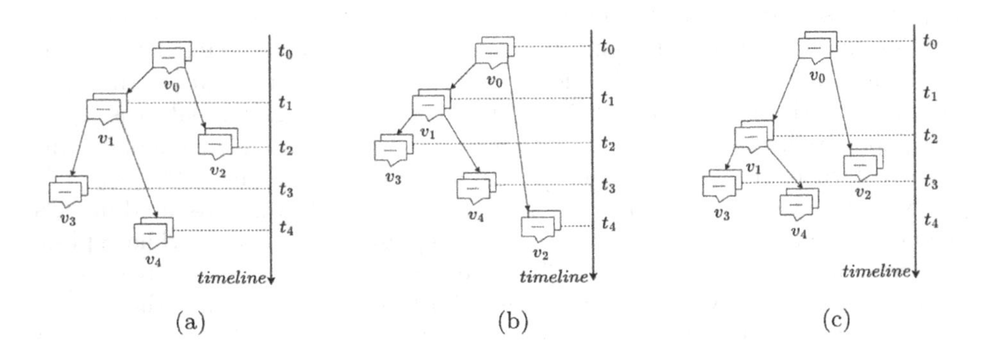

Fig. 1. Example of three news propagation graphs.

consists of three main components: (1) a *local sub-graph encoding module*, (2) a *self-attention-based global temporal evolution module*, and (3) a *neural Hawkes process-based self-exciting module*.

The local sub-graph encoding module encodes a local sub-graph, which models the interaction between a graph node and its neighbouring nodes. Each local sub-graph represents a single temporal stage in the graph evolution process. By encoding the full sequence of local sub-graphs, SEAGEN learns fine-grained news propagation patterns that can then be used for fake news detection.

A Transformer-based global temporal evolution module then integrates the sequence of local sub-graphs and captures the overall temporal evolution (i.e., news propagation) process. It uses a self-attention mechanism to re-weight the (encoded) local sub-graphs based on their content and timestamp of interactions.

A neural Hawkes process-based self-exciting module models the self-exciting phenomenon using a neural network and Hawkes process. We establish fake news detection and Hawkes intensity prediction to capture the self-exciting nature of social media-based fake news.

To summarise, this paper makes the following contributions:

- we propose a novel model named SEAGEN for fake news detection based on a sub-graph sequence-based approach to model temporal news propagation patterns in social networks;
- to learn local propagation patterns, we propose a time-aware encoder to encode local sub-graphs that model fine-grained user interaction events;
- to learn the overall propagation patterns, we use a self-attention-based global temporal evolution module and a neural Hawkes process module, where the former module integrates patterns learned from the local sub-graphs and the latter captures the self-exciting phenomenon;
- we perform extensive experiments on social media datasets. The results show that SEAGEN outperforms state-of-the-art models with an overall accuracy exceeding 93% for fake news detection, including a detection accuracy increase in early stage.

2 Problem Formulation and Data Structure

We consider a fake news detection dataset from social media, which consists of a set of claims $\mathcal{C} = \{C_1, C_2, \ldots, C_{|\mathcal{C}|}\}$. Each claim $C_i = \{v_0^i, v_1^i, v_2^i, \ldots, v_{n_i}^i, G_i\}$ corresponds to a news item, where v_0^i is a source post (e.g., a source tweet). There should be n_i ($n_i \geq 0$) responding posts (e.g., retweets or replies) $\{v_1^i, v_2^i, \ldots, v_{n_i}^i\}$ listed in chronological order.

Graph $G_i = \langle V_i, E_i, T_i \rangle$ is a temporal graph (a tree) representing the propagation pattern of the posts, where v_0^i is the root node (cf. the top left graph in Fig. 2). The set of nodes $V_i = \{v_0^i, v_1^i, v_2^i, \ldots, v_{n_i}^i\}$ represents the source and the responding posts. The set of edges $E_i = \{e_{st}^i | s, t = 0, \ldots, n_i\}$ represents the responding relationships between the posts, where s and t represent the subscripts of a post and a response to it, respectively. When $s = 0$, it refers to the source post v_0^i. The set $T_i = \{t_{st}^i | s, t = 0, \ldots, n_i\}$ represents the occurrence times of the edges in E_i. In this graph representation, each node $v_j^i \in V_i$ is represented by a vector $\mathbf{x}_j^i$, which is a text embedding of the post content. We use text embeddings from a pre-trained BERT-base text encoder [3] for simplicity.

Given graph G_i, our goal is to classify if it represents a fake or a true news.

We note that replies and retweets represent two different interactions on Twitter. Replies are comments with textual contents, while retweets re-post source posts and usually express a supportive attitude. To take advantage of the textual content of replies and the supportive information of retweets, we represent the replies with their own textual contents and retweets with the source post's textual contents, respectively.

3 Proposed Model

As shown in Fig. 2, SEAGEN consists of three modules: a *local sub-graph encoding* (LSGE) module, a *Transformer-based global evolution capturing* (GEC) module, and a *neural Hawkes process* module. The LSGE module extracts a sequence of sub-graphs from G_i and learns a representation for each sub-graph encoding both the structural and the temporal information (Sect. 3.1). The learned embeddings of the sequence of sub-graphs are fed into the GEC module that re-weights the embeddings and prepares for the neural Hawkes process (Sect. 3.2). The weighted embedding sequence is given to a neural Hawkes process to establish the interaction intensities (Sect. 3.3) and to a feedforward neural network (FFN) to determine whether G_i represents a fake news item. The output of the neural Hawkes process model and the FFN are used to compute the loss function for model training (Sect. 3.4).

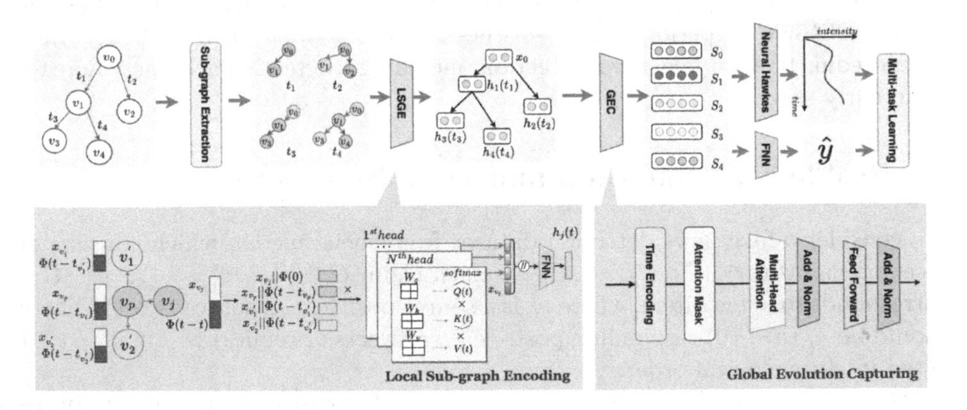

Fig. 2. Architecture of SEAGEN.

3.1 Local Sub-graph Encoding Module

Given graph G_i, our LSGE module first extracts a sequence of local sub-graphs. At each time t_j that corresponds to the occurrence of node $v_j^i \in G_i$ (i.e., post w_j^i) except for the source post (i.e., $j > 0$), we extract a sub-graph of G_i. The sub-graph is formed by v_j^i, v_j^i's parent node v_p^i, and v_p^i's neighbor nodes $\mathcal{N}(v_i^i; t_j)$ before time t_j. For example, as shown in Fig. 2, as time t_3, node v_3 occurs. A local sub-graph is formed by v_3, v_1 (i.e., v_3's parent node), and v_0 (i.e., a neighbour of v_1 that occurs before t_3.), in which the latter occurring node v_4 is not included. For simplification, in what follows, we omit the superscript i for the nodes in graph G_i, and we simply use t instead of t_j to represent the occurrence time of v_j.

Each local sub-graph forms a small conversation represented by the topic of discussion when v_j occurs. Then, the sub-graph is embedded by multi-head attention to form a new embedding $h_j(t)$ for node v_j, as detailed below.

When the interaction between parent node v_p and child node v_j occurs at time t, we consider v_p's neighbourhood $\mathcal{N}(v_p; t) = \{v_1^{'}, ..., v_N^{'}\}$ which takes place prior to time t. The node features of v_j, v_p, and v_p's neighbours are input together into the sub-graph encoder to produce a sub-graph representation $\mathbf{h}_j(t)$. For any node v_k, we use the text representation $\mathbf{x}_k \in \mathbb{R}^{d_0}$ to denote its node features, where d_0 is the dimension of the text representation. Every interaction between node v_k and its parent node is also associated with a posting time t_k. To compute the sub-graph representation $\mathbf{h}_j(t)$, a multi-head attention (MHA) mechanism [20] is utilised to integrate features from two interacting nodes and their neighbours. The query $\mathbf{Q}(t)$ is defined from the child node v_j which initiates the interaction. The key $\mathbf{K}(t)$ and the value $\mathbf{V}(t)$ are defined from the parent node v_p and its neighbours. Formally,

$$\mathbf{h}_j(t) = \mathbf{FFN}(\mathbf{x}_j || \hat{\mathbf{h}}(t)) \tag{1}$$

$$\hat{\mathbf{h}}(t) = \mathbf{MHA}(\mathbf{Q}(t), \mathbf{K}(t), \mathbf{V}(t)) \tag{2}$$

$$\mathbf{Q}(t) = \mathbf{x}_j || \Phi(0) \tag{3}$$

$$\mathbf{K}(t) = \mathbf{V}(t) = [\mathbf{x_p} || \Phi(t - t_p), \mathbf{x_1^{'}} || \Phi(t - t_1^{'}), ..., \mathbf{x_N^{'}} || \Phi(t - t_N^{'})] \tag{4}$$

Here, $\mathbf{h}_j(t)$ is the sub-graph representation of the current interaction and also the computed hidden representation for node v_j, $\Phi(\cdot)$ represents the generic time encoding [21], $||$ is the concatenation operator, and t_p, $\{t_1^{'}, ..., t_N^{'}\}$ are the posting times of node v_p and its neighbours $\mathcal{N}(v_p; t)$. FFN is a feedforward network and MHA is a multi-headed attention layer. FNN and MHA contain all trainable parameters of the LSGE.

After LSGE, the representation of the full temporal graph G_i consists the embeddings of the source node and the sub-graphs (each corresponding to a responsive node in G_i): $\mathbf{H} = \{\mathbf{x}_0, \mathbf{h}_1(t_1), ..., \mathbf{h}_k(t_k), ..., \mathbf{h}_N(t_N)\}$.

3.2 Global Evolution Capturing Module

Next, we employ a Transformer [20] encoder based self-attention module, Global Evolution Capturing (GEC) module, to capture the global evolution process. The output of LSGE, which is the sub-graph enhanced representation of all nodes in G_i, is fed into GEC in chronological order.

The original Transformer only considers element-wise relative positions via Positional Encoding [20]. In our case, we need to be aware the interaction absolute occurrence times for later neural Hawkes process module. Therefore, an adaptive temporal-aware Positional Encoding is formulated as:

$$[\mathbf{z}(t_j)]_k = \begin{cases} \cos(t_j/10000^{\frac{k-1}{M}}), \text{if } k \text{ is odd} \\ \sin(t_j/10000^{\frac{k}{M}}), \text{if } k \text{ is even} \end{cases} \tag{5}$$

where $\mathbf{z}(t_j) \in \mathbb{R}^M$ is the time encoding of the j-th element in the sequence, and M is the dimension of the encoding. Then the output of LSGE, i.e., $\mathbf{H}$, is first enhanced with the time encoding as follows:

$$\mathbf{H}' = \mathbf{H} + \mathbf{Z} \tag{6}$$

After adding the time encoding, $\mathbf{H}'$ is passed to a self-attention module:

$$\mathbf{S} = Softmax(\frac{\mathbf{QK}^\top}{\sqrt{M_K}})\mathbf{V} \tag{7}$$

$$\mathbf{Q} = \mathbf{H}'\mathbf{W}^{\mathbf{Q}}, \mathbf{K} = \mathbf{H}'\mathbf{W}^{\mathbf{K}}, \mathbf{V} = \mathbf{H}'\mathbf{W}^{\mathbf{V}}. \tag{8}$$

Here, $\mathbf{Q}$, $\mathbf{K}$, ad $\mathbf{V}$ are the query, key, and value of the self-attention [20] transformed from $\mathbf{H}'$, while $\mathbf{W}^{\mathbf{Q}}$, $\mathbf{W}^{\mathbf{K}}$, and $\mathbf{W}^{\mathbf{V}}$ are the weights of the linear transformations. Note that multi-head self-attention is also implemented.

The output $\mathbf{S}$ will be used in the neural Hawkes process module to compute the continuous conditional intensity which describe the dynamics of the news propagation process.To prevent leftward information flow (i.e., the neural Hawkes process predicts intensities by inferring future events' timestamps), attention mask is implemented like [20]. The veracity prediction is also computed based on $\mathbf{S}$, as defined by the following equation.

$$\hat{y} = \mathbf{FFN}(MeanPooling(\mathbf{S})) \tag{9}$$

3.3 Neural Hawkes Process Module

A Hawkes process [5] is a self-exciting point process. It can simulate news propagation by modelling the generation of social media posts (tweets) over a continuous time domain. The frequency of posts and responses/retweets generated is determined by an underlying intensity function which considers the impact of past posts. The intensity function $\lambda(t)$ models the self-exciting nature by summation of the impact of past posts, which is defined as:

$$\lambda(t) = \mu_0(t) + \sum_{t_k<t}^{t} \phi(t - t_k) \tag{10}$$

where $\mu_0(t)$ is the base intensity, ϕ is a so-called kernel function that is used to modulate the effect of previous events k on the intensity $\lambda(t)$.

The traditional Hawkes process oversimplifies the dynamics of point processes, and assumes that all previous events have positive impact on the occurrence of the current event. However, a user's behaviour can contribute to the spread of rumors on social media and/or curb them. To overcome this limitation, we adopt a neural Hawkes based process [22] to model the complicated self-exciting phenomenon.

The output of GEC $\mathbf{S} = \{\mathbf{s}_1, ..., \mathbf{s}_N\}$ is fed forward into the neural Hawkes process module, to compute the continuous user interaction intensity. Given a self-attentive sequence of interactions with timestamps before time t: $\mathcal{H} = \{(\mathbf{s_k}, t_k) : t_k < t\}$, the continuous intensity at time t, $\lambda(t|\mathcal{H}_t)$, is computed by:

$$\lambda(t|\mathcal{H}_t) = f(\alpha\frac{t - t_j}{t_j} + w^\top \mathbf{s_j} + b) \tag{11}$$

$$f(x) = \beta \log(1 + \exp(x/\beta)) \tag{12}$$

where α, β, and w are learnable parameters, $\mathbf{s}_j$ is the self-attentive hidden state for corresponding responsive post that occurs just before time t.

3.4 Model Training

For a sequence $\mathbf{S}$ over observation interval $[t_1, t_L]$ with a continuous conditional intensity function given as $\lambda(t|\mathcal{H}_t)$, the log-likelihood can be computed as:

$$\mathcal{L}_S = \sum_{j=1}^{L} \log \lambda(t|\mathcal{H}_t) - \int_{t_1}^{t_L} \lambda(t|\mathcal{H}_t)dt \tag{13}$$

where the left part is the event log-likelihood and the right part is the non-event log-likelihood. To calculate the integral non-event log-likelihood, Monte Carlo integration [14] is utilised.

Meanwhile, the veracity prediction loss is computed as a cross-entropy loss:

$$\mathcal{L}_C = -y \log(\hat{y_1}) - (1 - y) \log(\hat{y_0}) \tag{14}$$

where the $\hat{y} = [\hat{y_0}, \hat{y_1}]$ denotes the probability of a given piece of news to be true or false. The final loss function is the sum of L_S and L_C as weighted by γ.

$$\mathcal{L} = \mathcal{L}_C + \gamma \cdot \mathcal{L}_S \tag{15}$$

Table 1. Statistics of the Datasets.

Statistic	#source tweets	#users	#fake news	#true news	Avg. time length
Twitter	1,147	29,858	578	569	158 h
FakeNewsNet	4,168	45,109	2,079	2,089	1,951 h

4 Experiment

4.1 Datasets

We use public Twitter datasets Twitter [9] and FakeNewsNet [16]. The former in particular is formed by true-rumours and false-rumours from the Twitter15 [9] and Twitter16 [9] datasets, named as true news and fake news, respectively.

FakeNewsNet consists of two classes of data: true news and fake news. Table 1 summarises the datasets.

4.2 Baseline Methods

We compare with state-of-the-art fake news detection models including:

- **RvNN** [10] uses a recurrent neural network to learn discriminative features from post contents by following their non-sequential propagation structure.
- **Sta-PLAN** [7] uses a self-attention mechanism and position encoding [20] to extract textual features for sequence embedding learning.
- **STS-NN** [6] jointly models the spatial and the temporal structures of the message propagation process using a gated recurrent unit (GRU).
- **Bi-GCN** [1] represents social media posts as nodes in a graph and utilises a graph convolutional network (GCN)-based model to encode the graph.
- **Dy-GCN** [2] takes snapshots of the message evolution process, builds a graph for every snapshot, and then encodes the graph snapshots by a GCN.
- **GACL** [19] enhances Bi-GCN by generating adversarial training samples and training based on contrastive learning.

4.3 Experiment Setting

We run the baseline models with the default settings as reported in their original papers. We implement our model SEAGEN in Python 3.8 and run it on a NVIDIA A100 GPU. Datasets are split into training and test sets with a split ratio of 8:2 without overlapping. A pre-trained BERT model is used to compute textual embeddings as the initial graph node features of SEAGEN. The weighting parameter γ in the final loss function is set as $5e-5$ via a grid search in $\{5e-3, 5e-4, 5e-5, 5e-6\}$. The model parameters are optimised using the Adam algorithm, and the model performance is evaluated by a 5-fold cross validation. The average accuracy and F1 scores are reported as the evaluation metrics. We will release our code in: https://github.com/gszswork/SEAGEN

Table 2. Fake News Detection Performance on `Twitter` and `FakeNewsNet` (F-F1: F1 score for fake news detection; T-F1: F1 score for true news detection).

Method	`Twitter`			`FakeNewsNet`		
	Acc	F-F1	T-F1	Acc	F-F1	T-F1
RvNN	0.805	0.803	0.807	0.828	0.801	0.829
Sta-PLAN	0.780	0.780	0.779	0.800	0.794	0.801
STS-NN	0.834	0.834	0.833	0.858	0.857	0.858
Bi-GCN	0.864	0.865	0.863	0.889	0.889	0.890
Dy-GCN	0.873	0.872	0.873	0.896	0.894	0.896
GACL	0.878	0.875	0.880	0.905	0.906	0.902
`SEAGEN`	**0.908**	**0.910**	**0.906**	**0.930**	**0.929**	**0.931**
gain	+3%	+3.5%	+2.6%	+2.5%	+2.3%	+2.9%

4.4 Performance Comparison

Table 2 shows the overall model performance results. On the two datasets, our model `SEAGEN` significantly outperforms all the competitors – the performance gain is up to 3%. This confirms the effectiveness of `SEAGEN` and using the temporally evolving graph embeddings for fake news detection.

Further, we observe that the methods using graph neural networks (Bi-GCN, Dy-GCN, GACL) outperform sequence-based methods (RvNN, Sta-PLAN, and STS-NN), which confirms the effectiveness of graph based methods.

Meanwhile, the dynamic graph method Dy-GCN learns from snapshots of static graphs at different time points. It only yields a marginal improvement over Bi-GCN. Its performance is limited by its coarse-grained graph encoding because the temporal graph information between snapshots cannot be captured. GACL which is an enhanced Bi-GCN model achieves the second best performance. We attribute this to the adversarial training samples and contrastive learning.

4.5 Ablation Study

To analyse the impact of each module in `SEAGEN`, we implement the following variants:

- **w/o LSGE**: Removing the local sub-graph encoder and feeding the node features in temporal order directly into the Global Evolution Capturing module.
- **w/o GEC**: Removing the global evolution capturing module.
- **w/o Hawkes**: Removing the neural Hawkes process module and deactivating joint training.

The comparative performance patterns on the F1 scores are similar and hence are omitted due to space limit. Same below. Each variant is trained on the same datasets as the full model `SEAGEN`. The experimental results are shown in

Table 3. We see that the full model SEAGEN outperforms all model variants, confirming that each module contributes to the overall model performance. Among these variants, the w/o LSGE has the worst performance; this is because the SEAGEN will degrade to a pure sequence model without the LSGE kernel. We can also see that the GEC and Hawkes module do have contributions to the model performance by capturing evolution sequential and self-exciting features.

4.6 Early Detection Performance

Table 3. Fake News Detection Accuracy on Twitter and FakeNewsNet.

Variant	Twitter	FakeNewsNet
w/o LSGE	0.866	0.884
w/o GEC	0.878	0.915
w/o Hawkes	0.899	0.918
SEAGEN	0.908	0.930

To mitigate the negative impact of fake news, it is crucial to detect fake news as early as possible. Therefore, we further study our model performance on the early detection of fake news. We take the first 20%, 40%, 60%, 80%, and 100% (in terms of the response amount of a news) subsets of the two benchmark datasets to train and test the models and show the accuracy in Fig. 3. Here, we only show the most competitive baselines Dy-GCN and GACL to simplify the figure.

It can be observed that, with the change of subset amounts, our model SEAGEN consistently outperforms both Dy-GCN and GACL. As longer time spans are considered, the performance gaps increases because more detailed temporal information is provided. The performance gain of SEAGEN over GACL increases around 2% when only the first 20% replies/retweets of a source post is available, which can result from the temporal feature extraction in our modules. This confirms the effectiveness of SEAGEN in the early detection of fake news.

4.7 Case Study

We next showcase two samples from FakeNewsNet which are mis-classified by Bi-GCN, GACL and Dy-GCN but successfully classified by SEAGEN. The true news item A (Jennifer Aniston on a Friends Reunion: Anything Is a Possibility...I Mean, George Clooney Got Married.) and the fake news item B (Kristen Stewart On Dating Robert Pattinson: The Public Were The Enemy.) share almost the

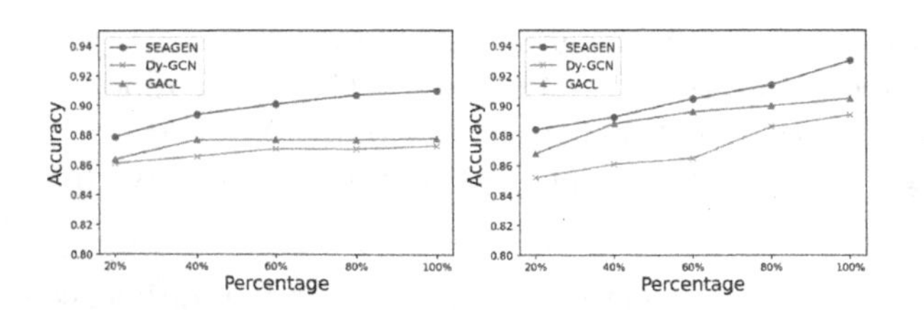

Fig. 3. Early Fake News Detection Performance on Twitter (left) and FakeNewsNet (right).

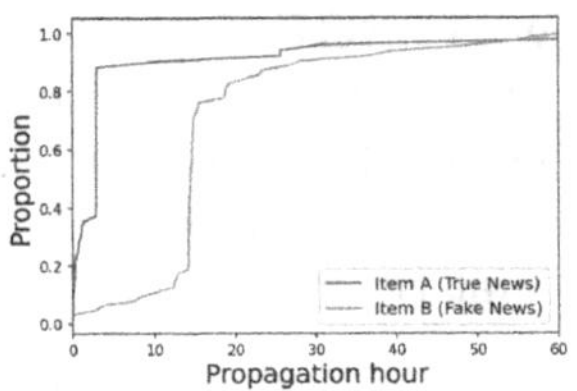

Fig. 4. ecdf plot for item A, B.

same propagation structure where most relies and retweets straightly interact with the source post. Examples A and B also have similar propagation time spans (around 60 h) and propagation size (around 70 responses), and hence they are difficult to be distinguished by static graph methods (Bi-GCN, GACL). However, their propagation speeds over time can vary. The propagation speeds of these two examples in the first 60 h are visualised by their empirical Cumulative Distribution Function (ecdf) in Fig. 4. Compared to that of the true news item A, the propagation of fake news item B keeps mild until a viral spread hits at around 15 h, which can be captured by `SEAGEN`'s GEC and the neural Hawkes module. In contract, Dy-GCN takes snapshots at equally spaced timestamps (e.g., 20 h, 40 h, 60 h), thus failing to capture the sudden increase at early stage, which exemplifies the weakness of coarse-grained graph snapshot-based methods.

5 Related Work

Existing approaches for fake news detection can be broadly divided into *content-based*, *social context-based*, and *environment-based*.

Content-based approaches learn content or style features from the text or media content of news [4]. They may also leverage external knowledge for fact checking [15]. Social context-based approaches detect fake news through user features [17] or propagation analysis. The propagation analysis is a hot topic and develops from sequence modelling [7,8] to graph modelling [1]. Propagation temporal features are also exploited recently, such as Choi et al. [2] encode the propagation as graph snapshots, Song et al. [18] utilises TGN [21] to encode the propagation graph, and [11] detects fake news through self-exciting difference. Environment-based approaches [12] mainly zooms out and considers the association across multiple news articles to debunk fake news.

In this paper, our approach focuses on social propagation graph, utilises graph and sequential features in a hybrid way. Self-exciting features is also captured by joint learning.

6 Conclusion

In this paper, we proposed a novel social media fake news detection model `SEAGEN`. It models the news propagation via self-exciting aware sub-graph sequence encodings. Extensive experiments on two benchmark datasets demonstrated the superiority of our model, including in detection in early stage. In the future work we intend to extend current work to user based temporal graph, which will be more suitable to be deployed as an online learning framework.

Acknowledgement. This research is supported by University of Melbourne and CSIRO.

References

1. Bian, T., Xiao, X., Xu, T., Zhao, P., Huang, W., Rong, Y., Huang, J.: Rumor detection on social media with bi-directional graph convolutional networks. In: AAAI (2020)
2. Choi, J., Ko, T., Choi, Y., Byun, H., Kim, C.: Dynamic graph convolutional networks with attention mechanism for rumor detection on social media. PLOS One. **16**(8), e0256039 (2021)
3. Devlin, J., Chang, M.W., Lee, K., Toutanova, K.: BERT: Pre-training of deep bidirectional transformers for language understanding. In: NAACL (2019)
4. Feng, S., Banerjee, R., Choi, Y.: Syntactic stylometry for deception detection. In: ACL (2012)
5. Hawkes, A.G.: Spectra of some self-exciting and mutually exciting point processes. Biometrika **58**(1), 83–90 (1971)
6. Huang, Q., Zhou, C., Wu, J., Liu, L., Wang, B.: Deep spatial-temporal structure learning for rumor detection on twitter. Neural Computing and Applications (2020)
7. Khoo, L.M.S., Chieu, H.L., Qian, Z., Jiang, J.: Interpretable rumor detection in microblogs by attending to user interactions. In: AAAI (2020)
8. Ma, J., et al.: Detecting rumors from microblogs with recurrent neural networks. In: IJCAI (2016)
9. Ma, J., Gao, W., Wong, K.F.: Detect rumors in microblog posts using propagation structure via kernel learning. In: ACL (2017)
10. Ma, J., Gao, W., Wong, K.F.: Rumor detection on twitter with tree-structured recursive neural networks. In: ACL (2018)
11. Naumzik, C., Feuerriegel, S.: Detecting false rumors from retweet dynamics on social media. In: WWW (2022)
12. Nguyen, V.H., Sugiyama, K., Nakov, P., Kan, M.Y.: Fang: Leveraging social context for fake news detection using graph representation. In: Proceedings of the 29th ACM International Conference on Information and Knowledge Management, pp. 1165–1174 (2020)
13. Nie, H.R., Zhang, X., Li, M., Dolgun, A., Baglin, J.: Modelling user influence and rumor propagation on twitter using Hawkes processes. In: DSAA (2020)
14. Robert, C.P., Casella, G., Casella, G.: Monte Carlo Statistical Methods, vol. 2. Springer, New York (1999). https://doi.org/10.1007/978-1-4757-4145-2
15. Samarinas, C., Hsu, W., Lee, M.L.: Improving evidence retrieval for automated explainable fact-checking. In: NAACL (2021)
16. Shu, K., Mahudeswaran, D., Wang, S., Lee, D., Liu, H.: Fakenewsnet: a data repository with news content, social context, and spatiotemporal information for studying fake news on social media. Big Data **8**(3), 171–188 (2020)
17. Shu, K., Wang, S., Liu, H.: Beyond news contents: the role of social context for fake news detection. In: WSDM (2019)
18. Song, C., Shu, K., Wu, B.: Temporally evolving graph neural network for fake news detection. Inf. Process. Manage. **58**(6), 102712 (2021)
19. Sun, T., Qian, Z., Dong, S., Li, P., Zhu, Q.: Rumor detection on social media with graph adversarial contrastive learning. In: WWW (2022)
20. Vaswani, A., et al.: Attention is all you need. In: NeurIPS (2017)
21. Xu, D., Ruan, C., Korpeoglu, E., Kumar, S., Achan, K.: Inductive representation learning on temporal graphs. In: ICLR (2020)
22. Zuo, S., Jiang, H., Li, Z., Zhao, T., Zha, H.: Transformer Hawkes process. In: ICML (2020)

Improving Machine Translation and Summarization with the Sinkhorn Divergence

Shijie Li[1], Inigo Jauregi Unanue[1,2], and Massimo Piccardi[1(✉)]

[1] University of Technology Sydney, Ultimo, NSW, Australia
Shijie.Li@student.uts.edu.au, Inigo.Jauregi@rozettatechnology.com, Massimo.Piccardi@uts.edu.au

[2] RoZetta Technology, Sydney, NSW, Australia

Abstract. Important natural language processing tasks such as machine translation and document summarization have made enormous strides in recent years. However, their performance is still partially limited by the standard training objectives, which operate on single tokens rather than on more global features. Moreover, such standard objectives do not explicitly consider the source documents, potentially affecting their alignment with the predictions. For these reasons, in this paper, we propose using an Optimal Transport (OT) training objective to promote a global alignment between the model's predictions and the source documents. In addition, we present an original implementation of the OT objective based on the Sinkhorn divergence between the final hidden states of the model's encoder and decoder. Experimental results over machine translation and abstractive summarization tasks show that the proposed approach has been able to achieve statistically significant improvements across all experimental settings compared to our baseline and other alternative objectives. A qualitative analysis of the results also shows that the predictions have been able to better align with the source sentences thanks to the supervision of the proposed objective.

Keywords: Natural Language Processing · Natural Language Generation · Neural Text Generation · Optimal Transport

1 Introduction

Natural language generation (NLG), a key field for the natural language processing (NLP) community, lends itself to a wide range of applications such as machine translation, text summarization, dialogue systems, and others [14]. In these tasks, attention-based sequence-to-sequence (seq2seq) models [23] are dominant, together with the conventional maximum-likelihood estimation (MLE), which is also known as *teacher forcing* in the area of recurrent neural networks (RNN). This approach maximizes the generation probability of the current

Supplementary Information The online version contains supplementary material available at https://doi.org/10.1007/978-3-031-33383-5_12.

© The Author(s), under exclusive license to Springer Nature Switzerland AG 2023
H. Kashima et al. (Eds.): PAKDD 2023, LNAI 13938, pp. 149–161, 2023.
https://doi.org/10.1007/978-3-031-33383-5_12

target word conditioned on all the previous ground-truth inputs and all the source words. However, it has been widely criticized for both its conditioning of the predictions on ground-truth information, unavailable at inference time, and its inability to capture sentence-level features by only operating at token level [19].

Several attempts have been made to address the limitations of standard MLE by adopting sentence-level objectives. For example, Ranzato et al. [19] have trained their models by directly optimizing sentence-level evaluation metrics such as the BLEU [16] and ROUGE [13] scores. However, these two metrics compute sentence similarity primarily based on surface matches of n-grams. While they can provide sentence-level information to a certain extent, they struggle to reward context-preserving lexical equivalence. Additionally, they are typically based on *hard* predictions, i.e., sequences of labels, which make the metrics non-differentiable as they are flat and subject to change discontinuously in the parameter space. For this reason, optimizing them often requires resorting to slow and high-variance gradient estimation techniques such as policy gradient.

An efficient alternative to optimizing the above n-gram-based metrics is to optimize embedding-based ones. Several such metrics have been proposed in recent years, including the Word Mover's Distance [10], MoverScore [26], and BERTScore [25]. The matching schemes in these advanced sentence-level metrics better preserve semantically-relevant information, especially when combined with contemporary pretrained language models such as BERT [5] and BART [11]. More importantly, they straightforwardly support optimization as they are based on continuous quantities. For instance, Jauregi Unanue et al. [8] have proposed fine-tuning machine translation models by using BERTScore as the training objective, reporting consistent improvements over a variety of language pairs. However, most of these methods only focus on measuring the similarity between predictions and references, and rarely pay attention to the source documents. While the source information can be covered by the inner cross-attention mechanism to some extent, the attention mechanism itself has been criticized in recent years. For example, in the absence of constraints, some of the source tokens may be rarely attended to. To amend this, some approaches have started to include explicit coverage terms in the models [7,17].

To address the above issues, in this paper, we focus on providing text generation models with sentence-level supervision directly from the source text. To achieve this goal, we propose a novel training objective based on the minimization of the recently-proposed Sinkhorn divergence (SD) [6] between the hidden states of the encoder and decoder. The Sinkhorn divergence is a variant of the general optimal transport (OT) problem, which can be used to optimally align two arbitrary sets of weighted elements. The proposed objective only utilizes the contextualized source information already learned by the encoder, without introducing any additional module or memory footprint. In addition, the inference remains unchanged and its run time is unaffected. Overall, our paper makes the following main contributions:

- A novel training objective for conditional text generation models such as machine translation and document summarization providing sentence-level supervision directly from the source text.

- An original implementation of the objective leveraging the context-aware hidden states of the encoder and the decoder, and the Sinkhorn divergence – a performing variant of OT distance.
- Experimental results on machine translation and abstractive summarization showing marked improvements in both text quality and word alignment over an MLE baseline and all other compared objectives.

2 Related Work

Sentence-Level Supervision. The sentence-level supervision used in early research was typically performed with the non-differentiable metrics used for evaluation, such as the BLEU and ROUGE scores used in [19]. With the increases in model capacity and training data size, advanced language models such as BERT [5] and BART [11] have shown their ability to learn context-aware representations of the input sentences. As a result, researchers have started to leverage these pretrained representations as sentence-and document-level signals. For example, Zhang et al. [25] have utilized BERT as a sentence-level evaluation metric, and Chen et al. [3] have focused on distilling knowledge learned by a large BERT model for training smaller, student models. Typically, these context-aware representations are extracted from the last layer of the language models, and we follow this line in our implementation.

Coverage of Source Information. Approaches for explicitly covering the source-side information are an important component of statistical machine translation, and also the founding idea behind the attention mechanism in contemporary NLG models [24], which adaptively focus on different parts of the source sentence at each generation step. In addition, networks such as the copying net [28] have directly allowed copying content from the source text to the predictions, leveraging the homogeneity of the source information and the output in NLG tasks. More recently, Garg et al. [7] have jointly trained an explicit alignment module for source and target sentences when training machine translation models and Parnell et al. [17] have proposed a reinforcement learning reward for multi-document summarization to even out the individual contributions of the source documents.

Optimal Transport. Optimal transport (OT) was first introduced in NLP by Kusner et al. [10] as a way to measure the distance between two documents. Since then, OT has been widely used in several other applications. For instance, Alqahtani et al. [1] have utilized it as an objective for word alignment while Zhao et al. [26] have used it as an evaluation metric. In terms of NLG tasks, Chen et al. [2] and Wang et al. [12] have demonstrated improved performance by minimizing the OT distance between the references and sentences generated with teacher forcing (TFOT) and student forcing (SFOT), respectively. In turn, Nguyen et al. [15] have used the OT distance for knowledge distillation and shown improvements in cross-lingual text summarization. However, none of these OT-based methods has paid explicit attention to context-aware source information.

3 Methodology

In this section, we first briefly recap the standard seq2seq training, then provide the basics of OT optimization, and finally introduce the proposed approach.

3.1 Sequence-to-Sequence Model Training

The seq2seq framework is essentially an encoder-decoder architecture, where the encoder is responsible for mapping a source sentence $X_{1:N} = (x_1, \dots, x_N)$ to a sequence of hidden vectors, or states, $H_{1:N} = (h_1, \dots, h_N)$, and the decoder is responsible for eventually mapping these hidden vectors to a target sentence, $Y_{1:M} = (y_1, \dots, y_M)$. In the original seq2seq model [23], only the last hidden state of the encoder, h_N, was fed into the decoder, limiting the source information available to the decoder. However, this limitation was removed by the attention mechanisms [14], which leverage all the encoder's hidden states. The standard MLE training objective of seq2seq models is to minimize the negative log-likelihood of the target sentence, $Y_{1:M}$, conditioned on $X_{1:N}$:

$$\mathcal{L}_{\text{MLE}} = -\log P_\theta(Y_{1:M}|X_{1:N}) = -\sum_{m=1}^{M} \log P_\theta(y_m|y_{<m}, X_{1:N}) \tag{1}$$

3.2 The Proposed Approach: A Contextual Sinkhorn Divergence

Optimal transport aims to determine the best linear assignment between the elements of two sets under given marginal constraints. To formally describe the proposed approach, we first introduce a cost matrix, C, such that $C_{ij} = c(x_i, y_j)$ is a distance between the vectorized token x_i and token y_j, which are denoted as h_n^S and h_m^T, respectively. Additionally, we introduce two discrete marginal distributions:

$$\Phi = \sum_{n=1}^{N} \phi_n \delta_{h_n^S} \quad ; \quad \Psi = \sum_{m=1}^{M} \psi_m \delta_{h_m^T} \tag{2}$$

where ϕ and ψ are individual weights with respect to each token in $X_{1:N}$ and $Y_{1:M}$ and δ_π is the Dirac function centred on the vector π. The weight vectors are discrete distributions, with their values lying in the simplex (i.e., $\phi_n, \psi_m \geq 0 \; \forall n, m; \sum_{n=1}^{N} \phi_n = \sum_{m=1}^{M} \psi_m = 1$). Hence, optimal transport aims to find a transport matrix, T, achieving the following minimization:

$$O(\Phi, \Psi) = \min_{T \in \Delta(\Phi, \Psi)} \langle T, C \rangle \tag{3}$$

where $\langle \cdot \rangle$ is the Frobenius dot-product and $\Delta(\Phi, \Psi)$ is the set of joint distributions with respective marginals Φ and Ψ. To achieve this minimization, OT matches token pairs of minimum cost from $X_{1:N}$ and $Y_{1:M}$ in a many-to-many manner, respecting their individual weights. Since this convex optimization can be computationally expensive, Cuturi [4] has proposed the Sinkhorn distance, which is an entropy-regularized OT that can be expressed as:

$$O_\epsilon(\Phi, \Psi) = O(\Phi, \Psi) + \epsilon \cdot h(T) \tag{4}$$

where $h(T)$ is the entropy of the transport matrix T and ϵ is a positive regularization coefficient. Note that one of the main benefits of the Sinkhorn distance is that it can be computed efficiently using a dual form:

$$O_\epsilon(\Phi, \Psi) = \langle \Phi, f \rangle + \langle \Psi, g \rangle \tag{5}$$

While the Sinkhorn distance is computationally efficient, it generally leads to a biased solution for a positive ϵ since $O_\epsilon(\Phi, \Phi) \neq 0$, and thus may not perform ideally as a training objective. For this reason, in our work we have chosen to experiment with the recently-proposed *Sinkhorn divergence* [6], which can be formally defined as:

$$\mathcal{L}_{\mathrm{SD}}(\Phi, \Psi) = O_\epsilon(\Phi, \Psi) - \frac{1}{2} O_\epsilon(\Phi, \Phi) - \frac{1}{2} O_\epsilon(\Psi, \Psi) \tag{6}$$

Intuitively, this divergence normalizes the standard Sinkhorn distance by discounting two symmetric terms, $O_\epsilon(\Phi, \Phi)$ and $O_\epsilon(\Psi, \Psi)$, that reflect the intrinsic "hardness" of its arguments, leaving only the alignment contribution in focus. The Sinkhorn divergence in Eq. 6, too, can be expressed concisely in dual form by simply subtracting the symmetric terms:

$$\mathcal{L}_{\mathrm{SD}}(\Phi, \Psi) = \langle \Phi, (f - f') \rangle + \langle \Psi, (g - g') \rangle \tag{7}$$

where f', g' are the solutions of the respective symmetric problems.

In our implementation, to cater for the information from the source sentence, we compute the Sinkhorn divergence "contextually" by setting the vectors h_n^S and h_m^T in Eq. 2 to the hidden states of the encoder and the decoder. Finally, we compose the seq2seq loss in Eq. 1 and the Sinkhorn divergence in Eq. 6 into our final training objective, which can be expressed as:

$$\mathcal{L} = \mathcal{L}_{\mathrm{MLE}} + \lambda \cdot \mathcal{L}_{SD} \tag{8}$$

where λ is a hyperparameter that controls the magnitude of the OT component. The objective shows that the proposed approach can be seamlessly incorporated into any contemporary encoder-decoder architecture.

4 Experiments

4.1 Datasets

Machine Translation. For this task, we have evaluated our approach on two standard datasets, IWSLT 2014 German↔English (De↔En)[1] and IWSLT

[1] We remark that there are a few misaligned sentence pairs in the official release of this dataset, which end up affecting the test BLEU score. For more details, please refer to https://github.com/pytorch/fairseq/issues/4146. Herein, we report the BLEU scores on the corrected dataset.

2015 English↔Vietnamese (En↔Vi), and one large-scale dataset ($\approx$ 4M parallel sentences), WMT 2014 English→German (En→De). For IWSLT De↔En and WMT En→De, we perform the same data pre-processing steps as in the Fairseq library[2]. For IWSLT En↔Vi, we use the publicly available dataset[3] with TED tst2012 and tst2013 as validation and test sets, respectively. For the two IWSLT datasets, we have tokenized sentences using the tokenizer and vocabulary from the pretrained mBERT base model, as distributed by Hugging Face.[4]. For WMT En→De, we have tokenized the dataset using the byte pair encoding (BPE) of Sennrich et al. [20], with 40K subword merge operations. For evaluation, we report the case-sensitive [22] detokenized sacreBLEU score, as suggested by Post et al. [18], and also the recently proposed BERTScore (F_{BERT}) [25] which nicely complements the BLEU score as it is based on embeddings rather than n-grams.

Abstractive Summarization. For this task, we have trained our models on the English Gigaword dataset provided by Hugging Face[5]. However, the default dataset contains roughly 190K documents in the validation set, which makes the validation process exceedingly slow. For the sake of efficiency, we have instead used the modified dataset provided by Zhou et al. [27]. For tokenization, we have used the same tokenizer of WMT En→De. For evaluation, we report the ROUGE-1, ROUGE-2 and ROUGE-L on both the original and modified test sets for a comprehensive comparison.

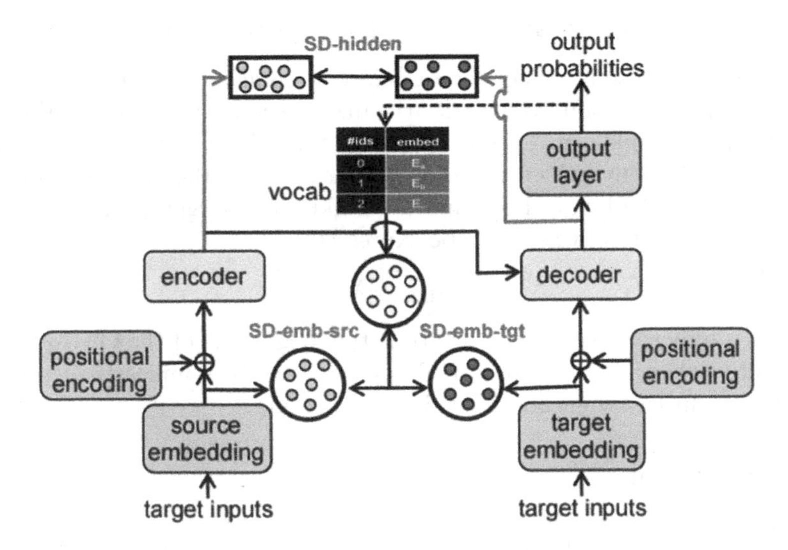

Fig. 1. Illustration of the three different training schemes. The dotted line indicates the path requiring smoothing techniques.

[2] https://github.com/pytorch/fairseq/tree/main/examples/translation.
[3] https://nlp.stanford.edu/projects/nmt/data/iwslt15.en-vi.
[4] https://huggingface.co/bert-base-multilingual-cased.
[5] https://huggingface.co/datasets/gigaword.

4.2 Models and Training

We have used Fairseq's *transformer_iwslt_de_en* configuration for the two IWSLT tasks and *transformer_wmt_en_de* configuration for the WMT task. For the MLE training, we have used the label-smoothed negative log-likelihood with smoothing parameter 0.1. For our combined training objective, we have explored a wide range of values for the hyperparameter λ in Eq. 8 using the IWSLT De→En dataset, and set it to 0.1 for all tasks based on the best performance on the validation set. A sensitivity analysis is presented in Sect. 4.3. During training, we have batched sentences with a maximum number of 8192 tokens and employed the inverse-sqrt learning rate scheduler and the Adam optimizer with $(\beta_1, \beta_2) = (0.9, 0.98)$. At inference time, we have used beam search decoding with a beam size of 5. For measuring the statistical significance with respect to the baseline model, we have used the paired bootstrap resampling test [9] with 2000 resamples.

The proposed approach minimizes the Sinkhorn divergence between the hidden states of the predicted and source sentences, and we therefore note it as SD-hidden hereafter. In order to single out the respective contributions of the contextual representation and the source text, we have also trained two further models for comparison. In the former, we have computed the same Sinkhorn divergence, but between the word embeddings of the predicted tokens and the source tokens (SD-emb-src), and in the latter, those of the predicted tokens and the target tokens (SD-emb-tgt). It is important to note that, in order to use conventional word embeddings in the two compared models, the predictions from the output layer of the decoder need to be "softened". In our experiments, we

Table 1. BLEU and F_{BERT} scores for the IWSLT 2014 De↔En and IWSLT 2015 En↔Vi translation tasks. (†) p-value < 0.05. (‡) p-value < 0.01.

Model	De→En				En→De			
	dev		test		dev		test	
	BLEU	F_{BERT}	BLEU	F_{BERT}	BLEU	F_{BERT}	BLEU	F_{BERT}
Transformer	35.14	67.41	33.44	65.94	29.90	64.27	28.22	63.24
+ SD-emb-src	35.11	67.22	33.39	65.78	29.92	64.41	28.34	63.36
+ SD-emb-tgt	35.27	67.56	33.67†	66.04	29.89	64.44†	28.36	**63.53**‡
+ SD-hidden	**35.59**‡	**67.70**‡	**34.05**‡	**66.31**‡	**30.19**†	**64.66**‡	**28.78**‡	63.46†
Model	Vi→En				En→Vi			
	dev		test		dev		test	
	BLEU	F_{BERT}	BLEU	F_{BERT}	BLEU	F_{BERT}	BLEU	F_{BERT}
Transformer	24.62	57.96	27.48	61.19	26.59	85.14	29.85	86.63
+ SD-emb-src	24.37	57.90	27.43	61.14	27.10†	85.22	30.28	**86.70**
+ SD-emb-tgt	24.27	57.61	27.57	61.01	26.68	**85.26**†	29.63	86.62
+ SD-hidden	**25.03**	**58.14**	**28.23**†	**61.26**	**27.14**†	**85.26**†	**30.41**†	86.61

have used the probability-averaged embedding as suggested by TFOT [2]. The three different training schemes are summarized in Fig. 1.

4.3 Results and Discussion

Machine Translation. We first report the results for the machine translation task over the two IWSLT datasets in Table 1 and the WMT dataset in Table 2. For the two IWSLT translation tasks, both SD-emb models show little or even no improvement compared to the baseline model. However, our SD-hidden model shows consistent improvements over almost all datasets and evaluation metrics, of up to 0.75 pp in BLEU and 0.39 pp in F_{BERT}. Also over the WMT 2014 En→De dataset, our SD-hidden model has performed the best, achieving increases of up to 0.45 pp in BLEU and 0.42 pp in F_{BERT}. Notably, the SD-emb models also show noticeable improvements in this dataset.

For comparison with the literature, we also include other reported sacreBLEU scores over the same datasets.

Table 2. BLEU and F_{BERT} scores for the WMT 2014 En→De translation task. (†) p-value < 0.01. (‡) p-value < 0.001. (⋆) from [21]. (-) not available.

Model	En→De			
	dev		test	
	BLEU	F_{BERT}	BLEU	F_{BERT}
Other Reported Results				
Transformer⋆	-	-	26.5	-
Rel-Transformer⋆	-	-	26.8	-
Our Implementations				
Transformer	29.99	61.47	26.61	63.27
+ SD-emb-src	30.22‡	61.69‡	26.97	63.54
+ SD-emb-tgt	30.20‡	61.65‡	26.95	63.46
+ SD-hidden	**30.3**‡	**61.75**‡	**27.06**†	**63.69**‡

Abstractive Summarization. Table 3 shows the results for the abstractive summarization task. For a fair comparison, where available, we report results for both the original and the modified test set. On the original test set, our SD-hidden model performed the best, achieving 0.22 pp, 0.27 pp and 0.23 pp improvements in ROUGE-1, ROUGE-2 and ROUGE-L scores, respectively. Also on the modified test set, it has achieved marked improvements over the baseline of 0.58 pp, 0.32 pp and 0.65 pp in ROUGE-1, ROUGE-2 and ROUGE-L scores, respectively. Yet, in this case, the SD-emb-src approach has slightly outperformed the proposed approach in two metrics. In all cases, all these results confirm the importance of attending to the source information in the training objective.

Comparison with the Standard OT. We have also compared the performance of the Sinkhorn divergence in Eq. 6 with the standard OT distance of Eq. 4 over the same hidden states. For simplicity, we have limited this comparison to the IWSLT 2014 En→De translation task. The results, reported in Table 4, show that the Sinkhorn divergence has outperformed the standard OT distance in all cases.

Table 3. ROUGE-1, ROUGE-2 and ROUGE-L scores for the Gigaword summarization task. (†) p-value < 0.05. (‡) p-value < 0.01. (∘) from [24]. (•) from [27]. (◇) from [28]. (-) not available.

Model	English Gigaword								
	dev			test			test*		
	RG-1	RG-2	RG-L	RG-1	RG-2	RG-L	RG-1	RG-2	RG-L
Other Reported Results									
Transformer∘	-	-	-	-	-	-	37.57	18.90	34.69
SEASS•	-	-	-	46.86	24.58	43.53	36.15	17.54	33.63
SeqCopyNet◇	-	-	-	47.27	25.07	44.00	35.93	17.51	33.35
Our Implementations									
Transformer	48.00	25.46	44.65	48.35	26.28	44.86	37.90	19.01	35.13
+ SD-emb-src	47.98	25.30	44.51	**49.07‡**	**26.79**	**45.51‡**	37.68	18.64	34.72
+ SD-emb-tgt	47.93	25.31	44.37	48.67	26.39	44.95	37.67	18.59	34.61
+ SD-hidden	**48.42‡**	**25.65**	**45.10‡**	48.93†	26.60	**45.51†**	**38.12**	**19.28**	**35.36**

* original test set provided by the Hugging Face library.

Table 4. BLEU and F_{BERT} scores for the standard OT (Sink. Dis. in the table) and the proposed Sinkhorn divergence (Sink. Div. in the table) in the IWSLT 2014 En→De translation task.

Method	En→De			
	dev		test	
	BLEU	F_{BERT}	BLEU	F_{BERT}
Transformer	29.90	64.27	28.22	63.24
+ Sink. Dis	29.81	64.53	28.58	63.40
+ Sink. Div	**30.19**	**64.66**	**28.78**	**63.46**

Performance Sensitivity to the Value of λ. Table 5 shows the performance with variable values of λ in the IWSLT 2014 En→De translation task. The key observation is that the results seem reasonably stable, and that values of 0.01 and 0.1 (and, likely, any values in between) have clearly outperformed the baseline (column 0) on the validation set. As a reassuring indication of stability, the same values have also outperformed the baseline on the test set. For convenience, we have used the best value over this validation set (i.e., 0.1) for all tasks.

Qualitative Analysis of the Word Alignments. To further investigate the proposed approach, in Fig. 2 we visualize the optimal transport matrices between the word embeddings of the source and the reference obtained with the different models for a sample from the IWSLT 2014 De→En dataset. For the baseline model, most words are wrongly aligned, especially for the first few tokens (e.g., "also" means "so" in German, "es" means 'it' etc.). This shows that the internal attention mechanism of the transformer is not particularly effective at aligning word embeddings. Conversely, the proposed model shows a remarkable performance, even when compared to the two SD-emb models that directly seek to align word embeddings during training. SD-emb-src has the second-best perfor-

Table 5. Performance sensitivity to the value of regularization parameter λ for the IWSLT 2014 De→En translation task.

dataset	λ						
	0	0.001	0.01	0.1	0.2	0.5	1
Valid BLEU	35.14	35.22	35.34	**35.59**	35.33	35.13	34.92
Valid F_{BERT}	68.81	68.52	69.03	**69.14**	68.76	68.38	68.39
Test BLEU	33.44	33.38	33.82	**34.05**	33.66	33.61	33.44
Test F_{BERT}	67.39	66.98	67.63	**67.80**	67.58	67.22	67.24

mance, yet it has still failed to correctly align the first three tokens. We have also observed that this behaviour has been even more pronounced in the case of long sentences. For these sentences, our SD-hidden model has displayed a consistent ability to align along the diagonal axis, which is correct in first approximation, while all the other models have predominantly reported very scattered alignments.

It is also noteworthy that the best alignments between word embeddings have not been obtained by the SD-emb objectives that explicitly optimize them. We speculate that a reason for this may be the impact of subword tokenization. For instance, the word *circle* in Fig. 2 is an intact token in the English corpus while it has been tokenized into two subwords, *kr* and *##eis*, in the German corpus. This may somehow break the "equilibrium" in the alignments, as two tokens need to match only one. Since subwords are given the same weight as intact tokens, the unmatched weight may be forcefully assigned to other tokens. We assume that this may be one of the reasons why optimizing the transport directly between word embeddings may lead to poorer alignments. Conversely, optimizing the transport over the hidden states of the transformer may afford more degrees of freedom to mollify this behavior. We leave further investigation to future work.

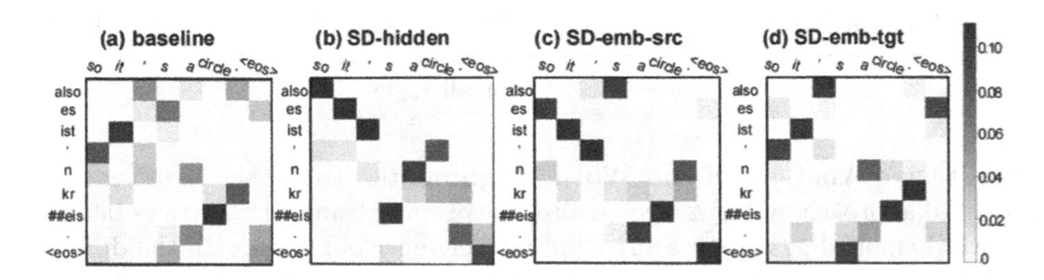

Fig. 2. OT matrices of a sample from the IWSLT 2014 De→En translation task.

Table 6. Examples of generated text for the IWSLT 2014 De→En translation task and English Gigaword Summarization task. This table uses color coding to highlight some correct and incorrect phrases. Red: incorrect phrases; Green: correct phrases.

Translation Examples	
Reference	he just looked up at the sky, and he said, "excuse me, can you not see that i'm driving?"
Baseline	he just looked up in heaven and said, you can't see i'm driving? "
SD-emb-src	he just looked up in heaven and said, "so, really, can you see that i'm driving?"
SD-emb-tgt	he just looked up into heaven and said, "forgive me, can't you see i'm driving cars?"
SD-hidden	he just looked up into the sky and said, "excuse me, can't you see that m driving?"
Reference	now these decisions vary in the number of choices that they offer per decision.
Baseline	now these decisions are different in the number of choices that you make when you make choices.
SD-emb-src	now these choices are different from the number of choices that they offer per choice.
SD-emb-tgt	now these choices are different from the number of choices they make.
SD-hidden	now these decisions are different in the number of choices they offer per decision.
Summarization Examples	
Reference	credit agricole announces 1.1-billion-euro bid for greek bank emporiki
Baseline	credit agricole launches offer to buy rest of greek bank
SD-emb-src	credit agricole bids for greek bank
SD-emb-tgt	credit agricole launches 1.1-bln-euro offer for greek bank
SD-hidden	credit agricole bids 1.1 bln euros for emporiki bank
Reference	palestinian official urges arabs to invest in jerusalem
Baseline	palestinian official calls for holy war for east jerusalem
SD-emb-src	palestinian official calls for arab investment in east jerusalem
SD-emb-tgt	palestinian official calls for arab investment in east jerusalem
SD-hidden	palestinian official urges arabs to invest in east jerusalem

Examples of Generated Text. In Table 6, we show a few examples from the IWSLT 2014 De→En translation task and the English Gigaword summarization task. Overall, it is easy to appreciate the higher quality of the generated sentences provided by the proposed approach for these samples. In the first translation example, the proposed approach has been the only one that was able to retrieve "sky" (metaphorical) instead of "heaven" (metaphysical) and the "excuse me," opener. In the second translation example, the proposed approach has been the only one to correctly nuance "different in" and "per decision". Also in the summarization examples, the proposed model seems to have been the most faithful to the reference. For example, in the second summarization example, the MLE baseline has returned a major mistake by predicting the incorrect phrase "calls for holy war".

5 Conclusion

In this work, we have proposed a novel training objective for NLG tasks that minimizes the Sinkhorn divergence between the contextual representations of the predictions and the source text. The proposed objective shares the computational efficiency of the well-known Sinkhorn distance and is, in principle, applicable to any of the seq2seq models in common use. The experimental results over various translation and summarization datasets have shown that the proposed approach has been able to achieve statistically-significant improvements over our MLE baseline and two, alternative OT objectives. A qualitative analysis of selected samples has shown that the proposed approach has led to word embeddings that can more effectively align the source and target, even over those of other OT-trained models which are explicitly trained to align word embeddings. In addition, a few examples of generated text have shown that the attention devoted to the source does not come at a price of fluency and adequacy of the generated text.

Acknowledgements. The first author is funded by the China Scholarship Council (CSC) from the Ministry of Education of the P. R. of China.

References

1. Alqahtani, S., Lalwani, G., Zhang, Y., Romeo, S., Mansour, S.: Using optimal transport as alignment objective for fine-tuning multilingual contextualized embeddings. In: EMNLP (2021)
2. Chen, L., et al.: Improving sequence-to-sequence learning via optimal transport. In: ICLR, pp. 1–16 (2019)
3. Chen, Y.C., Gan, Z., Cheng, Y., Liu, J., Liu, J.: Distilling knowledge learned in bert for text generation. In: ACL (2020)
4. Cuturi, M.: Sinkhorn distances: lightspeed computation of optimal transport. In: NIPS, pp. 2292–2300 (2013)
5. Devlin, J., Chang, M.W., Lee, K., Toutanova, K.: Bert: pre-training of deep bidirectional transformers for language understanding. In: NAACL (2019)
6. Feydy, J., Séjourné, T., Vialard, F.X., Amari, S., Trouvé, A., Peyré, G.: Interpolating between optimal transport and mmd using sinkhorn divergences. In: AISTATS (2019)
7. Garg, S., Peitz, S., Nallasamy, U., Paulik, M.: Jointly learning to align and translate with transformer models. In: EMNLP (2019)
8. Jauregi Unanue, I., Parnell, J., Piccardi, M.: Berttune: fine-tuning neural machine translation with bertscore. In: ACL/IJCNLP (2021)
9. Koehn, P.: Statistical significance tests for machine translation evaluation. In: EMNLP (2004)
10. Kusner, M.J., Sun, Y., Kolkin, N.I., Weinberger, K.Q.: From word embeddings to document distances. In: ICML (2015)
11. Lewis, M., et al.: Bart: denoising sequence-to-sequence pre-training for natural language generation, translation, and comprehension. arXiv abs/1910.13461 (2020)
12. Li, C., et al.: Improving text generation with student-forcing optimal transport. In: EMNLP, pp. 9144–9156 (2020)
13. Lin, C.Y.: Rouge: a package for automatic evaluation of summaries. In: ACL 2004 (2004)
14. Luong, T., Pham, H., Manning, C.D.: Effective approaches to attention-based neural machine translation. In: EMNLP (2015)

15. Nguyen, T., Luu, A.T.: Improving neural cross-lingual summarization via employing optimal transport distance for knowledge distillation. arXiv abs/2112.03473 (2021)
16. Papineni, K., Roukos, S., Ward, T., Zhu, W.J.: Bleu: a method for automatic evaluation of machine translation. In: ACL (2002)
17. Parnell, J., Unanue, I.J., Piccardi, M.: A multi-document coverage reward for relaxed multi-document summarization. In: ACL (2022)
18. Post, M.: A call for clarity in reporting bleu scores. In: WMT (2018)
19. Ranzato, M., Chopra, S., Auli, M., Zaremba, W.: Sequence level training with recurrent neural networks. CoRR abs/1511.06732 (2016)
20. Sennrich, R., Haddow, B., Birch, A.: Neural machine translation of rare words with subword units. arXiv abs/1508.07909 (2016)
21. Shaw, P., Uszkoreit, J., Vaswani, A.: Self-attention with relative position representations. In: NAACL (2018)
22. Shi, X., Huang, H., Jian, P., Tang, Y.-K.: Case-Sensitive Neural Machine Translation. In: Lauw, H.W., Wong, R.C.-W., Ntoulas, A., Lim, E.-P., Ng, S.-K., Pan, S.J. (eds.) PAKDD 2020. LNCS (LNAI), vol. 12084, pp. 662–674. Springer, Cham (2020). https://doi.org/10.1007/978-3-030-47426-3_51
23. Sutskever, I., Vinyals, O., Le, Q.V.: Sequence to sequence learning with neural networks. In: NIPS (2014)
24. Vaswani, A., et al.: Attention is all you need. arXiv abs/1706.03762 (2017)
25. Zhang, T., Kishore, V., Wu, F., Weinberger, K.Q., Artzi, Y.: BERTscore: evaluating text generation with BERT. arXiv abs/1904.09675 (2020)
26. Zhao, W., Peyrard, M., Liu, F., Gao, Y., Meyer, C.M., Eger, S.: Moverscore: text generation evaluating with contextualized embeddings and earth mover distance. arXiv abs/1909.02622 (2019)
27. Zhou, Q., Yang, N., Wei, F., Zhou, M.: Selective encoding for abstractive sentence summarization. In: ACL (2017)
28. Zhou, Q., Yang, N., Wei, F., Zhou, M.: Sequential copying networks. In: AAAI (2018)

Dual-Detector: An Unsupervised Learning Framework for Chinese Spelling Check

Feiran Shao and Jinlong Li(✉)

University of Science and Technology of China, Hefei, China
sfr97333@mail.ustc.edu.cn, jlli@ustc.edu.cn

Abstract. The task of Chinese Spelling Check (CSC) is to detect and correct spelling errors in Chinese sentences. Since the scale of labeled CSC training set is quite small, we propose an unsupervised Chinese spelling correction framework based on detectors. Two kinds of detectors: Dec-Err and Dec-Eva, are proposed to leverage the contextual information to detect misspelled characters and evaluate the corrections respectively. Both detectors are fine-tuned with our proposed hybrid mask strategy. Dec-Eva is a transformer encoder based detector, of which we modify the attention connections to reuse the contextual information and parallel evaluate possible corrections. Compared with supervised and unsupervised state-of-the-art methods, experimental studies show that our method achieves competitive results. Further empirical studies reveal the efficiency and flexibility of our method.

Keywords: Chinese spelling check · Natural language process · Text process

1 Introduction

Spelling Check aims to detect and correct spelling errors in texts. Spelling check can work as pre-processing for downstream applications like search engine [10] and essay scoring [12] or post-processing for text generation tasks like ASR [1] and OCR [11]. For Chinese Spelling Check (CSC), every Chinese character that can be displayed on computer does exist in Chinese while typos in many alphabetic languages may not even exist in vocabulary. There are no word delimiters between Chinese characters, making it an obstacle to exploiting Chinese words for CSC. Besides, most Chinese spelling errors are related to phonological or visual similarity [8]. In short, all above factors make it challenging to correct Chinese spelling errors, which requires contextual, phonological and visual information.

Early work [15–17] follows the framework of error detection, candidate generation and candidate evaluation, and unsupervised *n-gram* model is employed for detection and evaluation. The candidates are generated by looking up in the candidate table. Recently, many works [2,9,18] achieve better performance by leveraging the power of pre-trained models like BERT [4] in the supervised

© The Author(s), under exclusive license to Springer Nature Switzerland AG 2023
H. Kashima et al. (Eds.): PAKDD 2023, LNAI 13938, pp. 162–173, 2023.
https://doi.org/10.1007/978-3-031-33383-5_13

Table 1. Statistics of SIGHAN dataset

Train Set	#Sent	Avg. length	#Errors
SIGHAN13	700	41.8	343
SIGHAN14	3437	49.5	5122
SIGHAN15	2339	31.3	3037
Test set	#Sent	Avg. length	#Errors
SIGHAN13	1000	74.3	1224
SIGHAN14	1062	50.0	771
SIGHAN15	1100	30.6	703

learning paradigm. But supervised learning requires enough labeled samples. So far, the commonly used CSC dataset are SIGHAN datasets, as illustrated in Table 1. SIGHAN datasets contain several hundred or thousands of erroneous sentences, which are quite small. Further more, fine-tuning pre-trained models on SIGHAN datasets may result in overfitting [7]. As a workaround for handling the absence of labeled data, unsupervised learning is considered. Li proposed an unsupervised CSC framework uChecker [7] that utilizes a confusion set to select the generated corrections. To exploit the potential of pre-trained models, previous methods fine-tuned them to generate corrections. Different from them, we leverage the pre-trained model to evaluate possible corrections from an automatically generated candidate table.

Customized mask strategy is commonly adopted as an unsupervised method to synthesize erroneous sentences for fine-tuning the CSC model. Unlike masking a character with "MASK" in BERT, CSC methods replace a character with one of its similar characters in the confusion set. However, the confusion set is handcrafted and may not be complete. To reduce the reliance on confusion set, we utilize pinyin[1] to automatically generate replacing characters and the candidate table for fine-tuning.

We propose an unsupervised Chinese spelling check framework Dual-Detector which contains two detectors Dec-Err and Dec-Eva. We first use Dec-Err to detect the misspelled characters and then get their candidates from a candidate table. After that, the Dec-Eva evaluates all the candidates parallel based on the context. Since it's costly to evaluate all the candidates by replacing them into the given context, we propose a novel transformer encoder based detector Dec-Eva, of which we modify the attention connections to reuse the contextual information and parallel evaluate all the candidates. Our contributions can be summarized as follows:

- We propose an unsupervised Chinese spelling check framework Dual-Detector that leverages contextual information and candidate table to correct spelling errors. Dual-Detector is flexible since we can correct new erroneous cases by just re-configuring the candidate table without training.
- To reduce the reliance on confusion set, we propose a hybrid mask strategy, which masks characters based on pinyin, to synthesize erroneous sentences and candidates for fine-tuning.

[1] Pinyin is the romanization spelling system for the sounds of Chinese characters.

- We propose the novel transformer encoder based detector Dec-Eva to parallel calculate candidates' misspelled probabilities. The speedup is roughly proportional to the number of candidates compared with sequential calculation.

Experiments show that our method achieves competitive results on SIGHAN14 and SIGHAN15 datasets with unsupervised methods. Our method even outperforms the supervised methods on SIGHAN13 dataset, indicating the effectiveness of our method in low-resource settings. We have conducted analyses to verify the flexibility and efficiency of our method as well.

2 Method

2.1 Overview

Chinese spelling check task aims to detect and correct spelling errors in the given sentence. Given a sentence $X = (x_1, x_2, \ldots, x_n)$, where n denotes the length of the sentence, CSC would generate a corrected sentence $\hat{Y} = (\hat{y}_1, \hat{y}_2, \ldots, \hat{y}_n)$. The ground truth is represented as $Y = (y_1, y_2, \ldots, y_n)$. Unlike grammatical error correction, the input and output of Chinese spelling check are of the length.

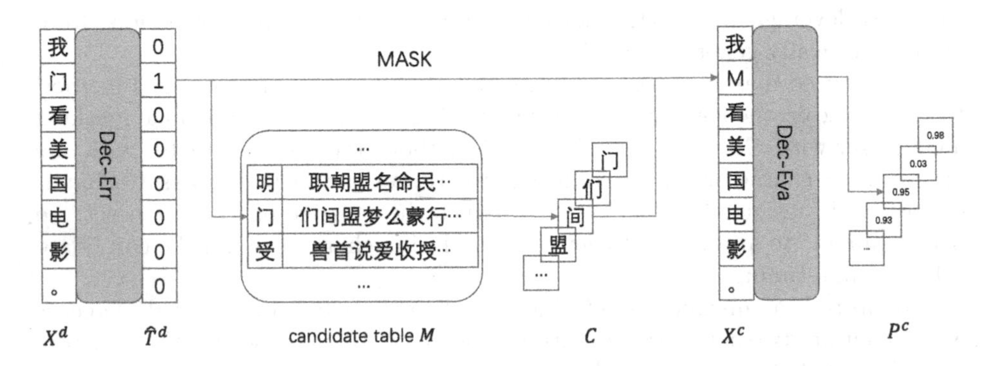

Fig. 1. Overview of our framework. The erroneous sentence "我门看美国电影" means "My door watches American movies". The correct sentence should be "我们看美国电影" " which means "We watch American movies". Sentence is first fed to the Dec-Err to get which characters are wrong. The output of "门" is 1 which indicates that it's misspelled. "MASK" will replace "门" as the denoised context. Then we look up "门" in candidate table to get its candidates (i.e., "们间盟梦么蒙行···"). The candidates are fed to the Dec-Eva simultaneously along with the denoised context. "们" is considered as correction as its misspelled probability is lowest.

As illustrated in Fig. 1, our framework is composed of a detector Dec-Err that detects misspelled characters, a candidate table and a detector Dec-Eva that evaluates all the candidates.

More specifically, the Dec-Err takes a sequence of characters X^d as input and outputs a sequence of labels $\hat{T}^d$. Characters with label 1 are considered misspelled. The candidate table will give lists of candidates for the misspelled characters. As the misspelled characters may mislead the contextual meaning, we replace the characters in misspelled positions with "MASK" token to denoise. The denoised sentence is used as the context X^c and fed into the Dec-Eva with the candidates C. The Dec-Eva evaluates all the candidates parallel and candidates with the lowest misspelled probabilities are selected as the correction. Superscript d and c are used to distinguish between variables in Dec-Err and Dec-Eva.

2.2 Hybrid Mask Strategy

To utilize the large amount of unlabeled corpus and reduce the reliance on phonetic confusion set, we propose the hybrid mask strategy that utilizes pinyin to synthesize data to build our CSC model. The hybrid mask strategy consists of four mask types: phonetic mask I, phonetic mask II, visual mask and random mask, as illustrated in Table 2.

Table 2. Four masks of the hybrid mask strategy. An example of phonetic mask II: character "消" is first converted to its pinyin "xiao", which is further converted to a similar pinyin "xiu". Then we randomly sample a character in the vocabulary with the pinyin "xiu", which gives us "休".

Mask Type	Proportion	Examples
Phonetic mask I(same pinyin)	30%	飞 → "fei" → 肥
Phonetic mask II(similar pinyin)	30%	消 → "xiao" → "xiu" → 休
Visual mask	10%	们 → 门
Random mask	30%	你 → 快

The phonetic mask first converts a character to its pinyin or similar pinyin. Then we replace the original character with a character from the vocabulary that has the corresponding pinyin. The visual mask replaces a character with a character in the shape confusion set. The random mask replaces a character with a random character. When we apply the hybrid mask strategy to a character, it will be masked by one of these four types according to the proportion in Table 2. As most spelling errors are related to phonetic similarity [8], we set the proportions empirically.

For the Dec-Err, the hybrid mask strategy is used to synthesize erroneous sentences from error-free sentences. The Dec-Err is fine-tuned by identifying these masked characters. For the Dec-Eva, we randomly select several characters from error-free sentence as detected misspelled. The candidates for these characters, other than themselves, are generated by sampling from the hybrid mask strategy. The Dec-Eva learns to select the original characters from the candidates.

2.3 Detector Dec-Err

The Dec-Err aims to detect spelling errors in sentences. The input of Dec-Err is sentence $X^d \triangleq (x_1, x_2, \ldots, x_n)$ and the outputs of Dec-Err are detected labels $\hat{T}^d = (\hat{t}_1^d, \hat{t}_2^d, \ldots, \hat{t}_n^d)$ and misspelled positions $Pos = (pos_1, pos_2, \ldots, pos_m)$, where $\hat{t}_i^d \in \{0, 1\}$ and m is the number of detected spelling errors.

Firstly, we calculate the misspelled probabilities P^d as follows:

$$\begin{aligned} P^d &= (p_1^d, p_2^d, \ldots, p_n^d) \\ &= DecErr(X^d) \end{aligned} \tag{1}$$

where $DecErr$ is the Dec-Err, p_i^d is the misspelled probability of ith character. Models like ELECTRA [3] and Bi-LSTM can be used as Dec-Err.

Then we convert the misspelled probabilities to labels $\hat{T}^d = (\hat{t}_1^d, \hat{t}_2^d, \ldots, \hat{t}_n^d)$ as follows:

$$\hat{t}_i^d = \begin{cases} 1, & p_i^d > 1 - p_i^d \\ 0, & otherwise \end{cases} \tag{2}$$

where label $\hat{t}_i^d$ indicates whether ith character misspelled, $\hat{t}_i^d = 1$ represents that it's detected as misspelled.

In addition, we get the misspelled position Pos which is defined as:

$$Pos = \{i | \hat{t}_i^d = 1, i = 1, 2, \ldots, n\} \tag{3}$$

Then we sort the Pos and denote it as $Pos = (pos_1, pos_2, \ldots, pos_m)$, where $m = \sum_i^n \hat{t}_i^d$ is the number of detected misspelled characters.

2.4 Candidate Table

The candidate table M contains the Chinese characters and their possible corrections which we refer to as candidates. The table can be considered as a mapping $\mathcal{M} : x \rightarrow (c_1, c_2, \ldots)$ from character to candidates.

Given the misspelled position pos_i and the character x_{pos_i}, we query the misspelled character from the table to get its candidates $\boldsymbol{c_i}$, denoted by:

$$\boldsymbol{c_i} = \mathcal{M}(x_{pos_i}) = (c_{i,1}, c_{i,2}, \ldots) \tag{4}$$

where $\boldsymbol{c_i}$ are candidates of ith misspelled character x_{pos_i}, $|\boldsymbol{c_i}|$ is the number of candidates of character x_{pos_i}. Note that the number of candidates for different characters may vary.

2.5 Detector Dec-Eva

The detector Dec-Eva detects whether the candidates are spelling errors in the given context. We use it to parallel evaluate the candidates and the most likely candidates are selected as corrections. The Dec-Eva stacks 12 Transformer encoder blocks. The detailed architecture of the Dec-Eva is illustrated in Fig. 2.

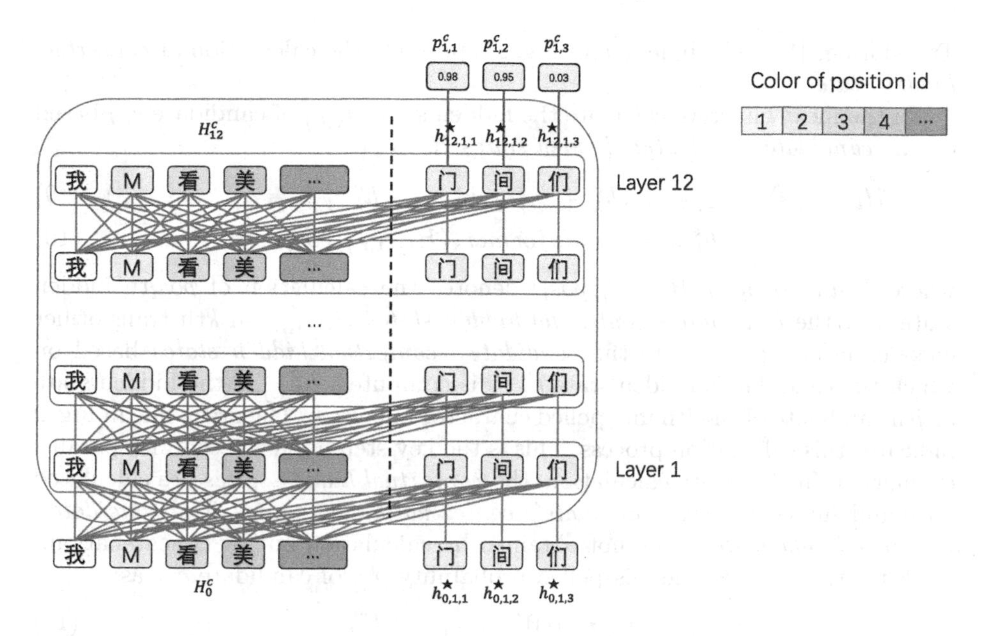

Fig. 2. Structure of the Dec-Eva. The blue line, green line and red line indicate Eq. (8), Eq. (10) and Eq. (11) respectively. The position ids of characters and candidates are indicated by their colors. (Color figure online)

First, we get the denoised context $X^c = (x^c_1, x^c_2, \ldots, x^c_n)$ by replacing misspelled characters with "MASK":

$$x^c_i = \begin{cases} x_i, & \hat{t}^d_i = 0 \\ MASK, & \hat{t}^d_i = 1 \end{cases} \tag{5}$$

Then we get the input of the Dec-Eva, which are the embeddings of context and candidates:

$$h^c_{0,i} = E(x^c_i, i) \tag{6}$$

$$h^\star_{0,i,j} = E(c_{i,j}, pos_i) \tag{7}$$

where $E(x^c_i, i)$ denotes the embedding of character x^c_i in position i, $h^c_{0,i}$ is the embedding of character x^c_i, $h^\star_{0,i,j}$ is the embedding of jth candidate of ith misspelled character.

The *contextual hidden states* H^c_{k-1} in layer $k-1$ are used as input of kth transformer encoder layer to get the *contextual hidden states* H^c_k in layer k:

$$\begin{aligned} H^c_k &= (h^c_{k,1}, h^c_{k,2}, \ldots, h^c_{k,n}) \\ &= Transformer_k(H^c_{k-1}) \end{aligned} \tag{8}$$

where $Tranformer_k(\cdot)$ is the kth transformer encoder layer, $h^c_{k,i}$ is ith hidden state in H^c_k. The detailed implementation of $Transformer_k(\cdot)$ is introduced in

Transformer [13]. The blue line in Fig. 2 represents the calculation of *contextual hidden states.*

In the meanwhile, we calculate the hidden state $h^{\star}_{k,i,j}$ of candidate $c_{i,j}$ based on the *candidate's contextual hidden states* $\mathcal{H}_{k-1,i,j}$:

$$\mathcal{H}_{k-1,i,j} \triangleq (h^{c}_{k-1,1}, \ldots, h^{c}_{k-1,pos_i-1}, h^{\star}_{k-1,i,j}, h^{c}_{k-1,pos_i+1}, \ldots, h^{c}_{k-1,n}) \quad (9)$$

$$h^{\star}_{k,i,j} = Transformer_k(\mathcal{H}_{k-1,i,j}; pos_i) \quad (10)$$

where $Transformer_k(\mathcal{H}_{k-1,i,j}; pos_i)$ denotes the calculation of pos_ith hidden state over the *candidate's contextual hidden states* $\mathcal{H}_{k-1,i,j}$ in kth transformer encoder layer, $\mathcal{H}_{k-1,i,j}$ are the *candidate's contextual hidden states* based on which the candidate's hidden state $h^{\star}_{k,i,j}$ is computed, $h^{\star}_{k,i,j}$ is the hidden state of jth candidate of the ith misspelled character in layer k. The green line in Fig. 2 indicates this calculation process. This is the key step in our method for parallel evaluation. In the entire calculation, the *contextual hidden states* are calculated once and shared to every *candidate's contextual hidden states.* The *candidate's contextual hidden states* are not obtained by calculation but by concatenation.

After that, we get the misspelled probability $p^{c}_{i,j}$ of candidate $c_{i,j}$ as:

$$p^{c}_{i,j} = \sigma(W^{c}(h^{\star}_{12,i,j}) + b^{c}) \quad (11)$$

where $h^{\star}_{12,i,j}$ is the hidden state of jth candidate at ith misspelled position in last layer, W^{c} and b^{c} are parameters of dense layer that projects hidden state to a logit value, σ represents the *sigmoid* function that converts the logit value to misspelled probability.

The correction y^{c}_{i} of ith misspelled character is defined as:

$$y^{c}_{i} = c_{i,j^*}, \; j^* = \arg\min_{j} p^{c}_{i,j} \quad (12)$$

Eventually, we get the corrected sentence $\hat{Y} = (\hat{y}_1, \hat{y}_2 \ldots \hat{y}_n)$ as:

$$\hat{y}_i = \begin{cases} x_i, & i \notin Pos \\ y^{c}_{j}, & i \in Pos \; and \; i = pos_j \end{cases} \quad (13)$$

2.6 Training

We train the Dec-Err and Dec-Eva separately. The loss function of the Dec-Err is defined as:

$$L^{d} = -\sum_{i=1}^{n}(t^{d}_{i} \cdot \log p^{d}_{i} + (1 - t^{d}_{i}) \cdot \log(1 - p^{d}_{i})), \; t^{d}_{i} = \begin{cases} 0, & x_i = y_i \\ 1, & otherwise \end{cases} \quad (14)$$

where t^{d}_{i} is the ground truth of detection label.

The loss of Dec-Eva is computed as:

$$L^{c} = -\sum_{i=1}^{m}\sum_{j=1}^{|c_i|}(t^{c}_{i,j} \cdot \log p^{c}_{i,j} + (1 - t^{c}_{i,j}) \cdot \log(1 - p^{c}_{i,j})), \; t^{c}_{i,j} = \begin{cases} 0, & c_{i,j} = y_{pos_i} \\ 1, & otherwise \end{cases} \quad (15)$$

where $t^{c}_{i,j}$ is the ground truth of candidate's label.

3 Experiments

To verify the effectiveness of our method, we have compared it with state-of-the-art methods. And then we have conducted analyses to verify its flexibility and efficiency.

3.1 Datasets and Settings

Datasets. The wiki2019zh[2] corpus contains one million wiki pages and Wang's hybrid dataset [14] consists of 270k sentences. These two datasets are considered to be error-free and are used to fine-tune our models. To utilize these datasets, we generated sentences with spelling errors from them. For wiki2019zh, we synthesize erroneous sentences with our hybrid mask strategy. For Wang's hybrid dataset, they generated erroneous sentences in a hybrid approach of OCR and ASR. In addition, the SIGHAN test sets, as illustrated in Table 1, are used for evaluation.

Settings. The detector Dec-Err is first tuned on wiki2019zh which is noised with hybrid mask strategy with 5% characters, and then is fine-tuned on Wang's hybrid dataset. As the wiki2019zh is huge, we only fine-tuned the Dec-Err for 150k steps. For the Dec-Eva, we selected 10% characters as positive detection results and mask them with "MASK" as context. We sampled 10 similar characters from the hybrid mask strategy along with the character itself as candidates. We fine-tuned Dec-Eva on the wiki2019zh data for 75k steps. In the inference, we used the erroneous cases in Wang's hybrid dataset as our candidate table to correct errors in SIGHAN test sets. This candidate table, which contains more than 30,000 erroneous cases generated by ASR and OCR approaches, is considered to be more realistic than the hand-crafted confusion set.

We used ELECTRA[3] to initialize the Dec-Err and Dec-Eva. We set the batch size to 128 and the learning rate to 5e-5. Adam optimizer [6] is adopted with the default settings.

3.2 Main Results

We compare our method with supervised methods FASPell [5], SpellGCN [2], MLM-phonetics [18], PLOME-Finetune [9] and unsupervised methods uChecker [7] and PLOME-Pretrain. Following the previous work, the precision, recall and F1 scores on sentence-level are reported. A sentence is considered as a correct correction only if all the misspelled characters are corrected. The detection capability of our method is evaluated based on the correction result following the previous two-stages method [18].

Table 3 illustrates the detection and correction results. Our method are competitive with the unsupervised model uChecker on SIGHAN15 dataset. The performance of the pre-trained CSC model PLOME drops significantly when it's not

[2] https://github.com/brightmart/nlp_chinese_corpus.
[3] https://github.com/ymcui/Chinese-ELECTRA.

Table 3. The performance of our method and supervised and unsupervised methods on SIGHAN test sets.

Dataset	Model	detection level			correction level		
		Pre	Rec	F1	Pre	Rec	F1
SIGHAN13	Supervised Methods						
	FASPell(2019)	76.2	63.2	69.1	73.1	60.5	66.2
	SpellGCN(2020)	80.1	74.4	77.2	78.3	72.7	75.4
	MLM-phonetics(2021)	82.0	78.3	80.1	79.5	77.0	78.2
	Unsupervised Methods						
	uChecker(2022)	75.4	73.4	74.4	72.6	70.8	71.7
	Ours	87.4	80.1	83.6	86.6	79.4	**82.9**
SIGHAN14	Supervised Methods						
	FASPell(2019)	61.0	53.5	57.0	59.4	52.0	55.4
	SpellGCN(2020)	65.1	69.5	67.2	63.1	67.2	65.3
	MLM-phonetics(2021)	66.2	73.8	69.8	64.2	73.8	68.7
	Unsupervised Methods						
	uChecker(2022)	61.7	61.5	61.6	57.6	57.5	57.6
	Ours	64.3	60.0	62.1	60.9	56.9	**58.8**
SIGHAN15	Supervised Methods						
	FASPell(2019)	67.6	60.0	63.5	66.6	59.1	62.6
	SpellGCN(2020)	74.8	80.7	77.7	72.1	77.7	75.9
	MLM-phonetics(2021)	77.5	83.1	80.2	74.9	80.2	77.5
	PLOME-Finetune(2021)	77.4	81.5	79.4	75.3	79.3	77.2
	Unsupervised Methods						
	PLOME-Pretrain(2021)	41.8	47.5	44.5	34.2	38.9	36.4
	uChecker(2022)	75.4	72.0	73.7	70.6	67.3	**68.9**
	Ours	72.7	69.0	70.8	70.6	66.9	68.7

fine-tuned on SIGHAN datasets, indicating that the supervised method may not perform well in the low-resource setting. On the SIGHAN14, our method gets 1.2 points improvement compared with uChecker. On the SIGHAN13, our method achieves better performance than previous state-of-the-art models, which is 11.2 points improvement compared with the unsupervised method uChecker and 4.7 points improvement compared with the supervised methods. Considering that the SIGHAN13 train set only contains 700 sentences, the supervised methods gain limited improvement on it. Our method attempts to exploit the unlabeled corpus and achieves better performance. This reveals the effectiveness of our method in low-resource settings. And in practical scenarios, it is difficult to obtain labeled sentence pairs, thus our unsupervised learning method is considered to be a more effective option.

3.3 Analysis

Flexibility. In practical scenarios, it's likely to encounter new erroneous cases unseen before. A flexible CSC model should be able be correct these cases by simple configuration. While the supervised methods have to be trained on the entire updated dataset or may suffer from catastrophic forgetting if trained incrementally, our method corrects the unseen cases simply by configuring the candidate table. To empirically verify the flexibility of our method, we simulated such situations by creating three candidate tables of different sizes with inclusion relationships. As our candidate table contains all the erroneous cases in Wang's hybrid synthetic dataset, we first randomly selected 90% of these erroneous cases as candidate table CT90 and the rest was considered to be the unseen erroneous cases. Then, by adding parts of the unseen cases to CT90, we got CT95 and CT100. We repeated this procedure with different random seeds for 4 times and measured their performance on SIGHAN test sets, as showed in Table 4.

Table 4. Average performance of different candidate table sizes.

Dataset	Candidate table	correction level		
		Pre	Rec	F1
SIGHAN13	CT90	82.6	69.5	75.5
	CT95	85.1	75.1	79.8
	CT100	86.6	79.4	82.9
SIGHAN14	CT90	58.8	52.1	55.2
	CT95	59.7	53.6	56.5
	CT100	60.9	56.9	58.8
SIGHAN15	CT90	68.0	59.9	63.7
	CT95	69.9	63.2	66.4
	CT100	70.6	66.9	68.7

As we add new erroneous cases to CT90, the precision, recall and F1 score increase gradually. The improvement is due to the corrections over the added cases. Since the pre-trained model learns the contextual representation, the Dec-Eva leverages the contextual representation to evaluate the candidates. By providing the candidates, we don't need to train on the new erroneous cases.

Efficiency. To verify the efficiency of Dec-Eva, we conducted a comparison experiment. We first measured the time cost of parallel evaluating all the candidates in Dec-Eva. Then, by replacing the detected erroneous character with their candidates and feeding them with context to the Dec-Err one by one, we measured the time cost of this sequential evaluation. Table 5 illustrates the comparison results on SIGHAN test sets.

Table 5. Time costs comparison.

Test set	Dec-Eva(ms/sent)	Sequential(ms/sent)	Avg. len	Speedup	Avg. #cand
SIGHAN13	63.2	927.9	74.3	14.7	15.7
SIGHAN14	55.6	647.4	50.0	11.6	12.2
SIGHAN15	50.6	558.3	30.6	11.0	12.0

We list the average time costs of evaluating all the candidates in each sentence of the two approaches. The average length of sentences and the average number of candidates per sentence are listed as Avg. len and Avg. #cand. The speedups are 14.7, 11.6 and 11.0 on three test sets respectively. We can see that the speedup is roughly proportional to the number of candidates. Theoretically, the time complexity of sequential evaluation is approximately $O(n * |C|)$ and the time complexity of Dec-Eva is $O(n + |C|)$, where n is the length of sentence and $|C| = \sum_{i=0}^{m} |\boldsymbol{c_i}|$ is the sum number of candidates for all the detected misspelled characters. The experimental results are consistent with the theoretical analysis.

4 Conclusion

We propose an unsupervised Chinese spelling check framework Dual-Detector that leverages the contextual information to evaluate possible corrections from the candidate table. In order to efficiently evaluate all possible corrections, we modify the attention connections of transformer encoder to parallel calculate the misspelled probabilities of the corrections. We introduce a hybrid mask strategy that exploits unlabeled data to fine-tune the models. Experimental results show it's feasible to conduct Chinese spelling check in the detection and evaluation pipeline framework.

The detection and evaluation framework leverage the power of pre-trained models and many more powerful pre-trained models are emerging. In future, we will verify the performance when initializing the detectors with these pre-trained models. As candidate table is essential in the framework, we will further explore generating the candidate table automatically.

References

1. Chao, Y.C., Chang, C.H.: Automatic spelling correction for ASR corpus in traditional Chinese language using Seq2Seq models. In: 2020 International Computer Symposium (ICS), pp. 553–558. IEEE (2020)
2. Cheng, X., et al.: SpellGCN: incorporating phonological and visual similarities into language models for Chinese spelling check. In: Proceedings of the 58th Annual Meeting of the Association for Computational Linguistics, pp. 871–881. Association for Computational Linguistics (2020). https://doi.org/10.18653/v1/2020.acl-main.81. https://aclanthology.org/2020.acl-main.81
3. Clark, K., Luong, M.T., Le, Q.V., Manning, C.D.: ELECTRA: pre-training text encoders as discriminators rather than generators. In: ICLR (2020). https://openreview.net/pdf?id=r1xMH1BtvB
4. Devlin, J., Chang, M.W., Lee, K., Toutanova, K.: BERT: pre-training of deep bidirectional transformers for language understanding. In: Proceedings of the 2019 Conference of the North American Chapter of the Association for Computational Linguistics: Human Language Technologies, Volume 1 (Long and Short Papers), Minneapolis, Minnesota, pp. 4171–4186. Association for Computational Linguistics (2019). https://doi.org/10.18653/v1/N19-1423. https://aclanthology.org/N19-1423

5. Hong, Y., Yu, X., He, N., Liu, N., Liu, J.: FASPell: a fast, adaptable, simple, powerful Chinese spell checker based on DAE-decoder paradigm. In: Proceedings of the 5th Workshop on Noisy User-Generated Text (W-NUT 2019), pp. 160–169 (2019)
6. Kingma, D.P., Ba, J.: Adam: a method for stochastic optimization. In: Bengio, Y., LeCun, Y. (eds.) 3rd International Conference on Learning Representations, ICLR 2015, San Diego, CA, USA, 7–9 May 2015, Conference Track Proceedings (2015). http://arxiv.org/abs/1412.6980
7. Li, P.: uChecker: masked pretrained language models as unsupervised Chinese spelling checkers. In: Proceedings of the 29th International Conference on Computational Linguistics, pp. 2812–2822. International Committee on Computational Linguistics, Gyeongju, Republic of Korea (2022). https://aclanthology.org/2022.coling-1.248
8. Liu, C.L., Lai, M.H., Chuang, Y.H., Lee, C.Y.: Visually and phonologically similar characters in incorrect simplified Chinese words. In: Coling 2010: Posters, pp. 739–747 (2010)
9. Liu, S., Yang, T., Yue, T., Zhang, F., Wang, D.: Plome: pre-training with misspelled knowledge for Chinese spelling correction. In: Proceedings of the 59th Annual Meeting of the Association for Computational Linguistics and the 11th International Joint Conference on Natural Language Processing (Volume 1: Long Papers), pp. 2991–3000 (2021)
10. Martins, B., Silva, M.J.: Spelling correction for search engine queries. In: Vicedo, J.L., Martínez-Barco, P., Muńoz, R., Saiz Noeda, M. (eds.) EsTAL 2004. LNCS (LNAI), vol. 3230, pp. 372–383. Springer, Heidelberg (2004). https://doi.org/10.1007/978-3-540-30228-5_33
11. Nguyen, T.T.H., Jatowt, A., Coustaty, M., Doucet, A.: Survey of post-OCR processing approaches. ACM Comput. Surv. (CSUR) **54**(6), 1–37 (2021)
12. Ramesh, D., Sanampudi, S.K.: An automated essay scoring systems: a systematic literature review. Artif. Intell. Rev. 1–33 (2021)
13. Vaswani, A., et al.: Attention is all you need. In: Proceedings of the 31st International Conference on Neural Information Processing Systems, NIPS 2017, pp. 6000–6010. Curran Associates Inc., Red Hook (2017)
14. Wang, D., Song, Y., Li, J., Han, J., Zhang, H.: A hybrid approach to automatic corpus generation for Chinese spelling check. In: Proceedings of the 2018 Conference on Empirical Methods in Natural Language Processing, pp. 2517–2527 (2018)
15. Xie, W., et al.: Chinese spelling check system based on n-gram model. In: Proceedings of the Eighth SIGHAN Workshop on Chinese Language Processing, pp. 128–136 (2015)
16. Yeh, J.F., Li, S.F., Wu, M.R., Chen, W.Y., Su, M.C.: Chinese word spelling correction based on n-gram ranked inverted index list. In: Proceedings of the Seventh SIGHAN Workshop on Chinese Language Processing, pp. 43–48 (2013)
17. Yu, J., Li, Z.: Chinese spelling error detection and correction based on language model, pronunciation, and shape. In: Proceedings of The Third CIPS-SIGHAN Joint Conference on Chinese Language Processing, pp. 220–223 (2014)
18. Zhang, R., et al.: Correcting Chinese spelling errors with phonetic pre-training. In: Findings of the Association for Computational Linguistics: ACL-IJCNLP 2021, pp. 2250–2261 (2021)

QA-Matcher: Unsupervised Entity Matching Using a Question Answering Model

Shogo Hayashi[1(✉)], Yuyang Dong[2], and Masafumi Oyamada[2]

[1] BizReach, Inc., Tokyo, Japan
shogo.hayashi@bizreach.co.jp
[2] NEC Corporation, Tokyo, Japan
{dongyuyang,oyamada}@nec.com

Abstract. Entity matching (EM) is a fundamental task in data integration, which involves identifying records that refer to the same real-world entity. Unsupervised EM is often preferred in real-world applications, as labeling data is often a labor-intensive process. However, existing unsupervised methods may not always perform well because the assumptions for these methods may not hold for tasks in different domains. In this paper, we propose QA-Matcher, an unsupervised EM model that is domain-agnostic and doesn't require any particular assumptions. Our idea is to frame EM as question answering (QA) by utilizing a trained QA model. Specifically, we generate a question that asks which record has the characteristics of a particular record and a passage that describes other records. We then use the trained QA model to predict the record pair that corresponds to the question-answer as a match. QA-Matcher leverages the power of a QA model to represent the semantics of various types of entities, allowing it to identify identical entities in a QA-like fashion. In extensive experiments on 16 real-world datasets, we demonstrate that QA-Matcher outperforms unsupervised EM methods and is competitive with supervised methods.

Keywords: entity matching · question answering

1 Introduction

Entity matching (EM), also referred to as entity resolution and record linkage, is a fundamental problem in data integration [3]. The task is to find a set of records that refer to the same real-world entity as depicted in Fig. 1. Typically, EM is carried out in blocking and matching steps. In the blocking step, a blocker filters out obvious non-matches from a set of record pairs. Then, in the matching step, a matcher classifies the record pairs into "matches" or "non-matches". With the recent significant advances in deep learning and natural language processing, supervised EM methods have shown significant improvements in performance [15,17].

S. Hayashi—This work was conducted while the author was affiliated with NEC.

© The Author(s), under exclusive license to Springer Nature Switzerland AG 2023
H. Kashima et al. (Eds.): PAKDD 2023, LNAI 13938, pp. 174–185, 2023.
https://doi.org/10.1007/978-3-031-33383-5_14

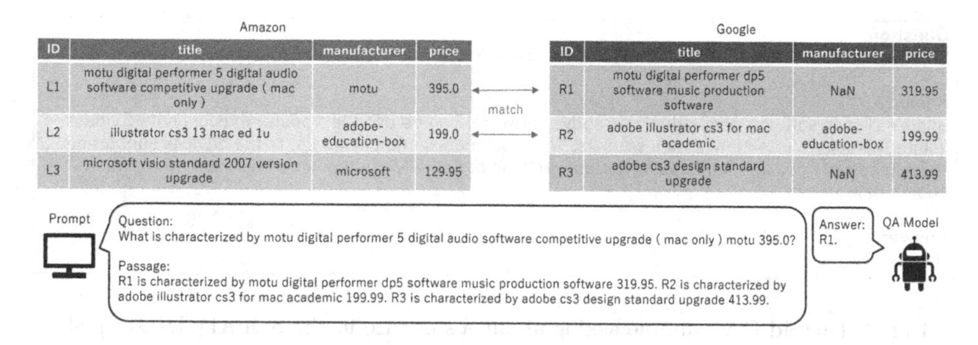

Fig. 1. (upper) Record examples from the Amazon-Google dataset. (lower) The idea of the proposed QA-Matcher that solves EM as QA using a trained QA model.

Unsupervised EM methods have been eagerly studied because collecting labeled data is often a labor-intensive task. However, unsupervised methods may not work well in some domains because their assumptions do not always hold in various domains. For example, a commonly used clustering-based method, ZeroER [23], generates similarity vectors of record pairs through similarity functions specified by users, and partitions the similarity vectors into match and non-match clusters using a Gaussian mixture model. However, selecting suitable similarity functions varies across different domains, and it requires a profound comprehension of a particular domain. For example, in product matching, semantic similarities are favored for product descriptions, while string similarities are preferred for product IDs. Therefore, it cannot be assumed that the selected similarity functions are always appropriate, and similarity vectors form clusters.

In this paper, we propose QA-Matcher, an unsupervised EM model that works for tasks in various domains without particular assumptions. Our idea is to transform EM into question answering (QA) and solve it using a trained QA model, as illustrated in Fig. 1. First, given two sets of records $\{L1, L2, L3\}$ and $\{R1, R2, R3\}$, we generate a question that asks which record has the characteristics (i.e., attribute values) of $L1$, and a passage (context) that describes the attribute values of $\{R1, R2, R3\}$. Then, we answer the question using a trained QA model, and consider a record pair, $(L1, R1)$, a match if the answer is $R1$, and a non-match otherwise. QA-Matcher can identify identical entities in a QA fashion by leveraging a QA model to represent the semantics of various types of entities. Further, QA-Matcher is easy to use with only one hyperparameter k, which is the number of records in a passage, and it is not limited to a certain QA model. We can easily access, use, and change more advanced QA models via the Transformers library [22]. We conduct extensive empirical studies using 16 benchmark datasets from various domains. The results show that QA-Matcher outperforms existing unsupervised methods in almost all datasets and even has competitive performance compared to supervised methods.

Question:
What's the name of the software used to manage music and other media on Apple devices?

Passage:
Apple's iTunes software (and other alternative software) can be used to transfer music, photos, videos, games, contact information, e-mail settings, Web bookmarks, and calendars, to the devices supporting these features from computers using certain versions of Apple Macintosh and Microsoft Windows operating systems.

Answer:
iTunes

Fig. 2. Factoid QA example asking about a software in the SQuAD dataset [18].

2 Preliminaries

We briefly introduce a QA framework based on retriever and reader modules.

2.1 Question Answering

QA [18] is a task that involves answering questions, typically factoid questions [9]. Factoid questions require finding facts that a certain token sequence is an answer. An example of a factoid question in the SQuAD dataset [18] is shown in Fig. 2, which asks which software has a particular characteristic. The dataset contains many types of entities, such as person, music, and companies. Open-domain QA [20] is a typical QA framework that involves retriever and reader modules. The retriever module is responsible for finding a relevant passage or context for a given question from a set of passages using nearest neighbor search. Sparse retrievers use BoW (Bag-of-Words model) of binary, TF-IDF, or BM25 scores to represent vectors of texts. Dense retrievers such as DPR [11] use dense vector representations to encode the semantics of texts. After the retrieval, the reader module answers the question using the retrieved passage. For example, a reader based on BERT [5] represents a question and passage as a single input token sequence and computes the hidden vectors of the tokens. For a question q and a passage p, the score of a span from token position i to j, $\mathrm{score}_{\mathrm{QA}}(i, j \mid q, p)$, is defined as

$$\mathrm{score}_{\mathrm{QA}}(i, j \mid q, p) = s^\top h_i + e^\top h_j, \tag{1}$$

where $h_i \in \mathbb{R}^H$ is the hidden vector corresponding to the i-th token and $s, e \in \mathbb{R}^H$. Finally, the answer is determined as the token span whose start and end positions maximize the score, i.e., $(i^*, j^*) = \arg\max_{i,j} \mathrm{score}_{\mathrm{QA}}(i, j \mid q, p)$.

3 Proposed Method

We propose QA-Matcher, a novel unsupervised EM model that can work for tasks in various domains without making any particular assumptions. QA-Matcher leverages a trained QA model to represent the semantics of entities. Additionally, it is easy to use, because it has only one hyperparameter, and trained QA models are easily accessed through libraries such as Transformers without the need for optimizing model parameters.

3.1 Idea: Solving Entity Matching as Question Answering

Our idea is to solve EM as QA. Specifically, we convert a record r into a factoid question $q(r)$ asking which record has the characteristics or attribute values of r. When a trained QA model answers r' to $q(r)$ in reference to a passage describing some records except for r, the record pair (r, r') can be considered most similar in the records or a match, a non-match otherwise. Based on the answers by the QA model, QA-Matcher predicts labels of record pairs.

By reframing EM as QA, where a factoid question seeks to identify the record with characteristics similar to the target record, we can leverage QA models to identify matching records. Moreover, given that QA datasets comprise various types of entities (e.g., person, software, music), QA models trained on these datasets can effectively comprehend the semantics of records from diverse domains in an unsupervised manner.

One may think sentence embedding methods, such as Sentence-BERT [19], are beneficial for EM by representing records as sentences. Nevertheless, they are trained on general sentences, whereas QA models are trained on factoid questions that require an understanding of specific entities. Therefore, QA models are expected to perform better on EM tasks, which will be demonstrated in the experiments.

3.2 Problem Setting

We represent a record r as a set of key-value pairs, $r = \{(a_i, v_i)\}_{i=1}^{L}$. Here, a_i and v_i denote an attribute name and a corresponding value (typically, a string), respectively. Let R and R' denote sets of records, between which matching records are identified. This paper focuses on the matching step: two sets of records R, R' and a set of record pairs after blocking, $P = \{(r_i, r'_i)\}_{i=1}^{n} \subset R \times R'$, are given. Our goal is to classify record pairs in P into matches or non-matches.

3.3 Framework

Figure 3 shows the overall framework of QA-Matcher. For a record pair $(r, r') \in P$, a retriever retrieves two subsets of k records relevant to r and r', denoted as $R'_k(r) \subset R' \backslash \{r'\}$ and $R_k(r') \subset R \backslash \{r\}$, respectively. Then, factoid questions $q(r)$ and $q(r')$ and passages $p(R'_k(r) \cup \{r'\})$ and $p(R_k(r') \cup \{r\})$ are generated based on a question and passage prompts. The record pair is classified by a QA classification module based on the answers provided by a trained QA model for the question-passage pairs $(q(r), p(R'_k(r) \cup \{r'\}))$ and $(q(r'), p(R_k(r') \cup \{r\}))$. After the classification of all pairs in P, a reclassification module reclassifies the pairs based on their scores obtained from the QA model.

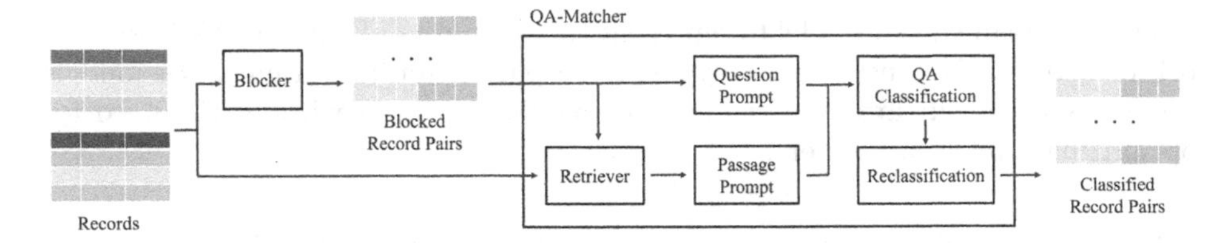

Fig. 3. Framework of QA-Matcher.

3.4 Retriever

A retrieval module is involved to select a limited number of relevant records for generating a passage. First, records in R and R' are converted into texts by concatenating their attribute values: for $r = \{(a_i, v_i)\}_{i=1}^{L}$,

$$s(r) = \text{“}v_1\ v_2\ \ldots v_L\text{”}. \tag{2}$$

Next, we retrieve k records for a query r from $R' \backslash \{r'\}$, denoted as $R'_k(r)$. Similarly, $R_k(r')$ is retrieved from $R \backslash \{r\}$ for a query r'. We can choose to use either a sparse retriever or a dense retriever. Sparse retrievers are preferred when the query terms match with the retrieved records, while dense retrievers are preferred for capturing semantic similarity between the query and the target terms. A retriever performs k-nearest neighbor vector search, and its speed can be improved using techniques such as an inverted index or an existing approximate nearest neighbor[1].

The number of records for a passage k is chosen based on the trade-off between hit ratio and computational cost. Retrieving more records results in a higher hit ratio but requires a higher computational cost, and vice versa. In practice, we found a sparse retriever with $k = 20$ performs well.

3.5 Question and Passage Prompts

A question prompt converts r to a factoid question $q(r)$ asking which is r using Eq. (2), represented as

$$q(r) = \text{“What is characterized by } s(r)\text{?”}. \tag{3}$$

A passage prompt converts $R'_k(r) \cup \{r'\} = \{r'_1, r'_2, \ldots, r'_{k+1}\}$ to a passage $p(R'_k(r) \cup \{r'\})$, represented as

$$\begin{aligned} p(R'_k(r) \cup \{r'\}) &= \text{“}t(r'_1)\ t(r'_2)\ \ldots t(r_{k+1})\text{”}, \\ t(r'_i) &= \text{“ID}_{r'_i} \text{ is characterized by } s(r'_i)\text{.” } (i = 1, \ldots, k+1), \end{aligned} \tag{4}$$

where $\text{ID}_{r'_i}$ is an arbitrary name of r'_i identifiable in $R'_k(r) \cup \{r'\}$.

[1] http://ann-benchmarks.com/.

3.6 QA Classification

A trained QA model predicts an answer denoted by $\mathrm{QA}(q(r), p(R'_k(r) \cup \{r'\}))$ for the question and passage. The prediction is also applied to the other question-passage pair $(q(r'), p(R_k(r') \cup \{r\}))$. If the answers to both questions about r and r' are each other, they are mutually most similar among R, R' and predicted as a match. Formally, the label of a record pair (r, r') given $R_k(r'), R'_k(r)$ is predicted as

$$\begin{aligned} f(r, r') = & \mathbb{1}\left[\mathrm{QA}\left(q(r), p(R'_k(r) \cup \{r'\})\right) = \mathrm{ID}_{r'}\right] \\ & \cdot \mathbb{1}\left[\mathrm{QA}\left(q(r'), p(R_k(r') \cup \{r\})\right) = \mathrm{ID}_r\right], \end{aligned} \tag{5}$$

where $\mathbb{1}[\cdot]$ takes 1 if the argument is true and 0 otherwise. In Eq. (5), we ensure that the QA model outputs only record names, by limiting answer candidates to $\{\mathrm{ID}_{r'_i} \mid r'_i \in R'_k(r) \cup \{r'\}\}$ or $\{\mathrm{ID}_{r_i} \mid r_i \in R_k(r) \cup \{r\}\}$.

3.7 Reclassification

In the predicted record pairs $\{(r_i, r'_i, f(r_i, r'_i))\}_{i=1}^n$ after applying Eq. (5) to P, there may exist false positives because Eq. (5) may predict a pair (r, r') as a match even when r does not match any record in R'. False negatives can occur if there are multiple matching records (e.g., both different records r_1 and r_2 in R match $r' \in R'$) in a passage.

To classify the record pairs correctly, reclassification is performed based on the score provided by the QA model. First, using Eq. (1), the average of scores that r and r' are answers to each other's questions, denoted by $\mathrm{score}_{\mathrm{av}}(r, r')$, is calculated as

$$\begin{aligned} \mathrm{score}_{\mathrm{av}}(r, r') = & \frac{1}{2}\Big(\mathrm{score}_{\mathrm{QA}}\left(\mathrm{ID}_{r'} \mid q(r), p(R'_k(r) \cup \{r'\})\right) \\ & + \mathrm{score}_{\mathrm{QA}}\left(\mathrm{ID}_r \mid q(r'), p(R_k(r') \cup \{r\})\right)\Big). \end{aligned} \tag{6}$$

Here, we abbreviate the span of $\mathrm{ID}_{r'}$ in the token sequence. By applying Eq. (5) and (6) to P, we obtain a set of score-label pairs $\tilde{\mathcal{D}} = \{(\mathrm{score}_{\mathrm{av}}(r_i, r'_i), f(r_i, r'_i))\}_{i=1}^n$. Subsequently, a one-dimensional linear classifier $h : \mathbb{R} \to \{0, 1\}$ is trained using $\tilde{\mathcal{D}}$ by maximizing likelihood. The trained classifier h reclassifies the record pairs as $\{(r_i, r'_i, h(\mathrm{score}_{\mathrm{av}}(r_i, r'_i)))\}_{i=1}^n$. Because false positives (false negatives) are expected to have low (high) scores, their classification results in Eq. (5) will be corrected by h.

4 Experiments

We demonstrate the effectiveness of QA-Matcher by comparing it to existing supervised and unsupervised methods using benchmark datasets.

Table 1. Summary of benchmark datasets.

Type	Dataset	Domain	Size	# Attr
Structured (two record sets share the same schema or attributes)	BeerAdvo-RateBeer	beer	91	4
	iTunes-Amazon$_1$	music	109	8
	Fodors-Zagats	restaurant	189	6
	DBLP-ACM$_1$	citation	2473	4
	DBLP-Scholar$_1$	citation	5742	4
	Amazon-Google	software	2293	3
	Walmart-Amazon$_1$	electronics	2049	5
Dirty (some attribute values are injected in wrong attributes)	iTunes-Amazon$_2$	music	109	8
	DBLP-ACM$_2$	citation	2473	4
	DBLP-Scholar$_2$	citation	5742	4
	Walmart-Amazon$_2$	electronics	2049	5
Textual (attribute values are long texts)	Abt-Buy	product	1916	3
	Company	company	22503	1
Heterogeneous (two record sets do not share the same schema)	Walmart-Amazon$_3$	electronics	2049	(4, 5)
	Walmart-Amazon$_4$	electronics	2049	(4, 4)
	Walmart-Amazon$_5$	electronics	2049	(4, 4)

4.1 Experimental Settings

Datasets: We use 16 datasets with structured, dirty, textual, and heterogeneous types, as summarized in Table 1. These datasets are originally from the ER Benchmark datasets [14] and the Magellan data repository [4] and preprocessed, including blocking, by Mudgal et al. [17]. Heterogeneous datasets are created from Walmart-Amazon$_1$ as Fu et al. [7] did. In order to make a fair comparison with supervised methods, we evaluate F1 scores for the methods on test sets.

Comparing Methods: We compare QA-Matcher with an unsupervised EM method, ZeroER[2] [23]. Since ZeroER ZeroER uses similarity functions for homogeneous schemas, it cannot be directly applied to heterogeneous datasets. Therefore, we concatenate attribute values of heterogeneous records and apply ZeroER to the concatenated values as Fu et al. [7] did. To demonstrate that QA-Matcher can produce results comparable to supervised methods without label information, we compare it with two supervised methods, Ditto [15,16] and DeepMatcher [17].

Configurations of QA-Matcher: We use a retriever based on BoW with binary scoring and set the number of retrieved records to generate a passage to $k = 20$ (sensitivity analysis of k will be performed later). We use a standard extractive QA model trained using the SQuAD dataset, bert-large-uncased-whole-

[2] https://github.com/chu-data-lab/zeroer.

word-masking-finetuned-squad from the Transformers library[3] [22]. This paper investigates the feasibility of utilizing QA models for EM in a fundamental setting; therefore, an extensive comparison of different QA models, retrievers, and question-passage formats is beyond our scope.

Table 2. F1 scores. The best results among the unsupervised methods are shown in bold, and the second best results are underlined. "–" indicates the computation did not end within a week.

Dataset	Unsupervised					Supervised	
	QA-Matcher	Sentence-BERT (QA-Matcher w/o QA)	QA-Matcher w/o retriever	QA-Matcher w/o reclas.	ZeroER	Ditto	Deep Matcher
BeerAdvo-RateBeer	**0.9333**	0.8966	0.4667	**0.9333**	0.7407	0.9437	0.7880
DBLP-Scholar$_1$	0.7276	0.5758	0.3355	0.6671	**0.7921**	0.9560	0.9470
DBLP-ACM$_1$	0.9754	0.9865	0.3844	**0.9888**	0.9541	0.9899	0.9845
Fodors-Zagats	**1.0000**	0.9333	0.2178	0.7586	0.9767	1.0000	1.0000
iTunes-Amazon$_1$	**0.9434**	0.6977	0.3971	0.8302	0.4800	0.9706	0.9120
Amazon-Google	0.6555	0.5266	0.2285	**0.6992**	0.2027	0.7558	0.7070
Walmart-Amazon$_1$	**0.6907**	0.4344	0.1722	0.6064	0.6645	0.8676	0.7360
DBLP-ACM$_2$	0.9832	0.9853	0.3771	**0.9921**	0.3974	0.9903	0.9810
iTunes-Amazon$_2$	**0.7727**	0.2500	0.4091	0.6957	0.4333	0.9565	0.7940
DBLP-Scholar$_2$	**0.7342**	0.5748	0.3379	0.6635	0.4169	0.9575	0.9380
Walmart-Amazon$_2$	**0.6433**	0.4855	0.1722	0.5808	0.2329	0.8569	0.5380
Abt-Buy	0.7218	0.3828	0.2205	**0.7569**	0.2110	0.8933	0.6280
Company	0.5197	0.4554	0.4308	**0.6131**	–	0.9385	0.9270
Walmart-Amazon$_3$	**0.6653**	0.4329	0.1722	0.5803	0.0000	0.8106	0.6710
Walmart-Amazon$_4$	**0.6653**	0.4329	0.1723	0.5803	0.0000	0.8211	0.6340
Walmart-Amazon$_5$	**0.6565**	0.4494	0.1725	0.5556	0.0000	0.8140	0.6650

4.2 Results

The results are shown in Table 2. QA-Matcher outperforms ZeroER on almost all datasets from various domains. In particular, it is competitive with Ditto and superior to DeepMatcher on several datasets such as the BeerAdvo-RateBeer and iTunes-Amazon$_1$ datasets, despite not accessing any label information. Figure 4 shows a matching record pair from iTunes-Amazon$_1$ and the corresponding question and passage generated by QA-Matcher. ZeroER misclassified the pair as a match, possibly due to the existence of many different tokens in the two records. In contrast, QA-Matcher correctly classified it by comprehending the meanings of the records in the QA format.

For the dirty datasets, where some attribute values are misplaced in other attributes, the results of QA-Matcher did not deteriorate significantly, unlike ZeroER, which is based on similarities between the same attribute values. For the heterogeneous datasets, QA-Matcher also produced competitive results with

[3] https://huggingface.co/.

id	Song_Name	Artist_Name	Album_Name	Genre	Price	CopyRight	Time	Released
1946	I Have Seen the Rain (Featuring James T. Moore)	P!nk featuring James T. Moore	I 'm Not Dead	Pop , Music , Electronic , R&B / Soul , Pop/Rock , Dance	$ 0.9 9	2006 RCA/JIVE Label Group , a unit of Sony Music Entertainment	3:29	4-Mar-06
23456	I Have Seen The Rain (Featuring James T. Moore) (Main Version)	P!nk	I 'm Not Dead [Explicit]	Dance & Electronic , Pop , R&B	$ 0.9 9	(C) 2009 Foo & Blu , LLC under exclusive License to Interscope Records	3:29	April 4 , 2006

Question:
What is characterized by I Have Seen the Rain (Featuring James T. Moore) P!nk featuring James T. Moore I 'm Not Dead Pop , Music , Electronic , R&B / Soul , Pop/Rock , Dance $ 0.99 2006 RCA/JIVE Label Group , a unit of Sony Music Entertainment 3:29 4-Mar-06?

Passage:
ID23456 is characterized by I Have Seen The Rain (Featuring James T. Moore) (Main Version) P!nk I 'm Not Dead [Explicit] Dance & Electronic , Pop , R&B $ 0.99 (C) 2009 Foo & Blu , LLC under exclusive License to Interscope Records 3:29 April 4 , 2006. ID52797 is characterized by Not Over You Caribou Uptempo Pop for Work Alternative Rock , Dance & Electronic $ 0.99 2006 The Leaf Label Ltd 3:38. ID52809 is characterized by On My Own Caribou Uptempo Pop for Work Alternative Rock , Dance & Electronic $ 1.29 2006 The Leaf Label Ltd 3:52. ···

Fig. 4. (upper) A matching record pair from iTunes-Amazon$_1$ that was correctly classified by QA-Matcher but not by ZeroER or Sentence-BERT (QA-Matcher without a QA model). (lower) The corresponding question and passage generated by QA-Matcher.

DeepMatcher. For the Abt-Buy dataset, QA-Matcher demonstrated superior performance in comparison to both ZeroER and DeepMatcher. However, it exhibited considerably lower performance compared to the supervised methods on the Company dataset, which may be caused by the lower performance of the retriever, as shown later.

4.3 Ablation Study

We next conduct ablation studies of QA-Matcher. In particular, we evaluate the performance of a QA model, a retriever, and reclassification, respectively. First, we compare a QA model and a sentence embedding method, Sentence-BERT [19] (sentence-transformers/paraphrase-MiniLM-L6-v2 from the Transformers library), to demonstrate that QA models are better at capturing the semantics of records that sentence embedding methods. A record pair (r, r') is classified as a match if r is most similar to r' in $R'_k(r) \cup \{r'\}$ with respect to the cosine similarity of vectors embedded by Sentence-BERT. Next, we compare a retriever and a random retriever to show that finding similar records is beneficial. Third, we compare results with and without reclassification to show that it enhances performance.

As shown in Table 2, QA-Matcher outperforms Sentence-BERT (QA-Matcher without a QA model) on almost all datasets, which may suggest that QA models are beneficial in identify entities. The performance of QA-Matcher without a retriever (with a random retriever) was poor because the random retriever does not ensure that the record pairs predicted as match are most similar to each other. QA-Matcher with reclassification consistently achieved high performance on all datasets by utilizing the scores provided by the QA model. In contrast, QA-Matcher without reclassification did not produce high performance consistently.

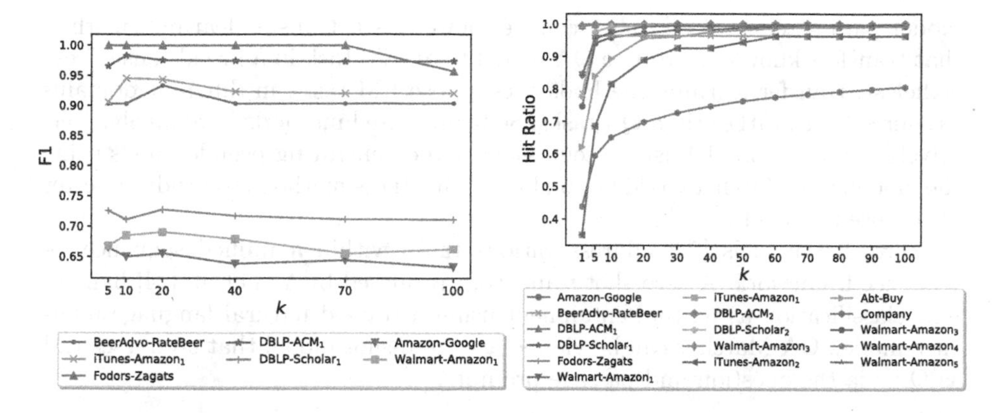

Fig. 5. F1 scores for different numbers of retrieved records k.

Fig. 6. Hit ratio of the retriever for different numbers of retrieved records k.

4.4 Sensitivity Analysis

We study the effect of the value of the hyperparameter, k, the number of retrieved records to be described in a passage. Figure 5 shows the results for $k \in \{5, 10, 20, 40, 70, 100\}$ on the structured datasets. The performance of QA-Matcher was not significantly affected by the value of k, possibly due to the high performance of the retriever with small values of k.

To further study the impact of the retriever, we evaluate the retriever in terms of hit ratio, the proportion of retrieved records that contain the matching records. As depicted in Fig. 6, we found that the retriever was able to accurately retrieve the matching records with small values of k for all datasets except for the Company dataset. This indicates that the retriever's poor performance might have resulted in the lower performance of QA-Matcher on the Company dataset.

5 Related Work

EM is a fundamental problem in data integration [3]. EM is typically formalized as supervised binary classification [13]. Recently, methods based on deep neural networks [15,17] have demonstrated significant performance. These methods represent records as real-valued vectors using word or sentence embedding methods [1,5] and classify them. However, they require large amounts of labeled data.

To reduce the labeling costs, unsupervised methods have been developed [6, 23]. ZeroER [23] applies clustering using a Gaussian mixture model to similarity vectors of record pairs generated by similarity functions given by users. However, it may not work if similarity functions are not appropriate to a particular domain.

Other techniques for a low resource setting include transfer learning and self-supervised learning. Transfer learning methods [10,12] involve training a model using large amounts of labeled data from source tasks and transferring the knowledge from the source data to a target task, for example, by sharing model parameters. QA-Matcher can be viewed as a transfer learning method that transfers knowledge from a QA model to an EM task in a broad sense. Nevertheless, transfer learning methods presuppose EM tasks in different domains as source tasks, rather than QA tasks. Self-supervised methods [2,8] involve iteratively training a model using pseudo-labels and generating pseudo-labels using the model. QA-Matcher could be combined with these methods to produce better initial pseudo-labels.

Zero-shot models [21] address various tasks within a unified sequence-to-sequence framework. A zero-shot framework proposed by Yin et al. [24] handles text classification as textual entailment using a trained natural language inference model. QA-Matcher can be viewed as a zero-shot model that addresses EM as QA via the question and passage prompts.

6 Conclusion

We proposed QA-Matcher, an unsupervised EM model that operates for tasks in diverse domains by solving EM as QA using a trained QA model. We demonstrated that QA-Matcher effectively performed with no label information across 16 datasets from various domains by utilizing the knowledge of the QA model to represent the semantics of records.

References

1. Bojanowski, P., Grave, E., Joulin, A., Mikolov, T.: Enriching word vectors with subword information. Trans. Assoc. Comput. Linguist. **5**, 135–146 (2017)
2. Cappuzzo, R., Papotti, P., Thirumuruganathan, S.: Creating embeddings of heterogeneous relational datasets for data integration tasks. In: Proceedings of the 2020 ACM SIGMOD International Conference on Management of Data, pp. 1335–1349 (2020)
3. Cohen, W.W., Richman, J.: Learning to match and cluster large high-dimensional data sets for data integration. In: Proceedings of the Eighth ACM SIGKDD International Conference on Knowledge Discovery and Data Mining, pp. 475–480 (2002)
4. Das, S., et al.: The Magellan data repository. https://sites.google.com/site/anhaidgroup/projects/data
5. Devlin, J., Chang, M., Lee, K., Toutanova, K.: BERT: pre-training of deep bidirectional transformers for language understanding. In: Proceedings of the 2019 Conference of the North American Chapter of the Association for Computational Linguistics: Human Language Technologies, Volume 1 (Long and Short Papers), pp. 4171–4186 (2019)
6. Fellegi, I.P., Sunter, A.B.: A theory for record linkage. J. Am. Stat. Assoc. **64**(328), 1183–1210 (1969)
7. Fu, C., Han, X., He, J., Sun, L.: Hierarchical matching network for heterogeneous entity resolution. In: Proceedings of the 29th International Joint Conference on Artificial Intelligence, pp. 3665–3671 (2020)
8. Ge, C., Wang, P., Chen, L., Liu, X., Zheng, B., Gao, Y.: CollaborEM: a self-supervised entity matching framework using multi-features collaboration. IEEE Trans. Knowl. Data Eng. 1 (2021)

9. Iyyer, M., Boyd-Graber, J., Claudino, L., Socher, R., Daumé III, H.: A neural network for factoid question answering over paragraphs. In: Proceedings of the 2014 Conference on Empirical Methods in Natural Language Processing, pp. 633–644 (2014)
10. Jin, D., Sisman, B., Wei, H., Dong, X.L., Koutra, D.: Deep transfer learning for multi-source entity linkage via domain adaptation. Proc. VLDB Endow. **15**(3), 465–477 (2021)
11. Karpukhin, V., et al.: Dense passage retrieval for open-domain question answering. In: Proceedings of the 2020 Conference on Empirical Methods in Natural Language Processing, pp. 6769–6781 (2020)
12. Kasai, J., Qian, K., Gurajada, S., Li, Y., Popa, L.: Low-resource deep entity resolution with transfer and active learning. In: Proceedings of the 57th Annual Meeting of the Association for Computational Linguistics, pp. 5851–5861 (2019)
13. Konda, P., et al.: Magellan: toward building entity matching management systems. Proc. VLDB Endow. **9**(12), 1197–1208 (2016)
14. Köpcke, H., Thor, A., Rahm, E.: Evaluation of entity resolution approaches on real-world match problems. Proc. VLDB Endow. **3**(1), 484–493 (2010)
15. Li, Y., Li, J., Suhara, Y., Doan, A., Tan, W.: Deep entity matching with pre-trained language models. Proc. VLDB Endow. **14**(1), 50–60 (2020)
16. Li, Y., Li, J., Suhara, Y., Wang, J., Hirota, W., Tan, W.: Deep entity matching: challenges and opportunities. J. Data Inf. Qual. **13**(1), 1:1–1:17 (2021)
17. Mudgal, S., et al.: Deep learning for entity matching: a design space exploration. In: Proceedings of the 2018 International Conference on Management of Data, pp. 19–34 (2018)
18. Rajpurkar, P., Zhang, J., Lopyrev, K., Liang, P.: SQuAD: 100, 000+ questions for machine comprehension of text. In: Proceedings of the 2016 Conference on Empirical Methods in Natural Language Processing, pp. 2383–2392 (2016)
19. Reimers, N., Gurevych, I.: Sentence-BERT: sentence embeddings using siamese BERT-networks. In: Proceedings of the 2019 Conference on Empirical Methods in Natural Language Processing (2019)
20. Voorhees, E.M.: The TREC-8 question answering track report. In: Proceedings of the Eighth Text Retrieval Conference, vol. 99, pp. 77–82 (1999)
21. Wei, J., et al.: Finetuned language models are zero-shot learners. arXiv preprint arXiv:2109.01652 (2021)
22. Wolf, T., et al.: Transformers: state-of-the-art natural language processing. In: Proceedings of the 2020 Conference on Empirical Methods in Natural Language Processing: System Demonstrations, pp. 38–45 (2020)
23. Wu, R., Chaba, S., Sawlani, S., Chu, X., Thirumuruganathan, S.: ZeroER: entity resolution using zero labeled examples. In: Proceedings of the ACM SIGMOD 2020 International Conference on Management of Data, pp. 1149–1164 (2020)
24. Yin, W., Hay, J., Roth, D.: Benchmarking zero-shot text classification: datasets, evaluation and entailment approach. In: Proceedings of the 2019 Conference on Empirical Methods in Natural Language Processing and the Ninth International Joint Conference on Natural Language Processing, pp. 3912–3921 (2019)

Multi-task Student Teacher Based Unsupervised Domain Adaptation for Address Parsing

Rishav Sahay(✉), Anoop Saladi, and Prateek Sircar

International Machine Learning, Amazon, Bengaluru, India
{rissahay,saladias,sircarp}@amazon.com

Abstract. In an e-commerce business, the ability to parse postal addresses into sub-component entities (such as building, locality) is essential to take automated actions at scale for successful delivery of shipments. The entities can be leveraged to build applications for logistics related operations, e.g. geocoding, assessing address completeness. Training an accurate address parser requires a significant number of manually labeled examples which is very expensive to create, especially when trying to build model(s) for multiple countries with unique address structure. To tackle this problem, in this paper, we present a novel Unsupervised Domain Adaptation (UDA) framework to transfer knowledge acquired by training a parser on labeled data from one country (source domain) to another (target domain) with unlabeled data. We specifically propose a multi-task student-teacher model comprising of three components: 1) specialized teachers trained on source data to create a pseudo labeled dataset, 2) consistency regularization, that uses a new data augmentation technique for sequence tagging data, and 3) boundary detection, leveraging signals in addresses like commas and text box boundaries. Multiple experiments on diverse address datasets (In this paper, we do not reveal the name of the e-commerce countries on which we evaluate our models due to business confidentiality. We also mask finer address details with (XX) to preserve customer's privacy.) demonstrate that our approach outperforms state-of-the-art UDA baselines for Named Entity Recognition (NER) task in terms of F1-score by 2–9%.

Keywords: Named Entity Recognition · Address parsing · Unsupervised Domain Adaptation

1 Introduction

Address is the most critical customer data to make successful and reliable deliveries of products in an e-commerce business. Some of the challenges we observe with respect to address quality such as misspellings, non-standard address connotations lead to delivery delays/failures, adversely impacting crucial business metrics. These problems are more pertinent to emerging countries where addresses

© The Author(s), under exclusive license to Springer Nature Switzerland AG 2023
H. Kashima et al. (Eds.): PAKDD 2023, LNAI 13938, pp. 186–197, 2023.
https://doi.org/10.1007/978-3-031-33383-5_15

are not standardized. Address parsing helps in identification of unique subcomponents which in turn can be leveraged for multiple downstream tasks, such as assessing completeness of addresses, predicting geocodes from address text, identifying high density communities in cities to pilot new services. For example, if you consider an address *karkarbagh colony, near sbi bank*, the parser can signal a missing building entity which can be used to power customer nudges to request additional information during address creation/order placement, thereby improving address quality. Address parsers are essentially NER models that require huge amounts (>50K) of labeled addresses with tags such as building name, locality, road and house unit information for model training. While address reference data from open-sources like *OpenAddresses* [6] and *OpenStreetMap* [7] are available, they differ vastly from the noisy/unstructured addresses commonly entered by customers and hence, lead to dismal performance in a practical setting. On the other hand, obtaining manually annotated address label data is expensive, time-consuming and prone to human errors. To tackle the labeled data scarcity issue, we leverage annotated data available for existing source countries when training an address parser for a new target country. There are various challenges involved while transferring knowledge from one country to another: 1) different countries have different address structures. For example, the most common address writing format in India (IN) is building name, road name followed by locality, while in the case of United Arab Emirates (UAE), it is road name followed by building name, and locality, 2) different countries may have addresses in different languages. Even if they are written in the same language (as in our experiments fixed to English), they have a very low vocabulary overlap. Hence, to solve these problems, we propose a novel multi-task student teacher based UDA architecture for address parsing that uses labeled data from source country to learn a target country address parser. Our approach involves three steps. Firstly, we perform domain adaptive pre-training for the base teacher model using both source and target addresses. We then train multiple teachers using source labeled data for address parsing to help deal with structural differences between countries. Finally, the student model is trained on two student-teacher tasks, consistency regularization and entity boundary detection. While consistency regularization task helps in making the student model robust to noise, boundary detection task provides additional information to the student about the target address structure, thereby improving overall model performance. To summarize, we make the following contributions in this work:

- We propose a novel multi-task student-teacher based UDA framework for address parsing that uses two teachers - one teacher is learned directly on source data while the other uses shuffled data.
- We present a new data augmentation technique applicable to sequence tagging data like addresses to perform consistency regularization on the student model.
- We introduce boundary detection as an additional task that leverages self-supervised signals in addresses like customer commas and address text box boundaries.

- We evaluate our approach on proprietary e-commerce data and external datasets, and demonstrate its superiority over state-of-the-art baselines. Ablation study further confirms the effectiveness of each of the proposed components.

2 Related Work

Address Parsing: In [11], the authors develop a multinational address parser using subword embeddings and recurrent neural network architecture. They build a single model capable of parsing addresses from multiple countries at the same time. Specifically, they use MultiBPEmb [4] to vectorize each word and the subword embeddings are fed into a BiLSTM encoder. The last hidden state of each word is fed to a feed-forward layer that is fed as input to a Seq2Seq module. [12] improves upon the above multi-national parser by adding an attention mechanism while label decoding, along with domain adversarial training for domain invariant features. Although the models are meant for multi-national parsing, the addresses on which they are trained with are structured addresses [10] that lack real life noise provided by customers for e-commerce delivery. Further, training a single parser model for multiple countries leads to deteriorated results for emerging countries, given the differences in address formats between countries. Also, since we do not have access to annotated data from multiple countries (~20 as assumed in the paper), we focus on single-source single-target address parsing in our experiments.

UDA-NER: As shown in [3], domain adaptive fine tuning using Masked language Modelling (MLM) on the target data and source data before fine tuning on the source NER data can help in boosting the zero shot performance on the target data (with a different distribution as compared to the source). In [9], the authors propose a teacher-student learning method for cross-lingual NER where the student model is trained using mean squared error loss with the teacher's output probability distribution (soft pseudo labels) as the ground truth. They also extend the methodology to multi-source cross lingual NER by weighing each of the source teachers' soft labels using the similarity between the source and target language vectors. Similarly, in [1], the authors use a student-teacher framework for cross-lingual NER, where a teacher is trained on source labeled data and, in parallel, a language discriminator and encoder are trained on the token-level adversarial task. All the above works on UDA/cross-lingual NER deal with open domain text. Addresses, on the other hand, are quite different as compared to open-domain text as they have a unique linguistic, and possess a notion of structure that can vary across countries. Hence, directly applying existing methods for our use-case is not optimal which is further validated in the results section.

3 Proposed Methodology

The core architecture of an address parser is that of a name entity recognition model, which is a token wise classifier on the top of an encoder $f(\theta)$ (RoBERTa-base [5]). For a given address text $x = \{x_i\}_{i=1}^{L}$, the encoder maps it into a set of hidden state vectors $h = \{f_\theta(x_i)\}_{i=1}^{L}$. For a token $x_i \in x$, its hidden state vector h_i is used to derive the probability distribution over the entity labels (see Sect. 4.1) using a linear classification layer and *softmax* function. We have access to a source labeled data D_{src}^{l} in which each source address has been assigned a label sequence, unlabeled source D_{src}^{ul} and unlabeled target D_{tgt}^{ul} addresses. We assume that source and target use the same set of entity labels.

3.1 Adaptive Pre-training Using MLM

We first adapt an open source RoBERTa-base[1] model to the addresses in source and target using MLM task [2]. We take 20K unlabeled addresses from both D_{src}^{ul} and D_{tgt}^{ul} after which we concatenate-shuffle them. The training procedure used here follows the same masking criteria and training settings as done in [3]. The adapted model $f(\theta_{add})$ helps to understand the linguistics within addresses and the vocabulary of both source and target data.

3.2 Student-Teacher Framework

Post adaptive pre-training, we use a student-teacher framework for unsupervised domain adaptation.

Multi-teacher Training. As discussed earlier, source and target can have very different structures. For e.g. in UAE, *road* occurs at the start for most of the addresses, while in India *building* occurs at the start. Training a teacher directly on source data will force the model to memorize such source-specific address structure patterns and produce noisy outputs on target data. This leads to a poor quality student model. To tackle this issue, we introduce an additional teacher trained on an entity level shuffled source data D_{shuf}^{l}. To create D_{shuf}^{l}, we randomly pick up annotated entity *ent* at index i from a source address with tags $\{x_{src}, y_{src}\} \in D_{src}^{l}$ and place it at another random position j $(j \neq i)$ to obtain a new shuffled address with tags $\{x_{shuf}, y_{shuf}\}$. Since the entities are shuffled, the source addresses will have higher variance in terms of structure, thereby enforcing the teacher model to pay attention to an entity itself without getting affected by its neighbours. Thus, we train two teacher models, main teacher $f(\theta_T)$ and shuffled-data teacher $f(\theta_{T_{shuf}})$ on D_{src}^{l} and D_{shuf}^{l} respectively using a cross-entropy loss function. The two teacher models during training are initialized using the weights of $f(\theta_{add})$ and the word embeddings layer is frozen to ensure that the models do not forget the target information that was learnt during MLM training.

Student-Teacher Tasks. Using each of the two teachers, for a given address x' where $x' \in D_{tgt}^{ul}$, we obtain two sets of pseudo labels for i-token x'_i: output probability distribution of the entity labels $P(x'_i; \theta_T)$ from the main teacher

[1] https://huggingface.co/docs/transformers/model_doc/roberta.

and $P(x_i'; \theta_{T_{shuf}})$ from the shuffled-data teacher. The student-teacher loss is formulated as the *mean squared error (MSE)* between the output distributions of the entity labels by the student model $f(\theta_S)$ and pseudo labels generated by the teacher as done in [9]. For each of the losses, those tokens are only considered on which the corresponding teacher has a maximum output probability more than a certain threshold t (set to 0.85). The shuffled-data teacher is a structure agnostic model whose pseudo labels when combined with the main teacher's pseudo labels enhance the quality of the student model.

Algorithm 1: Data Augmentation for CR

Input: Unlabeled target data D_{tgt}^{ul}

1 Main teacher model $f(\theta_T)$

Output: Augmented Target data D_{aug}^{ul}

2 $EntDic \leftarrow \{\}$

3 $Y_{HL} \leftarrow []$

4 **for** $x\prime$ *in* D_{tgt}^{ul} **do**

5 $y_{hl}, y_{probs} \leftarrow$ get hard labels with max prob. with $f(\theta_T)$

6 $types, starts, ends, probs \leftarrow$ find all entities with avg probs from y_{hard} and y_{probs}

7 **for** $ind \leftarrow 0$ **to** $len(types)$ **do**

8 **if** $probs[ind] >= 0.90$ **then**

9 $EntDic[types[ind]].append(x\prime[starts[ind] : ends[ind])$

10 $Y_{HL}.append(y_{hl})$

11 $D_{aug}^{ul} \leftarrow []$

12 **for** $(x\prime, y_{hl})$ *in* (D_{tgt}^{ul}, Y_{HL}) **do**

13 $types, starts, ends \leftarrow$ find type, start ind, end ind of all entities from y_{hl}

14 **for** $ind \leftarrow 0$ **to** $len(types)$ **do**

15 $type \leftarrow types[ind]$

16 $randEnt \leftarrow \phi$

17 **while** $(end[ind] - start[ind]) \neq len(randEnt)$ **do**

18 $randEnt \leftarrow$ sample an entity from $EntDic[type]$

19 $x\prime \leftarrow$ replace the entity at ind by $randEnt$

20 $x\prime_{aug} \leftarrow x\prime$

21 $D_{aug}^{ul}.append(x\prime_{aug})$

3.3 Consistency Regularisation Task

Consistency regularization (CR) [8] is a well-studied technique used in semi-supervised and self-supervised settings that encourages the prediction of the network to be similar in the vicinity of the observed training examples. We leverage this technique to make the student model more robust to noisy pseudo labels predicted by the main teacher $f(\theta_T)$. Here, we introduce a new data augmentation technique which allows us to create synthetic target addresses with the same labels as pseudo labels produced by $f(\theta_T)$ for the original target address. Basically, we first create a dictionary with all the high confidence (>0.90) entities predicted by $f(\theta_T)$ on D_{tgt}^{ul}. The dictionary contains an entity type mapped to list of confident predicted entities. The confidence of an entity is measured by the average maximum probability of the first sub-word of each of the entity tokens. Then for each pseudo-labeled target address $x' \in D_{tgt}^{ul}$, we replace every entity within the address with another random entity of the same entity type (also same length to ensure label consistency) in the dictionary to get x'_{aug}. See Algorithm 1 for the pseudo code to perform the data augmentation. The loss between the probability distribution of i token of x' denoted as $P(x'_i; \theta_S)$ and i token of x'_{aug} denoted as $P(x'_{iaug}; \theta_S)$ is formulated as a MSE loss. The synthetic address provides an entity level viewpoint to the student model. In other words, if a particular entity occurs at a differently in another address, it still refers to the same entity. The loss function for the above task is given by

$$L_{CR} = \sum_{x', x'_{aug} \in D_{tgt}^{ul}} \sum_{i=1}^{L} MSE(P(x'_i; \theta_S), P(x'_{iaug}; \theta_S)) \tag{1}$$

3.4 Boundary Detection Task

Address inputs have self-supervised signals like commas provided by customers while entering address text. In our case, we also have access to data in separate text fields (line 1, line 2 etc.) as entered during address creation, which automatically provides logical boundaries within the address. The key motivation behind introducing this module is that such implicit signals separating entities within addresses can potentially help the student model to identify correct entity spans in the target domain. Specifically, we sample an equal amount of addresses with boundary signals from the target address database as that of D_{tgt}^{ul}. Since these boundary signals can be noisy because of insufficient commas entered by customers, we only consider those addresses with more than two commas. We refer to this data as D_{bs}. Note that the separation between text fields is also converted to comma during pre-processing. We then define the boundary detection task on the student model as token level binary classification task to predict commas (labeled as 1) after a token in an address:

$$L_{BS} = \sum_{x^c \in D_{bs}} \sum_{i=1}^{L} BCE(P(x_i^c; \theta_{SC}), y_i^c)) \tag{2}$$

where x^c is an address text, x_i^c is the token at i index, $y_i \in \{0, 1\}$, θ_{SC} refers to the parameters of the student encoder model along with the dense layer of binary classification and BCE refers to binary cross entropy loss function. Note that both the NER tasks and the boundary detection task share a common encoder.

The student model is trained in a multi-task fashion on 4 losses - 1) student-teacher loss with main teacher pseudo labels, 2) student-teacher loss with shuffled-data teacher pseudo labels, 3) consistency regularization loss, and 4)

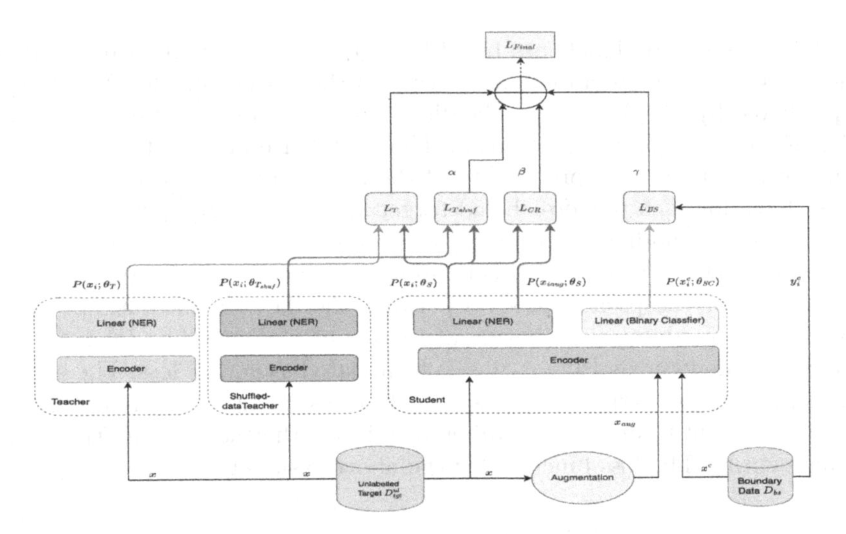

Fig. 1. Student training using multiple tasks

boundary detection task loss. We weigh the 3 new tasks with α, β and γ respectively. Figure 1 depicts the training of the student model using multiple tasks.

$$L_{Final} = L_T + \alpha * L_{T_{shuf}} + \beta * L_{CR} + \gamma * L_{BS} \quad (3)$$

4 Experiments, Data and Results

4.1 Data

We use proprietary e-commerce addresses from 3 countries, namely C1, C2 and C3 as well as external labeled addresses [10] from USA (US), Australia (AU) and Great Britain (GB), to evaluate our model. For the in-house datasets, the train and test data are manually annotated with 4 broad entity types, namely *bld* (building name, number and apartment names), *loc* (locality, sub-locality, community), *road* (road name, road number) and *unit* (apartment number, door number and floor number) in BIO format. We also have access to unlabeled raw customer addresses (with/without boundary signals as entered by the customers) in the e-commerce databases. We test our model under 6 different source to target transfer settings. The external addresses on the other hand are tagged with 6 entity types, namely *Unit*, *StreetNumber*, *StreetName*, *Province*, *PostalCode* and *Municipality*. We intentionally choose more structured countries here to validate the robustness of our approach across emerging as well as established marketplaces. Since we do not have boundary separated customer addresses for this dataset, we synthesized boundaries after each entity in the labeled addresses using the BIO format. Note that we only test on countries with English addresses. Dataset statistics for each of the countries are mentioned in the Table 1. We reveal only the approximate size of the e-commerce datasets due to confidentiality issues.

Table 1. Labeled, unlabeled and boundary data size (in K)

Country	C1	C2	C3	US	GB	AU
Train size	100	20	10	30	30	30
Test size	10	5	1	4	4	4
Unlabeled size	200	100	100	30	30	30
Boundary size	200	100	100	30	30	30

4.2 Experiment Setup

We use Hugging Face RoBERTa-Base as the backbone model. The maximum sequence length is fixed to 70 while batch size as 32; For MLM, the number of epochs is set to 3 while for student and teacher training, it is fixed to 7. We used early stopping with a patience of 2 to terminate the student-teacher training using the validation loss. For MLM, the learning rate is fixed at 3e−5, while for teacher a peak learning rate of 5e−5 and 1e−4 for student is used with a linear schedule and warm steps of 0.1 as in [9]. α, β and γ set to 1(default). A weight decay of 0.01 is used during training and dropout is fixed to 0.1 for student/teacher. AdamW is used as the optimizer and the model is trained on a single Nvidia Tesla T4 GPU. We use micro average F1-score across entity labels as our evaluation metric. Each number reported in Sect. 4.4 is an average of 5 runs with different seeds.

4.3 Baselines

We compare our approach against the following baselines:

- **Lower Bound (LB):** We train an open-source RoBERTa model on the source data and test it on target addresses in a zero-shot fashion.
- **DAPT** refers to [3]. Here, we pre-train the model using 20K samples from source and target domain.
- **SSTS:** [9] is the teacher-student learning approach for NER in a knowledge transfer setting
- **SSTS-DAPT:** We first run MLM on RoBERTa (DAPT) and then use it to initialize the teacher.
- **AdvP:** [1] is the recent state-of-the-art for UDA-NER.

4.4 Results, Ablation Studies, Parameter Study and Case Study

Performance Analysis. As it can be observed from Table 2, our approach (or its ablation) outperforms all the baselines on each of the 9 transfer settings. Noticeably, SOTA for UDA-NER like SSTS and AdvP are outperformed by DAPT. We do see that performing student-teacher training on a source/target domain adapted model (SSTS-DAPT) gives a performance boost, suggesting the

Table 2. Performance comparison of all the methods on 9 transfer settings from both e-commerce addresses and [10] addresses measured by micro-average F1-Scores.

Country Pair	E-commerce						DeepParse		
	C1-C2	C1-C3	C3-C1	C3-C2	C2-C1	C2-C3	US-GB	GB-US	AU-US
LB	0.398	0.314	0.320	0.479	0.428	0.393	0.390	0.601	0.354
DAPT	0.602	0.467	0.450	0.465	0.609	0.443	0.630	0.815	0.574
SSTS	0.434	0.391	0.383	0.490	0.453	0.432	0.372	0.598	0.364
AdvP	0.486	0.391	0.379	0.478	0.462	0.456	0.408	0.602	0.358
SSTS-DAPT	0.609	0.494	0.502	0.501	0.647	0.437	0.671	0.828	0.570
Our Approach	**0.666**	**0.511**	**0.540**	**0.527**	**0.693**	**0.512**	**0.762**	0.816	**0.604**
wo ST	0.644	0.495	0.522	0.522	0.676	0.483	0.649	**0.864**	0.567
wo CR	0.658	0.509	0.533	0.513	0.680	0.510	0.758	0.795	0.598
wo BS	0.653	0.491	0.528	0.523	0.677	0.505	0.735	0.818	0.589

Table 3. Performance of [11] on e-commerce datasets (F1-Scores)

Method	C1	C2	C3
[11]	0.15	0.23	0.12

Table 4. Entity level results for GB-US transfer.

Entity	Precision	Recall	F1-Score
Municipality	0.964	0.968	0.966
PostalCode	0.986	0.997	0.992
Province	0.969	0.982	0.975
StreetName	0.909	0.939	0.924
StreetNumber	0.991	0.993	0.992
Unit	0.952	0.970	0.961

benefits of adaptive pre-training. Our final proposed approach, however outperforms the best baselines by 2% to 9% which shows the advantages of adding the new tasks. We show the entity type level scores (precision, recall and F1-score) using our approach for one of the DeepParse addresses transfer settings (GB-US) in Table 4. The scores mentioned in the table are from the seed with a max overall F1-Score out of the 5 seed runs. Lastly, we test the performance of the multi-national parser mentioned in [11] on our e-commerce address dataset shown in the Table 3. Since the parser emitted different set of tags as in the e-commerce dataset, we created a mapping between the tag sets (like *StreetNumber*, *StreetName* mapped to *road*, *municipality* to *loc*, *unit* to *unit* and *bld* to *O*). The low F1-Scores for the datasets show the lack of generalization of the parser to real life customer addresses, thus justifying our claims in Sect. 2.

Ablation and Parameter Studies. We perform ablation studies on all the transfer settings shown in Table 2 by removing each of the proposed components - 1) shuffled-data teacher *wo ST*, 2) consistency regularization *wo CR*, and 3) boundary detection *wo BS*. We see a consistent drop in the F1 scores

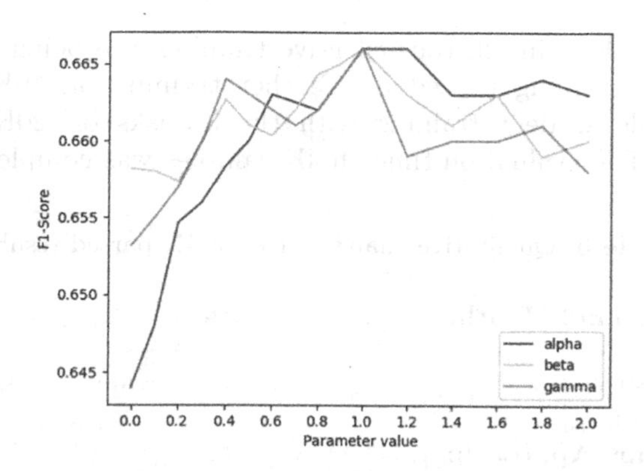

Fig. 2. F1 scores on the test data for C1-C2 when α, β or γ is varied from $[0, 2]$

for almost all of the datasets when removing a module. Specifically, the average drop in the scores for *wo ST* is 2.3%, *wo CR* is 0.8% and *wo BS* is 1.2% this. Thus, structural differences between the addresses of different countries can be handled well using shuffled-teacher and boundary detection, while consistency regularization helps dealing with noisy pseudo labels. We also study the impact of the 3 hyperparameters α, β and γ used in Eq. 3. We performed a grid search where we vary the values of α, β and γ in range [0, 2]. Here, we varied one of the hyper-parameters at a time while the others are set to default value of 1. As seen in Fig. 2, we observe an increasing trend in the F1-Score for C1 to C2 transfer with increase in parameter values till 1 and then a slight decreasing trend is observed.

Case Study. Finally as shown in Table 5, we perform a case study for UAE to IN transfer on 2 concrete examples where we compare the parsed outputs of SSTS and our method. In case 1, SSTS wrongly labels *13 XX* as *road* which suggests that the teacher memorized the address structure of UAE (where customers usually enter road names at the start). This was corrected by our approach that correctly labels it as *bld* which can be attributed to the structural in-variance brought in by the shuffled-teacher. Also, it detects *tarsali sussen road* partially while our approach fully recognized the entity. The boundary detection task helps here in detecting the full entity by signalling the student model right boundaries in a target address. In example 2, SSTS missed *marg* (hindi for road) which is a word used specifically in IN (not used in UAE) and mis-classifies it as *loc* while our approach correctly recognized it as *road*. This shows the importance of domain adaptive pre-training using MLM on source and target addresses.

4.5 Training/Inference Time

On a total 10K sample size of source and target addresses, MLM based domain adaptive pre-training took 30 mins/epoch. For every address, we created 10 augmentations as done in [2], thus effective training size being 100K. We used 4 GPUs for this training procedure. Teacher training on 10K samples took 1 min/epoch while student training with the 3 tasks on 20K samples took 6.10 min/epoch. The evaluation time on 4K samples was completed in 9 s.

Table 5. Qualitative analysis of UAE-IN parsed results

1	**Ground Truth:** [13 XX]$_{bld}$ [motinagar 2]$_{loc}$ [tarsali sussen road]$_{road}$ **SSTS:** [13 XX]$_{road}$ [motinagar 2]$_{loc}$ [tarsali]$_{loc}$ [sussen road]$_{road}$ **Our Approach:** [13 XX]$_{bld}$ [motinagar 2]$_{loc}$ [tarsali sussen road]$_{road}$
2	**Ground Truth:** [XX floor]$_{unit}$ [a764]$_{unit}$ [tulsi marg]$_{road}$ [sector19]$_{loc}$ uttar pradesh **SSTS:** [XX floor]$_{unit}$ [a764]$_{unit}$ [tulsi marg]$_{loc}$ [sector19]$_{loc}$ uttar pradesh **Our Approach:** [XX floor]$_{unit}$ [a764]$_{unit}$ [tulsi marg]$_{road}$ [sector19]$_{loc}$ uttar pradesh

5 Industrial Usecase

The address parser is catering to two use-cases in production for an emerging country C, namely 1) address quality scoring, and 2) community identification for launching new services. We trained an address parser for country C using labeled data from another country C1 and integrated parser based address completeness detection with the existing address quality scoring model for country C. This integration led to a 3% increase in the recall of the model for detecting junk addresses. The parser also assisted in identification of high-density communities for the launch of new value added e-commerce services. Our approach resulted in 67% and 133% more communities as compared to those identified earlier by operations team via manual process.

6 Conclusion and Future Work

In this paper, we propose a student-teacher based framework to transfer knowledge from training address parser on source country with labeled data to a target country with unlabeled data. Our approach uses multiple techniques like training shuffled-data teacher using shuffled source data, data augmentation for sequence tagging data for consistency regularization and learning from boundary signals to improve the target parser. Experiments on multiple e-commerce datasets and external data validate the effectiveness of our approach. In future, we plan to extend the solution to handle source and target countries with different languages and also, leverage multiple sources for training the model for the target country. Also, we wish to explore cases when we do have a limited amount of labeled data for the target country and how to include it in our training framework.

References

1. Chen, W., Jiang, H., Wu, Q., Karlsson, B., Guan, Y.: AdvPicker: effectively leveraging unlabeled data via adversarial discriminator for cross-lingual NER. In: Proceedings of the 59th Annual Meeting of the Association for Computational Linguistics. Association for Computational Linguistics, August 2021. https://doi.org/10.18653/v1/2021.acl-long.61. https://aclanthology.org/2021.acl-long.61
2. Devlin, J., Chang, M.W., Lee, K., Toutanova, K.: BERT: pre-training of deep bidirectional transformers for language understanding (2018). https://doi.org/10.48550/ARXIV.1810.04805. https://arxiv.org/abs/1810.04805
3. Han, X., Eisenstein, J.: Unsupervised domain adaptation of contextualized embeddings for sequence labeling (2019). https://doi.org/10.48550/ARXIV.1904.02817. https://arxiv.org/abs/1904.02817
4. Heinzerling, B., Strube, M.: BPEmb: tokenization-free pre-trained subword embeddings in 275 languages. In: Proceedings of the Eleventh International Conference on Language Resources and Evaluation, Miyazaki, Japan (2018)
5. Liu, Y., et al.: RoBERTa: a robustly optimized BERT pretraining approach (2019). https://doi.org/10.48550/ARXIV.1907.11692. https://arxiv.org/abs/1907.11692
6. openaddresses: Open addresses data (2019). https://openaddresses.io/
7. openstreetmap: Open street map data (2019). www.openstreetmap.org/#map=5/21.843/82.795
8. Sajjadi, M., Javanmardi, M., Tasdizen, T.: Regularization with stochastic transformations and perturbations for deep semi-supervised learning (2016). https://doi.org/10.48550/ARXIV.1606.04586. https://arxiv.org/abs/1606.04586
9. Wu, Q., Lin, Z., Karlsson, B.F., Lou, J.G., Huang, B.: Single-/multi-source cross-lingual NER via teacher-student learning on unlabeled data in target language (2020). https://doi.org/10.48550/ARXIV.2004.12440. https://arxiv.org/abs/2004.12440
10. Yassine, M., Beauchemin, D.: Structured Multinational Address Data (2020). https://github.com/GRAAL-Research/deepparse-address-data
11. Yassine, M., Beauchemin, D., Laviolette, F., Lamontagne, L.: Leveraging subword embeddings for multinational address parsing. CoRR abs/2006.16152 (2020). https://arxiv.org/abs/2006.16152
12. Yassine, M., Beauchemin, D., Laviolette, F., Lamontagne, L.: Multinational address parsing: a zero-shot evaluation. CoRR abs/2112.04008 (2021). https://arxiv.org/abs/2112.04008

Generative Sentiment Transfer via Adaptive Masking

Yingze Xie[1], Jie Xu[1](✉), Liqiang Qiao[1], Yun Liu[2], Feiran Huang[3], and Chaozhuo Li[4]

[1] School of Information Science and Technology, Beijing Foreign Studies University, Beijing, China
jxu@bfsu.edu.cn
[2] Department of Automation, Moutai Institute, Renhuai, Guizhou, China
[3] College of Cyber Security/College of Information Science and Technology, Jinan University, Guangzhou, China
[4] Microsoft Research Asia, Beijing, China

Abstract. Sentiment transfer aims at revising the input text to satisfy a given sentiment polarity while retaining the original semantic content. The nucleus of sentiment transfer lies in precisely separating the sentiment information from the content information. Existing explicit approaches generally identify and mask sentiment tokens simply based on prior linguistic knowledge and manually-defined rules, leading to low generality and undesirable transfer performance. In this paper, we view the positions to be masked as the learnable parameters, and further propose a novel AM-ST model to learn adaptive task-relevant masks based on the attention mechanism. Moreover, a sentiment-aware masked language model is further proposed to fill in the blanks in the masked positions by incorporating both context and sentiment polarity to capture the multi-grained semantics comprehensively. AM-ST is thoroughly evaluated on two popular datasets, and the experimental results demonstrate the superiority of our proposal.

1 Introduction

Sentiment transfer [6] aims at altering the sentiment polarity of a text while preserving its vanilla content meanings, which has been widely employed in a myriad of applications such as news sentiment transformation [3], passage editing [17] and data augmentation [12,28]. For example, when inputting a text sequence "stale food and poor service", the expected sentiment-transferred output would be "fresh food and good service", which modifies the sentiment polarity from negative to positive while maintaining the content information.

Existing sentiment transfer models could be roughly categorized into two categories. The first type of work implicitly disentangles content and sentiment [1–3,8,9,15,19] by learning the latent representations of content and sentiment respectively, and then combining the content representation and target sentiment signal to generate transferred sentences. Such implicit approaches

© The Author(s), under exclusive license to Springer Nature Switzerland AG 2023
H. Kashima et al. (Eds.): PAKDD 2023, LNAI 13938, pp. 198–209, 2023.
https://doi.org/10.1007/978-3-031-33383-5_16

generally employ GAN (Generative Adversarial Network) to remove sentiment attributes from content representation and generate text indistinguishable from real data. However, previous works [7] noted that implicit methods generally suffer from inferior performance on content preservation and low interpretability. Another way of sentiment transfer is decoupling content and sentiment explicitly [5,10,14,16,18,20,23,26], which first identifies sentiment-associated words and replaces them with words related to the target sentiment while keeping other words unchanged. Explicit methods benefit from the simplicity and explainability since the sentiment-tokens are explicitly uncovered and replaced.

Despite the promising performance of existing sentiment transfer models, they are still facing two crucial challenges. First, it is intractable to separate the sentiment style from the semantic content precisely. Existing implicit methods generally split the hidden representation of the input sequence into two vectors, which contain style and content information, respectively. However, such implicit methods generally perform unsatisfactorily on content preservation subtask [7], probably brought by the loss of content information in the process of disentangling. Explicit models are capable of identifying emotion-associated tokens explicitly, while they generally locate these tokens simply based on prior linguistic knowledge and rules. Such heuristic methods are incapable of ensuring the precise correlations between tokens in masking positions and sentiment signals, leading to an obscure disentanglement. Second, it is nontrivial to combine content and target sentiment to generate a target sentence effectively. Existing works usually assume that the overall emotional tendency of the generated sentence should be close to the target sentiment label. [23] leverages Attribute Conditional Masked Language Model (AC-MLM) to fill in masked positions. [14] retrieves new phrases associated with target attributes from the corpus and combines them with content information. However, these methods only focus on sentence-level sentiment labels and pay no attention to the word-level polarities, thus cannot capture such fine-grained semantic information and the connections between sentence-level sentiment and word-level polarity.

In this paper, we propose a novel **S**entiment **T**ransfer model AM-ST to handle the mentioned challenges based on **A**daptive **M**asking. First, we identify emotion-associated tokens and mask them using a trainable mask module, which is capable of adaptively learning the optimal mask positions. Then, a sentiment-aware masked language model is leveraged to fill in blanks in these masked positions, incorporating both context and sentiment polarity to improve transfer accuracy. Specifically, following the assumption of [23] that transferred sentence can be generated by simply replacing several emotional-related words, AM-ST adopts a mask classifier to identify the appropriate mask positions and then mask them with special tokens. In order to certify that sentiment information only appears in the masked tokens and not in the rest tokens, we design two types of losses: classification losses and adversarial losses, ensuring the clear separation of sentiment and content. After that, in the filling blanks phase, we adopt a sentiment-aware masked language model based on a reconstruction loss to predict both sentence- and token-level polarities in the masked positions.

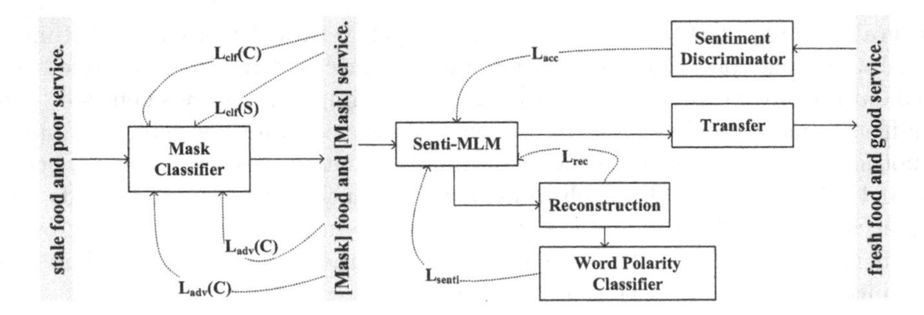

Fig. 1. Framework of the proposed AM-ST model.

Experimental results on two popular datasets demonstrate the superiority of our proposal. Our major contributions are summarized as follows:

- To alleviate the challenge of unclear disentanglement brought by low identifying quality, we propose an adaptive masking module, setting mask position as a trainable parameter to certify accuracy in identifying sentiment words.
- We further propose a sentiment-aware masked language model in the infilling blanks stage, which captures both semantic context and emotional signals.
- Experimental results on popular datasets indicate that the proposed AM-ST model consistently outperforms SOTA models.

2 Problem Definition

Given an input text sequence $X = \{x_1, x_2, ..., x_N\}$ (x_i indicates a single word), its source sentiment label l, and the target sentiment label $\widetilde{l}$, sentiment transfer aims to generate the target sentence $\widetilde{X}$ which maintains the content information of X while having the target sentiment label $\widetilde{l}$. Following the previous work [8], we select the binary sentiment label set $\{positive, negative\}$.

3 Methodology

3.1 Framework

Figure 1 demonstrates the framework of the proposed AM-ST model. Given the input sentence *stale food and poor service*, AM-ST first utilizes the adaptive masking module to separate the sentiment words from other content words, resulting the masked sentiment token set S and the content token set C. After masking the sentiment tokens with the special tokens, the sentence is converted to *[mask] food and [mask] service*. Then, based on the learned content text C and the target sentiment label $\widetilde{l}$, we further propose a generative module to predict tokens to infill words conforming target sentiment. Both the word-level and sentence-level sentiment polarities are properly incorporated to generate desirable sentiment-transferred sentences. The input text is transferred into the positive sentiment *"fresh food and good service"*.

3.2 Adaptive Sentiment Token Masking

In this stage, we aim to identify and mask sentiment-associated tokens to achieve pure content text. Previous works generally rely on prior linguistic knowledge (e.g., sentiment dictionary, co-occurrence frequency of words with a particular sentiment) to identify the sentiment tokens. However, such rule-based heuristic methods might not be optimal. Downstream datasets might contain unique sentiment tokens, which might be inconsistent with the general knowledge and ignored by previous works, leading to low generality and inferior performance. Instead of directly leveraging prior linguistic knowledge to discover and mask sentiment tokens, we view the mask positions as the learnable variables. Our model is expected to automatically locate the positions of the sentiment words to better align with the domain-specific knowledge.

Given the input text sequence X and the vanilla sentiment label l, we first need to decide whether token x_i should be masked. A mask classifier as shown in Fig. 2(a) is integrated to learn the masking probability of each token. The mask classifier is implemented as an attention-based classifier.

$$y = softmax(\alpha \cdot H) \tag{1}$$

where H denotes the hidden vectors generated by a Bi-LSTM model [4], and α denotes the attention weight vector generated based on the attention mechanism [21]. The output $y \in \mathbb{R}^{N \times 1}$ indicates the probability of x_i containing sentiment information and should be masked. If the predicted probability exceeds a threshold, we will mask the token in this position. We define S as the set of the tokens being masked, and C as the set of the rest of the tokens in X. The mask classifier will update its parameters adaptively during model training. In addition, we introduce extra constraints to facilitate disentangling between S and C. Intuitively, all tokens in S should contain sentiment information but no content information, while tokens in C should only include content information without any sentiment information. Such constraints are comprehensively satisfied by the following two types of losses.

Classification Loss. Classification losses ensure that the extracted token set C and S should capture the content and sentiment information, respectively. Thus, we design the following two classification losses for these two types of tokens.

Sentiment Classification Loss. The extracted set S should be associated with sentiment information, and we employ a pre-trained sentiment classifier $clf(S)$ implemented as a fully-connected layer with activation function Softmax to examine this. The input of the $clf(S)$ is S, and the loss function of $clf(S)$ is:

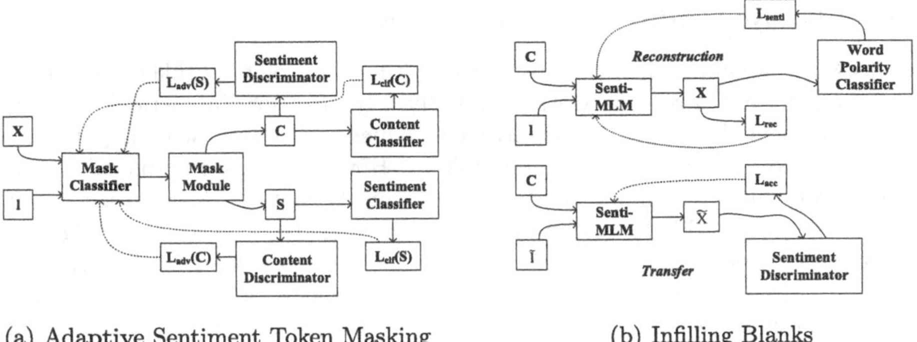

(a) Adaptive Sentiment Token Masking (b) Infilling Blanks

Fig. 2. Two phases of the proposed AM-ST model.

$$L_{clf}(S) = -\sum_{l \in \text{ labels}} t_s(l) \log y_{s(S)}(l) \tag{2}$$

where $t_s \in \mathbb{R}^{N \times 2}$ denotes the output of $clf(S)$. $y_{s(S)}(l)$ denotes the probability of emotional polarity of S being l. $L_{clf}(S)$ is a cross-entropy loss and approximates the distance between the predicted distribution $y_{s(S)}$ and ground-truth distribution t_s. It is worth noting that classifier $clf(S)$ is pre-trained and its parameters will not be updated during the training of our model, so the reduction of $L_{clf}(S)$ is not caused by the classifier's better predictive capacity, but brought by richer sentiment information encoded in the input text.

Content Classification Loss. To ensure the content information is captured by C, we design a content classification loss to measure the closeness between the vanilla input X and the content set C in terms of textual content. Following previous work [8], we use bag-of-words (BoW) to measure content completeness. To testify whether C preserves content well, we employ a Softmax content classifier $clf(C)$ to predict BoW distribution with the input of C. The loss function of $clf(C)$ is defined as the cross-entropy between ground-truth distribution t_c and predicted distribution $y_{c(C)}$:

$$L_{clf}(C) = -\sum_{w \in \text{ vocabulary}} t_c(w) \log y_{c(C)}(w) \tag{3}$$

where $t_c(w) = \frac{\text{count}(w,X)}{N}$ is the ground-truth BoW distribution, $\text{count}(w, X)$ is the frequency of word w in the vocabulary appearing in X. $y_{c(C)}(w)$ denotes the predicted probability of word w's appearance.

Adversarial Loss. To separate content and sentiment information, it is indispensable to examine whether C and S contain overlapping information [11,13]. We further introduce two adversarial losses to accomplish the objective.

Sentiment Adversarial Loss. To ensure that content set C contains as little sentiment information as possible, we adopt a two-step training paradigm. In the first step, we introduce a Softmax sentiment discriminator $dis(S)$ to predict the sentiment label of C. To improve the performance of $dis(S)$, we introduce $L_{dis}(S)$ to measure the distance between predicted distribution $y_{s(C)}$ and ground-truth distribution t_s:

$$L_{dis}(S) = -\sum_{l \in \text{ labels}} t_s(l) \log y_{s(C)}(l) \tag{4}$$

where $y_{s(C)}(l)$ represents the probability of the predicted label of C to be l. As $dis(S)$ has been trained to predict the sentiment label using C, in the second step, we introduce $L_{adv(S)}$ to punish the classification ability of $dis(S)$, which is implemented as the entropy of $y_{s(C)}$ and measures the accuracy of prediction. $L_{adv(S)}$ is maximized when $y_{s(C)}$ is evenly distributed, which means that sentiment labels are completely unpredictable for C:

$$L_{adv}(S) = -\sum_{l \in \text{ labels}} y_{s(C)}(l) \log y_{s(C)}(l) \tag{5}$$

Content Adversarial Loss. In order to remove content information in S, we introduce $L_{adv}(C)$, which is calculated in the similar manner of $L_{adv}(S)$. First, we adopt a content Softmax discriminator $dis(C)$ and optimize its ability to predict content BoW contribution $y_{c(S)}$ using S by minimizing $L_{dis}(C)$.

$$L_{dis}(C) = -\sum_{w \in \text{ vocabulary}} t_c(w) \log y_{c(S)}(w) \tag{6}$$

where t_c is the ground-truth distribution and $y_{c(S)}$ is the predicted distribution of $dis(C)$. Then, we punish the discernment of content discriminator. $L_{adv}(C)$ is the entropy of $y_{c(S)}$ and achieves its maximum when it is impossible to discern content distribution using S:

$$L_{adv}(C) = -\sum_{w \in \text{ vocabulary}} y_{c(S)}(w) \log y_{c(S)}(w) \tag{7}$$

Overall Objective Function. Based on the previous losses, the final objective function is formally designed as:

$$L_{total} = \lambda_1 L_{clf}(S) - \lambda_2 L_{adv}(S) + \lambda_3 L_{clf}(C) - \lambda_4 L_{adv}(C) \tag{8}$$

where λ_1, λ_2, λ_3 and λ_4 are the weights of the corresponding losses.

It is worth noting that although both the attention-based method in [23] and our model utilizes classifiers to find mask position, the former classifier is not updated synchronously during model training, but trained in advance. Therefore, the performance of [23] cannot be continuously improved as it ignores the feedback on whether masked tokens are related to sentiment. Instead, our mask position classifier adaptively updates its parameters using the above four losses and continues improving identifying accuracy.

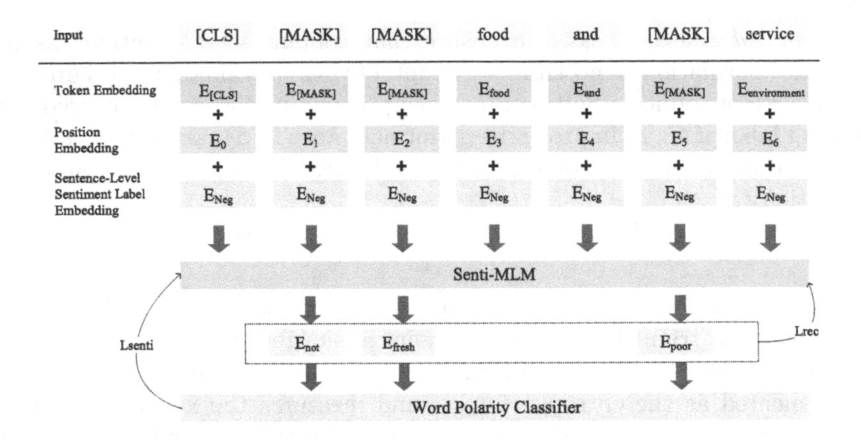

Fig. 3. Sentiment aware masked language model (Senti-MLM). Token embeddings, position embeddings and sentence-level sentiment embeddings are the inputs, and Senti-MLM is trained to predict tokens and word-level polarities in the masked positions.

3.3 Infilling Blanks

In this stage, our model will infill tokens in masked positions using a sentiment-aware masked language model (Senti-MLM) as shown in Fig. 2(b). Although MLM performs well in clozing tasks, it only considers contextual information and overlooks sentiment information. However, antonyms with opposite sentiment polarities tend to have different contexts. [24] proved that MLM performs well in learning features of domains and semantics, but tends to neglect opinion words and sentiments. Therefore, simply utilizing MLM to infill words in masked positions may not achieve desirable performance since tokens in these positions still contain rich sentiment information. To solve this problem, we adopt Senti-MLM, which is able to predict the tokens in the masked positions considering not only context but also sentiment information to facilitate transfer accuracy.

On the basis of the MLM of the pre-trained BERT model, we design a new paradigm to incorporate sentiment information. As shown in Fig. 3, the traditional segmentation embeddings are replaced by sentence-level sentiment label embeddings. The training objective is expected to predict tokens in masked positions and their word-level polarities. In the adaptive masking stage, we have proved that all masked tokens are associated with sentiment information. Therefore, the proposed Senti-MLM overcomes MLM's shortcoming of overlooking sentiment information and performs well in using words with proper sentiment polarity to fill in these blanks.

We employ a reconstruction task to train Senti-MLM. Given the set of content tokens C and the original label l, the objective aims to reconstruct the input sentence X using Senti-MLM. We measure the performance of the reconstruction task by L_{rec}, which reflects the content preservation ability of Senti-MLM:

$$L_{rec} = \sum_{t_i \in S} p\left(t_i \mid l; C\right) \tag{9}$$

To take word-level polarities into consideration, we train a word polarity classifier to predict word-level polarities using the hidden-state h generated by Senti-MLM, and its output is y_h. Then we introduce L_{senti} to measure Senti-MLM's ability to recover word-level polarities by comparing y_h and ground-truth word-level polarity t_{polar}.

$$L_{senti} = - \sum_{l \in \text{ labels}} t_{polar}(l) \log y_h(l) \tag{10}$$

Then, loss L_{rec} and L_{senti} are weighted combined to finetune Senti-MLM, and ϑ_1 and ϑ_2 denote the corresponding weights:

$$L_1 = \vartheta_1 L_{rec} + \vartheta_2 L_{senti} \tag{11}$$

After that, we leverage Senti-MLM to perform the transfer task. Content tokens C and target sentence label $\widetilde{l}$ are fed into Senti-MLM, and the transferred sentence $\widetilde{X}$ can be generated. To evaluate the transfer accuracy, we introduce $Lacc$ to measure whether $\widetilde{X}$ conforms with target label $\widetilde{l}$:

$$\widetilde{X} = \text{SentiMLM}(C, \widetilde{l}) \tag{12}$$

$$L_{acc} = -\log p(l \mid \widetilde{X}) \tag{13}$$

We continue finetuning Senti-MLM using L_{rec} and L_{acc}. The overall loss in this stage is $L_2 = \vartheta_3 L_{rec} + \vartheta_4 L_{acc}$, where ϑ_3 and ϑ_4 trade off between L_{rec} and L_{acc}.

4 Experiment

4.1 Experimental Settings

Dataset. Following previous works [14], we adopt two popular datasets, Yelp[1] and Amazon[2], to evaluate the performance of our proposal. Yelp contains business reviews in which each review is labeled with negative or positive sentiment. Similarly, Amazon dataset contains product reviews from Amazon, each of which is manually labeled as either negative or positive.

Baselines. We compare the proposed AM-ST model with the following popular baselines for verifying the performance.

- CrossAligned [19]: CrossAligned generates the original sentence back using a generator that combines content representation with the original label.

[1] https://www.yelp.com/dataset.

[2] https://www.kaggle.com/datasets/bittlingmayer/amazonreviews.

Table 1. Experimental results on Amazon and Yelp.

	Yelp		Amazon	
	ACC (%)	BLEU	ACC (%)	BLEU
CrossAligned [19]	73.1	3.1	74.1	0.4
StyleEmbedding [3]	8.7	11.8	43.3	10.0
MultiDecoder [3]	47.6	7.1	68.3	5.0
CycledReinforce [25]	85.2	9.9	77.3	0.1
DeleteAndRetrieval [14]	88.7	8.4	48.0	22.8
DisentangledRepresentation [8]	91.5	12.2	82.4	25.2
AC-MLM-Frequency [23]	95.1	11.6	64.5	27.2
AC-MLM-Fusion [23]	95.3	12.3	85.2	28.3
AM-ST	**97.1**	**12.9**	**86.4**	**29.7**

- StyleEmbedding [3]: Style embedding is fed into a decoder to generate text given different target sentiment signals.
- MultiDecoder [3]: MultiDecoder is a seq2seq model using multiple decoders. Each decoder independently generates a corresponding text style.
- CycledReinforce [25]: This model consists of a deemotionalizing module and an emotionalizing module, which extracts non-emotive semantic information and then emotionalizes neutral sentences.
- DeleteAndRetrieval [14]: DeleteAndRetrieval removes original attribute phrases and retrieves new terms related to the target attribute in the corpus.
- DisentangledRepresentation [8]: VAE with auxiliary multitask and adversarial objectives are used to learn content embeddings and style embeddings.
- AC-MLM-Frequency [23]: AC-MLM-Frequency converts the emotion transfer problem into a clozing task through a masked language model.
- AC-MLM-Fusion [23]: An extension of AC-MLM-Frequency which employs an attention mechanism to further filter retrieved sentiment words.

Evaluation Metrics. Following the previous work [14], we select Accuracy and BLEU as the evaluation metrics. Accuracy is calculated by how likely a transferred sentence conforms with the target sentiment, which is an indicator of transfer accuracy [22,27]. BLEU is computed by the similarity between human reference by [14] and the generated transferred sentence. A high BLEU score indicates that the model performs well in content preservation.

4.2 Quantitative Analysis

All sentiment transfer models are evaluated five times, and the average performance is reported in Table 1. For the StyleEmbedding model, the parameters

Table 2. Ablation Study.

	Yelp		Amazon	
	ACC (%)	BLEU	ACC (%)	BLEU
-$L_{clf}(S)$	96.1	12.7	84.7	29.0
-$L_{clf}(C)$	96.3	11.7	85.9	28.6
-$L_{adv}(S)$	96.4	12.4	85.8	29.4
-$L_{adv}(C)$	96.8	12.1	86.2	28.9
-L_{senti}	95.6	12.2	85.2	29.2
AM-ST	**97.1**	**12.9**	**86.4**	**29.7**

of the encoder and decoder are fixed when generating sentences of target sentiments, leading to inferior sentiment capability and thus achieving the worst performance. AC-MLMs are the strongest baselines because they take advantage of MLM's strong ability to predict masked tokens according to the semantic context. AC-MLM-Fusion achieves better performance since it further overcomes the deficiency of AC-MLM-Frequency by introducing an attention-based classifier to filter pseudo-sentiment words. One can clearly see that our proposal consistently outperforms baseline models on both datasets, verifying the effectiveness of the proposed adaptive masking mechanism and the token-level sentiment polarity.

4.3 Ablation Study

We further conduct an ablation study to verify the effectiveness of different components. Five ablation models are designed by removing different objective functions, namely sentiment classification loss in Eq. (2), content classification loss in Eq. (3), sentiment adversarial loss in Eq. (5) and content adversarial loss in Eq. (7). Table 2 presents the experimental results of different ablation models. One can easily observe that model performance significantly decreases after removing any components, verifying these modules are indispensable to a successful sentiment transfer. It is reasonable as the classification losses contribute to capturing the sentiment/content information, while the adversarial losses are capable of separating the sentiment and content information. The last module L_{senti} incorporates the token-level sentiment signals into the MLM process, facilitating the generation phase.

4.4 Parameter Sensitivity Analysis

Here we study the performance sensitivity of our proposal on eight core parameters: the weights λ_i in Eq. (8) and the weights ϑ_i of Formula (11). As the performance trends on the two datasets are similar, here we only report the results on the Yelp dataset. We first fix other hyper-parameters and then report the results by tuning the target hyper-parameter in the range of $[0, 0.5]$. Figure 4(a) and 4(b) presents the experimental results. One can see that with the increase of different hyper-parameters, the performance over all datasets first increases and then decreases, leading to a similar tendency. Thus, these hyper-parameters should be carefully tuned to achieve desirable performance.

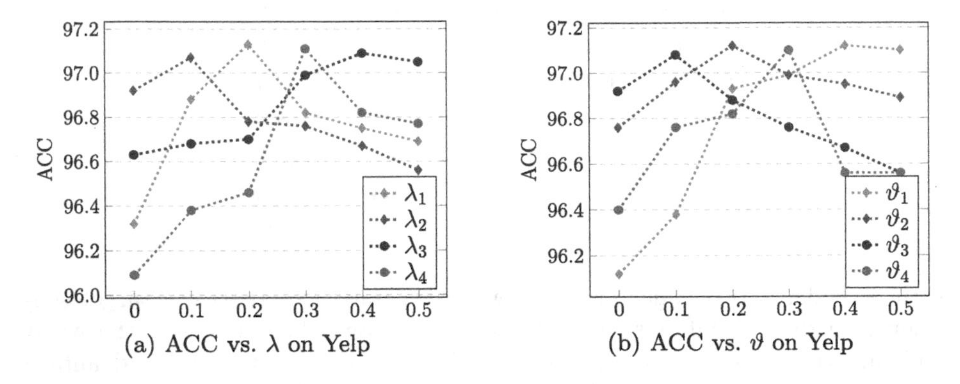

(a) ACC vs. λ on Yelp

(b) ACC vs. ϑ on Yelp

Fig. 4. Parameter sensitivity analysis on eight core hyper-parameters.

5 Conclusion

We present a novel model for sentiment transfer, which views the mask positions as trainable parameters to accurately identify and mask sentiment-related words. In addition, a sentiment-aware masked language model is adopted to infill blanks more efficiently by considering both context and word-level polarity. Experiments demonstrate that our model consistently outperforms SOTA models.

Acknowledgement. This work is supported in part by the National Natural Science Foundation of China (No. 62272200, 61932010, U22A2095) and the National Natural Science Foundation of China under Grant 62241205.

References

1. Chen, L., et al.: Adversarial text generation via feature-mover's distance. In: Advances in Neural Information Processing Systems, vol. 31 (2018)
2. Dathathri, S., et al.: Plug and play language models: a simple approach to controlled text generation. In: ICLR (2020)
3. Fu, Z., Tan, X., Peng, N., Zhao, D., Yan, R.: Style transfer in text: exploration and evaluation. In: AAAI (2018)
4. Graves, A., Schmidhuber, J.: Framewise phoneme classification with bidirectional LSTM networks. In: IJCNN, pp. 2047–2052 (2005)
5. Guu, K., Hashimoto, T.B., Oren, Y., Liang, P.: Generating sentences by editing prototypes. Trans. Assoc. Comput. Linguisti. **6**, 437–450 (2018)
6. He, R., McAuley, J.: Ups and downs: modeling the visual evolution of fashion trends with one-class collaborative filtering. In: WWW, pp. 507–517 (2016)
7. Hu, Z., Lee, R.K.W., Aggarwal, C.C., Zhang, A.: Text style transfer: a review and experimental evaluation. ACM SIGKDD Explor. Newsl. **24**(1), 14–45 (2022)

8. John, V., Mou, L., Bahuleyan, H., Vechtomova, O.: Disentangled representation learning for non-parallel text style transfer. In: ACL, pp. 424–434 (2019)
9. Krishna, K., Nathani, D., Samanta, B., Talukdar, P.: Few-shot controllable style transfer for low-resource multilingual settings. In: ACL, pp. 7439–7468 (2022)
10. Lee, J.: Stable style transformer: delete and generate approach with encoder-decoder for text style transfer. In: Proceedings of the 13th International Conference on Natural Language Generation, pp. 195–204 (2020)
11. Li, C., et al.: Adversarial learning for weakly-supervised social network alignment. In: AAAI (2019)
12. Li, C., et al.: PPNE: property preserving network embedding. In: Candan, S., Chen, L., Pedersen, T.B., Chang, L., Hua, W. (eds.) DASFAA 2017. LNCS, vol. 10177, pp. 163–179. Springer, Cham (2017). https://doi.org/10.1007/978-3-319-55753-3_11
13. Li, C., et al.: Distribution distance minimization for unsupervised user identity linkage. In: CIKM, pp. 447–456 (2018)
14. Li, J., Jia, R., He, H., Liang, P.: Delete, retrieve, generate: a simple approach to sentiment and style transfer. In: NAACL, pp. 1865–1874 (2018)
15. Liu, A., Wang, A., Okazaki, N.: Semi-supervised formality style transfer with consistency training. In: ACL, pp. 4689–4701 (2022)
16. Madaan, A., et al.: Politeness transfer: a tag and generate approach. In: ACL, pp. 1869–1881 (2020)
17. Parra, G.L., Calero, S.X.: Automated writing evaluation tools in the improvement of the writing skill. Int. J. Instr. **12**(2), 209–226 (2019)
18. Shang, M., et al.: Semi-supervised text style transfer: cross projection in latent space. In: EMNLP, pp. 4937–4946 (2019)
19. Shen, T., Lei, T., Barzilay, R., Jaakkola, T.: Style transfer from non-parallel text by cross-alignment. In: NeurIPS (2017)
20. Sudhakar, A., Upadhyay, B., Maheswaran, A.: Transforming delete, retrieve, generate approach for controlled text style transfer. In: EMNLP, pp. 3269–3279 (2019)
21. Vaswani, A., et al.: Attention is all you need. In: NeurIPS (2017)
22. Wang, Y., et al.: An adaptive graph pre-training framework for localized collaborative filtering. TOIS **41**(2), 1–27 (2022)
23. Wu, X., Zhang, T., Zang, L., Han, J., Hu, S.: Mask and infill: applying masked language model for sentiment transfer. In: IJCAI, pp. 5271–5277 (2019)
24. Xu, H., Shu, L., Yu, P., Liu, B.: Understanding pre-trained BERT for aspect-based sentiment analysis. In: COLING, pp. 244–250 (2020)
25. Xu, J., et al.: Unpaired sentiment-to-sentiment translation: a cycled reinforcement learning approach. In: ACL, pp. 979–988 (2018)
26. Zhang, Y., Xu, J., Yang, P., Sun, X.: Learning sentiment memories for sentiment modification without parallel data. In: EMNLP, pp. 1103–1108 (2018)
27. Zhao, J., et al.: Learning on large-scale text-attributed graphs via variational inference. arXiv preprint arXiv:2210.14709 (2022)
28. Zheng, X., Chalasani, T., Ghosal, K., Lutz, S., Smolic, A.: STaDA: style transfer as data augmentation. In: VISAPP (2019)

Unsupervised Text Style Transfer Through Differentiable Back Translation and Rewards

Dibyanayan Bandyopadhyay(✉) and Asif Ekbal

Department of Computer Science and Engineering, Indian Institute of Technology Patna, Bihta, Patna, India
{dibyanayan_2111cs02,asif}@iitp.ac.in

Abstract. In this paper, we propose an end-to-end system for unsupervised text style transfer (UTST). Prior studies on UTST work on the principle of disentanglement between style and content features, which successfully accomplishes the task of generating style-transferred text. The success of a style transfer system depends on three criteria, *viz.* Style transfer accuracy, Content preservation of source, and Fluency of the generated text. Generated text by disentanglement-based method achieves better style transfer performance but suffers from the lack of content preservation as the previous works suggest. To develop an all-around solution to all three aspects, we use a reinforcement learning-based training objective that gives rewards to the model for generating fluent style transferred text while preserving the source content. On the modeling aspect, we develop a shared encoder and style-specific decoder architecture which uses the Transformer architecture as a backbone. This modeling choice enables us to frame a differentiable back translation objective aiding better content preservation as shown through a careful ablation study. We conclude this paper with both automatic and human evaluation, showing the superiority of our proposed method on sentiment and formality style transfer tasks. Code is available at https://github.com/newcodevelop/Unsupervised-TST.

Keywords: Unsupervised Text Style Transfer · Back Translation · Reinforcement Learning

1 Introduction

Text style transfer is defined as converting input text into an output text which contains the same underlying semantics but a different style. Many of the style transfer datasets are non-parallel and thus standard supervised training fails in this regime. In the existing literature, there have been attempts to solve this problem using unsupervised methods [3,8,10].

Existing methods implement two lines of research, *viz. i)* one that focuses on disentanglement of style and content [8]. As reported in Subramanian et al. [17], the style information can still be recovered from the latent representation even

© The Author(s), under exclusive license to Springer Nature Switzerland AG 2023
H. Kashima et al. (Eds.): PAKDD 2023, LNAI 13938, pp. 210–221, 2023.
https://doi.org/10.1007/978-3-031-33383-5_17

after the model has been trained adversarially. This results in loss of content preservation in trade of better accuracy [10]. *ii)* The second line of research focuses on designing custom modeling choices to alleviate the shortcomings of the former methods, and thus achieves higher style transfer accuracy in trade of better content preservation.

To incorporate the best of both worlds, we make a modeling choice by designing an input text encoder that is shared between two style-specific decoders. Based on Style transformer [3], we also make use of Transformer [19] architecture to build both the encoder and decoder. These style-specific decoders are first trained via an auto-encoding objective where each decoder only decodes input text containing only a specific style. This approach helps to make the encoder learn style-independent features that are later utilized by the style-specific decoders for converting input sentences into their style-transferred form. This modeling choice also enables us to use a custom back-translation objective to train the network in an end-to-end fashion. Jointly training by this auto-encoding and back-translation objective was sufficient for the model to get better than random performance for style transfer. This shows a positive disentanglement between style and content, enabled by these modeling choices. Further, this results in higher content preservation as reflected in a carefully done ablation study.

While simply putting these objectives may work to some extent, there is more room for improvement. We observe that there are three necessities by which we should measure the performance of the trained model. Specifically, we want to *i)* Preserve content between the input source and generated output. *ii)* Transfer style and *iii)* Generate grammatical output in the process.

Previous works [7,8,11] mostly use the auto-encoding objective as the primary objective coupled with task-specific rewards, particularly, the style transfer reward is utilized by many recent works [10]. We incorporate rewards for all three dimensions discussed above.

Previous methods use BLEU [12] between source and generated text [21] as a reward for content preservation. Though it is a well-studied method to evaluate the similarity between the source and generated text, BLEU does not properly reflect human judgment. We propose to use a more robust semantic similarity-based metric as a reward to measure the distance between the source and generated text(c.f Sect. 3.5).

For generating grammatical sentences, we use negative perplexity obtained from the GPT2 [14] language model.

Prior research showed a trade-off between style transfer accuracy and content preservation and fluency. Hence, comparing models on all metrics is difficult. To compare models on an equal footing, we utilise the geometric mean of all measures following [8] as a composite score for content preservation, style transfer, and fluency. To summarize, the contributions are *three-fold: i)* We develop a shared encoder and style-specific decoder-based Transformer architecture which obtains superior content preservation and comparable style transfer accuracy compared to the existing methods, evaluated both automatically and manually. *ii)* We develop a differentiable back translation objective equipped with reinforcement learning. The ablation of different components shows the effect of various reward functions to train the model. *iii)* We perform case studies and error analysis to qualitatively illustrate the failure cases of our model and propose some ideas to overcome them.

2 Related Work

Past years have seen numerous methods developed as a part of text style transfer [6]. Many of the existing works for unsupervised text style transfer operate solely on the disentanglement of source content and style in order to generate the same content with a different style. Various styles [13] and content-oriented loss [16] functions are proposed to tackle this problem. After Subramanian et.al. (2018) [17] proposed that disentanglement is not always successful, Dai et.al. (2019) [3] proposed an architecture without any explicit disentanglement constraint.

Another line of work focuses on pipeline-based training, which contains three steps, i.e., delete, retrieve, and generate [9]. Sudhakar et.al. (2019) [18] used GPT as the generator in this pipeline.

More recent studies in [23] focus on exploring the word-level style relevance which is assigned by a pre-trained style classifier. The authors in [11] proposed a tag-and-generate pipeline, which firstly identifies style attribute markers from source texts, then replaces them with a special token, and generates the outputs based on the tags.

Compared to these methods, we propose an all-around objective for the model by using reward-based learning. The above models also suffer from poor content preservation, which is mostly alleviated in our proposed model by using the novel differentiable back-translation objective.

3 Methodology

3.1 Problem Definition

We assume we have two datasets $D_1 = \{d_1^1, d_1^2, d_1^3, ..., d_1^n\}$ and $D_2 = \{d_2^1, d_2^2, d_2^3, ..., d_2^n\}$ with two separate styles x and y, respectively. The goal is generating data (d_1^p, d_2^q) such that its original style (x, y) are respectively reversed to (y, x). We implement the proposed architecture using a transformer encoder shared between two style-specific transformer decoders.

3.2 Shared Encoding

Our proposed architecture uses a shared encoder (E) which encodes both input d_1^p and d_2^q coupled with styles x and y, respectively. Formally,

$$c = E(d_a^p; X) \text{ and } a \in \{1, 2\} \text{ and } X \in \{x, y\} \tag{1}$$

where c is a style-agnostic content encoding of the source input sentence.

3.3 Auto-Encoding

$G1$ and $G2$ are two transformer decoders that get the same encoder representation c. Autoencoding of inputs is implemented by these decoders, where G_1 reconstructs input d_1^p while G_2 reconstructs input d_2^q. This auto-encoding procedure can be explained by the following loss function ($\mathcal{L}_{\mathcal{R}}$) in Eq. 2.

$$\mathcal{L}_{\mathcal{R}} = \mathop{\mathbb{E}}_{d_1^p \sim D_1} [-log p_{G_1}(d_1^p|c)] + \mathop{\mathbb{E}}_{d_2^q \sim D_2} [-log p_{G_2}(d_2^q|c)] \tag{2}$$

To act as discriminators for styles, we force G_1 and G_2 to only train by maintaining the following constraints:

$$p_{G_1}(d_2^q|c) = 0; c = E(d_2^q; y) \text{ and } p_{G_2}(d_1^p|c) = 0; c = E(d_1^p; x) \tag{3}$$

3.4 Differentiable Back-Translation

We employ a differentiable back translation training objective for better content preservation of the source sentence. G_1 is used to generate style transferred version of input sentence d_2^q coupled with input style x, denoted as $\widetilde{d_2^q}$. The sampling process for generation is non-differentiable owing to the non-differentiability of the softmax probability distribution. Therefore, we use Gumbel Softmax [5], as a continuous approximation of the categorical distribution. If corresponding decoder's output logit for $\widetilde{d_2^q}$ is $\widetilde{l_2^q}$, then the corresponding soft probability distribution at time-step q would be $\widetilde{p_2^q} = Gumbel(\widetilde{l_2^q}, \tau)$, where $Gumbel$ is defined as:

$$Gumbel(\widetilde{l_2^q}, \tau) = \frac{exp(log\ \widetilde{l_2^q} + g_i)/\tau}{\sum_{q=1}^{n} exp(log\ \widetilde{l_2^q} + g_i)/\tau} \tag{4}$$

where g_i are independently sampled from $Gumbel(0, 1)$. In the forward pass, we use one-hot approximation of $\widetilde{p_2^q}$ denoted as $\hat{p_2^q}$. Note that the backward pass is still differentiable as the following straight-through estimation technique is used to generate hard one-hot samples.

$$\hat{p_2^q} = \hat{p_2^q} - \widetilde{p_2^q}.detach() + \widetilde{p_2^q} \tag{5}$$

detach function shown above strips $\widetilde{p_2^q}$ of its gradient to obtain re-parameterized value of $\hat{p_2^q}$.

In the back translation, these sampled $\widetilde{d_2^q}$ are fed into the encoder (E) and then the input d_2^q is reconstructed back via G_2. Thus, back-translation loss associated with decoder G_1 is :

$$\mathcal{L}_{\mathcal{BT}}(G_1) = \mathop{\mathbb{E}}_{d_2^q \sim D_2} [-log p_{G_2}(d_2^q|c)] \text{ and } c = E(\widetilde{d_2^q}) \tag{6}$$

By a similar argument, back-translation loss associated with decoder G_2 would be given by:

$$\mathcal{L}_{\mathcal{BT}}(G_2) = \mathop{\mathbb{E}}_{d_1^p \sim D_1} [-log p_{G_1}(d_1^p|c)] \text{ and } c = E(\widetilde{d_1^p}) \tag{7}$$

So, the final back translation loss would be

$$\mathcal{L}_{\mathcal{BT}} = \mathcal{L}_{\mathcal{BT}}(G_1) + \mathcal{L}_{\mathcal{BT}}(G_2) \tag{8}$$

3.5 Reinforcement Learning

We design various reward functions for *i) Content preservation*, *ii) Style Transfer* and *iii) Fluency* of the generated output.

Content Preservation. As content preservation reward (c_r), we use the harmonic mean of BLEU and BERTSCORE [22]. The harmonic mean is chosen because if any of the metrics is zero, their composition should also be zero. BLEU between the generated sentence $\widetilde{d_a^p}$ and input sentence d_a^p reflects the syntactic similarity between the two, whereas BERTSCORE reflects their semantic similarity.

$$c_r = \frac{2 * BLEU(\widetilde{d_a^p}, d_a^p) * BERTSCORE(\widetilde{d_a^p}, d_a^p)}{BLEU(\widetilde{d_a^p}, d_a^p) + BERTSCORE(\widetilde{d_a^p}, d_a^p)} \tag{9}$$

Style Transfer. We use a fine-tuned distilBERT[1] architecture as a reward model for style classification. Predicted style label from distilBERT would be $\hat{y}$, where $\hat{y} = distilBERT(\widetilde{d_a^p})$. The style transfer reward (s_r) is generated by the following Equation

$$s_r = \mathbf{1}_{\{x\}}(\hat{y}) \tag{10}$$

$\mathbf{1}$ is a modified indicator function, which outputs a positive value of 1 if and only if the input style (x) and the predicted style ($\hat{y}$) are not the same ($x \neq \hat{y}$), thereby generating a reward of +1 in the case of successful style transfer.

Fluency. We fine-tune a pre-trained GPT2 model on the respective style transfer dataset, which acts as a reward model for fluency. We define fluency reward f_r as

$$f_r = -ppl(y_{1:M}) = -\frac{2^{-\frac{1}{M}L(y_{1:M})} - 1}{c - 1}$$

where $L(y_{1:M})$ is the negative log likelihood loss associated with GPT2 output when $\widetilde{d_a^p}$ is given as an input to GPT2. To use perplexity ($ppl(y_{1:M})$) as a reward, we first clip it to c (a fixed integer with value 10) and then normalize it such that PPL values lies between 0 to 1; following the equation as mentioned above. We take the negative of PPL as the fluency reward f_r, as we want to minimize PPL to obtain more fluent sentences. *Note that both distilBERT and GPT2 are frozen while training our model.*

3.6 Learning Technique

The goal of the RL technique is to maximize the expected reward or equivalently minimize the negative of it, which can be expressed as:

$$L_{RL}(\theta) = - \mathop{\mathbb{E}}_{d^s \sim p_\theta} [r(d^s)]$$

[1] https://huggingface.co/docs/transformers/model_doc/distilbert.

p_θ is the probability distribution (parameterized by θ) of the decoder output from where d^s is sampled. Here, $d^s = \{d_i^s\}_{i=1}^t$, where t is the number of words sampled and d_i^s refers to ith word being sampled. s as a superscript refers to the sampling process. $r(d^s)$ can be obtained from summing up all the rewards : $r(d^s) = c_r + s_r + f_r$

To compute the gradient of $L(\theta)$, we use REINFORCE [20] algorithm. Specifically, we subtract a baseline reward $r(d^g)$ obtained from generating d^g in a greedy fashion, shown via the superscript g.

The full form of the gradient of the loss function becomes:

$$\nabla_\theta L_{RL}(\theta) = - \mathop{\mathbb{E}}_{d^s \sim p_\theta} [(r(d^s) - r(d^g))\nabla_\theta \ logp_\theta(d^s)] \tag{11}$$

We finally calculate the overall loss function of the model as: $L(\theta) = \alpha\mathcal{L}_\mathcal{R} + \beta(\mathcal{L}_{BT} + L_{RL})$, where $(\alpha, \beta) = (1, 0)$ before a certain training epoch threshold and after that they are given equal weightage of 0.5.

4 Datasets, Experiments and Results

4.1 Datasets

For the experiments and evaluation, we use two commonly used datasets, namely, *YELP* [9] and *GYAFC* [15]. These two datasets are sentiment and formality transfer datasets, respectively. The YELP dataset comprises sentiment-labelled sentences with either positive or negative polarity. The GYAFC dataset contains formality labels along with 4 human references. As our method is unsupervised, we do not use parallel human reference sentences for training. They are only used to evaluate the content preservation capability of our model at the test time. Dataset statistics for these methods are shown in Table 1.

Table 1. Dataset statistics for YELP and GYAFC datasets

Dataset	Style	Training	Testing
YELP	Neg	266k	500
	Pos	177K	500
GYAFC	Informal	52k	500
	Formal	52K	500

4.2 Baselines

To compare our methods to that of existing systems, we use the following baselines: i) *Generative Style Transformer (GST)* [18], ii) *Style Transformer (ST)* [3], iii) *Deep Latent Sequence Model (DLSC)* [4], iv) *Dual Reinforcement Learning Framework (DualRL)* [10], v) *DeleteAndRetrieve Framework (Del&Ret)* [9] and vi) *A tag and generate approach (Tag&Generate)* [11]

Table 2. Automatic Evaluation for YELP dataset

Models	**Acc**	**Content Preservation**						**PPL**	s_{GM}	r_{GM}
		wrt Source			wrt Reference					
		s-BLEU	s-METEOR	s-BSC	r-BLEU	r-METEOR	r-BSC			
GST	88.3	45.8	73.7	91.3	20.3	45.0	83.6	123	29.4	22.2
ST	**92.1**	53.1	76.5	91.1	21.5	46.7	83.4	372	24.7	18.3
DLSC	84.6	49.1	64.8	87.6	21.0	39.2	80.7	**67**	32.4	24.3
DualRL	88.3	59.0	78.6	92.7	24.8	47.9	84.1	143	30.6	22.8
Del&Ret	48.0	40.6	83.7	91.7	11.6	44.1	81.8	1461	16.0	10.7
Tag&Generate	86.4	47.8	66.9	89.3	19.1	39.5	81.7	119	29.1	21.4
Ours	80.4	**75.1**	**82.9**	**93.9**	**30.8**	**49.7**	**85.2**	74	**36.4**	**26.9**

4.3 Automatic and Human Evaluation

Automatic Evaluation. We evaluate our models on both YELP & GYAFC datasets. We evaluate our proposed model on three different dimensions. They are *i)* Content preservation evaluated on both source and human reference, *ii)* Style transfer accuracy and *iii)* Fluency as measured by perplexity (PPL). To measure the style transfer accuracy of our model and other models, we use a fine-tuned BERT trained on the YELP dataset as a style classifier.

In Table 2 and Table 3, we specify the accuracy obtained by our model and other state-of-the-art baseline models in *Acc* column. Note that although our model achieves lower style transfer accuracy on YELP compared to state-of-the-art models, it is very easy to fool the style classifier by simply pre-pending 'Terrible' to positive source sentences and 'Wonderful' to negative source sentences. Doing this leads to 87.8% style classifier performance with a trade-off of very low fluency. Thus to compare separate models, we have to compare them against all three metrics, *i)* style transfer accuracy *ii)* content preservation as well as *iii)* fluency.

Table 3. Automatic evaluation for GYAFC dataset. Note that we do not provide r-METEOR and r-BSC due to having multiple human references and these metrics are ill-defined for multiple human references.

Models	Acc	Content Preservation				PPL	s_{GM}
		wrt Source			wrt Reference		
		s-BLEU	s-METEOR	s-BSC	r-BLEU		
DualRL	62.9	57.2	69.5	89.6	33.0	3443	14.6
Del&Ret	55.2	32.37	62.61	83.41	14.4	321	19.6
Ours	**79.82**	**59.54**	**78.82**	**91.0**	**35.1**	**545**	**28.9**

To measure content preservation performance, we use three metrics, BLEU, METEOR [1] and BERTSCORE, denoted as BSC. For each of them, we prepend them with r or s depending on whether the predicted text is compared to either human reference or source text respectively. We observe that across all the content preservation metrics, our models obtain a state-of-the-art result. This can

Table 4. Human evaluation on YELP (left) and GYAFC (right) datasets.

YELP				GYAFC			
Model	STA	CP	Fluency	Model	STA	CP	Fluency
GST	2.52	2.32	2.78	Del&Ret	2.22	2.22	1.71
DLSC	**2.69**	2.58	**2.84**	DualRL	1.81	2.67	2.12
Ours	2.65	**2.86**	2.65	Ours	**2.31**	**2.71**	**2.34**
kappa	0.32	0.39	0.30	kappa	0.43	0.31	0.43

be attributed to using the composition of BLEU and BERTSCORE as rewards and the additional back translation objective.

To measure the performance of our models on grammatical accuracy, we use perplexity (PPL) obtained using a pre-trained GPT2 model. Our model achieves a perplexity score of 74 and 545, which are the second best among all the baseline models.

To compare all the models (across all three dimensions), we develop two combined metrics called s_{GM} and r_{GM}. s_{GM} refers to the geometric mean of s-BLEU, Acc, s-METEOR, s-BSC, and (1/PPL) where all the above metrics except PPL are converted into 0-1 range. Whereas, r_{GM} refers to the geometric mean of different content preservation metrics evaluated on human references, following the work of Jogn et al. [8]

Human Evaluation. We conduct Human evaluation by recruiting two in-house annotators on both the YELP and GYAFC datasets by choosing 25 candidate sentences from each dataset. Each of the generated candidates is rated for fluency, style transfer accuracy (STA), and content preservation (CP) (all rated between 1 to 3 in a discrete manner). We report the average score across all dimensions in Table 4. Although we observe there is no clear winner across all the metrics, our model consistently performs better for content preservation. We also report the Cohen kappa [2] score to assess inter-rater agreement, which shows modest agreement across the model outputs.

5 Analysis

5.1 Ablation Studies

In Table 5, we show the results of ablations for our model on the YELP dataset. Full refers to the proposed model. In the second row, we show the result for the

Table 5. Ablation studies on YELP dataset

Model	Acc	s-BLEU	r-BLEU	PPL
Full	80.4	**75.1**	**30.8**	**74**
w/o BT	**94.8**	0.6	0.5	14.29
w/o Fluency	84.1	71.60	30.59	305
w/o BT & Fluency	86.7	69.81	30.79	244

model without back translation (BT) loss. Note that despite more accuracy gain and very low PPL, the proposed model without a BT component fails to adhere to source or human references leading to a very low BLEU score. Similarly in the third row, we show the results for our model without optimizing it for fluency reward, resulting in a very high PPL of 305. This results in ungrammatical sentences being generated on the output side. Also, we observe a clear trade-off between style transfer accuracy and content preservation indicators like s-BLEU and r-BLEU. Finally, in the last row, we illustrate the results for our model both without optimizing it for fluency and without BT. This leads to high PPL and also a drop in s-BLEU and r-BLEU owing to disabling back-translation training objectives.

Table 6. Case study from our model and other existing SOTA baseline models on YELP dataset

Models	Positive to Negative (P2N)	Negative to Positive (N2P)
Input	the bread is definitely home made and i could probably eat it all day	however, the manager came back & told me my order was coming up
GST	the bread is made fresh and i could n ' t even eat it all day	well, the manager came and told me my order was right up
DLSC	the bread is not home made and i could probably eat it all day	Loved it!
Ours	the bread is not home made and i could probably not eat it all day	amazing, the manager came back & told me my order coming up
Input	while the menu is simple, what it does offer is truly first - rate	when i picked up the order, i was given another totally different price
GST	while the menu is simple, what it does offer is n ' t anything special	when i picked up the order, i was totally happy price
DLSC	while the menu is simple, what it does have gotten something about	great food
Ours	while the menu is simple, what it does offer is truly worst	when i picked up the order, i was given another amazing price
Input	this adds the last little touch to what was already an amazing place	my services were very rushed and she did n't even do a good job
GST	this was the only little spot to what was already an amazing place	my services were very good and she even did a good job
DLSC	this adds the whole little way to what was already an awful place	my services were very pleasant and she did a great job and good job
Ours	this pulled the last little touch to what was already an terrible place	my services were very amazing and she did superb good job

5.2 Case Study

Table 6 shows positive source input to negative transfer (column P2N) and negative source input to positive transfer (shown in column N2P). GST and DLSC, two state-of-the-art models, are compared to ours. We can infer the following from all examples: Our methodology preserves content by converting source text into style-transferred sentences while preserving grammatical structure. This produces natural, source-compliant sentences. As illustrated in Table 2, this lowers style transfer accuracy since modest changes in the input source may not always be adequate for the style classifier to identify output sentences from our model as opposite polarity to the source.

5.3 Error Analysis

Our proposed model has several shortcomings that remain undetected via automatic evaluation metrics. In this section, we discuss the primary failure cases of our model.

1. The style classifier considers the generated sentence a style transferred version of the input sentence because the model modifies the attribute that is not an adjective due to the same reward being applied to every time step. Consider the source sentence with positive polarity: 'super dragon is my favourite Chinese restaurant.' The model output is: 'super rude is my favourite Chinese restaurant.' This generated sentence is unnatural yet it gets classified as a negative sentence due to the presence of the word 'rude'.
2. In the RL setting, our model continuously tries to trick the style classifier, which is not adversarially robust, to retain as many source words in the target as possible while also changing style attributes. Hence, generated sentences can be fluent but strange. Some of the examples contain i) Prepending opposite polarity sentiment adjective at the beginning, ii) Flipping only a single style attribute in a sentence containing multiple style attributes.
3. Without back-translation training and with using fluency reward, the model learns to generate only single or two words having very low perplexity but successful transfer. As content preservation and style transfer are given the same quantity of reward, the model generates highly fluent style-transferred sentences without bothering about content preservation. The second row of Table 5 shows how the model obtains a low BLEU score when back translation training is disabled. This can be attributed due to improper reward scaling and usage of basic RL techniques like REINFORCE.

6 Conclusion and Future Works

In this paper, we propose a technique for unsupervised style transfer that combines reward-based learning with a differentiable back translation objective. Through detailed qualitative and quantitative analysis, we demonstrate that our proposed techniques are superior. In the error analysis, we highlight the lack of adversarial robustness of the style classifier and the occurrence of odd outputs from our models as a result of inappropriate reward scaling and reward hacking. Also, the gradient-based reinforcement learning technique used is REINFORCE, where the gradient estimate is very noisy, leading to erroneous updates. We would like to address these difficulties (problems with REINFORCE and providing appropriate rewards) in the future, as they are task-independent and would be advantageous for generic tasks of natural language generation (e.g. Unsupervised NMT).

Acknowledgement. The research reported in this paper is an outcome of the project "**HELIOS-Hate, Hyperpartisan, and Hyperpluralism Elicitation and Observer System**", sponsored by Wipro.

References

1. Banerjee, S., Lavie, A.: METEOR: an automatic metric for MT evaluation with improved correlation with human judgments. In: Proceedings of the ACL Workshop on Intrinsic and Extrinsic Evaluation Measures for Machine Translation and/or Summarization. pp. 65–72. Association for Computational Linguistics, Ann Arbor, Michigan (2005). https://aclanthology.org/W05-0909
2. Cohen, J.: A coefficient of agreement for nominal scales. Educ. Psychol. Meas. **20**(1), 37–46 (1960)
3. Dai, N., Liang, J., Qiu, X., Huang, X.: style transformer: unpaired text style transfer without disentangled latent representation. In: Proceedings of the 57th Annual Meeting of the Association for Computational Linguistics, pp. 5997–6007. Association for Computational Linguistics, Florence, Italy (2019). https://doi.org/10.18653/v1/P19-1601. https://aclanthology.org/P19-1601
4. He, J., Wang, X., Neubig, G., Berg-Kirkpatrick, T.: A probabilistic formulation of unsupervised text style transfer (2020). https://arxiv.org/abs/2002.03912
5. Jang, E., Gu, S., Poole, B.: Categorical reparameterization with gumbel-softmax. In: 5th International Conference on Learning Representations, ICLR 2017, Toulon, France, 24–26 April 2017, Conference Track Proceedings. OpenReview.net (2017). https://openreview.net/forum?id=rkE3y85ee
6. Jin, D., Jin, Z., Hu, Z., Vechtomova, O., Mihalcea, R.: Deep learning for text style transfer: a survey. Comput. Linguist. **48**(1), 155–205 (2022). https://doi.org/10.1162/coli_a_00426. https://aclanthology.org/2022.cl-1.6
7. Jing, Y., Yang, Y., Feng, Z., Ye, J., Song, M.: Neural style transfer: a review. CoRR abs/1705.04058 (2017). http://arxiv.org/abs/1705.04058
8. John, V., Mou, L., Bahuleyan, H., Vechtomova, O.: Disentangled representation learning for non-parallel text style transfer. In: Proceedings of the 57th Annual Meeting of the Association for Computational Linguistics, pp. 424–434. Association for Computational Linguistics, Florence, Italy (2019). https://doi.org/10.18653/v1/P19-1041. https://aclanthology.org/P19-1041
9. Li, J., Jia, R., He, H., Liang, P.: Delete, retrieve, generate: a simple approach to sentiment and style transfer. In: Proceedings of the 2018 Conference of the North American Chapter of the Association for Computational Linguistics: Human Language Technologies, Volume 1 (Long Papers), pp. 1865–1874. Association for Computational Linguistics, New Orleans, Louisiana (2018). https://doi.org/10.18653/v1/N18-1169. https://aclanthology.org/N18-1169
10. Luo, F., et al.: A dual reinforcement learning framework for unsupervised text style transfer. CoRR abs/1905.10060 (2019). http://arxiv.org/abs/1905.10060
11. Madaan, A., et al.: Politeness transfer: a tag and generate approach. In: Proceedings of the 58th Annual Meeting of the Association for Computational Linguistics, pp. 1869–1881. Association for Computational Linguistics, Online (2020). https://doi.org/10.18653/v1/2020.acl-main.169. https://aclanthology.org/2020.acl-main.169
12. Papineni, K., Roukos, S., Ward, T., Zhu, W.J.: Bleu: a method for automatic evaluation of machine translation. In: Proceedings of the 40th Annual Meeting of the Association for Computational Linguistics, pp. 311–318. Association for Computational Linguistics, Philadelphia, Pennsylvania, USA (2002). https://doi.org/10.3115/1073083.1073135. https://aclanthology.org/P02-1040

13. Prabhumoye, S., Quirk, C., Galley, M.: Towards content transfer through grounded text generation. In: Proceedings of the 2019 Conference of the North American Chapter of the Association for Computational Linguistics: Human Language Technologies, Volume 1 (Long and Short Papers), pp. 2622–2632. Association for Computational Linguistics, Minneapolis, Minnesota (2019). https://doi.org/10.18653/v1/N19-1269. https://aclanthology.org/N19-1269
14. Radford, A., Wu, J., Child, R., Luan, D., Amodei, D., Sutskever, I.: Language models are unsupervised multitask learners. OpenAI Blog **1**(8), 9(2019)
15. Rao, S., Tetreault, J.: Dear sir or madam, may I introduce the GYAFC dataset: Corpus, benchmarks and metrics for formality style transfer. In: Proceedings of the 2018 Conference of the North American Chapter of the Association for Computational Linguistics: Human Language Technologies, Volume 1 (Long Papers), pp. 129–140. Association for Computational Linguistics, New Orleans, Louisiana (2018). https://doi.org/10.18653/v1/N18-1012. https://aclanthology.org/N18-1012
16. Nogueira dos Santos, C., Melnyk, I., Padhi, I.: Fighting offensive language on social media with unsupervised text style transfer. In: Proceedings of the 56th Annual Meeting of the Association for Computational Linguistics (Volume 2: Short Papers), pp. 189–194. Association for Computational Linguistics, Melbourne, Australia (2018). https://doi.org/10.18653/v1/P18-2031. https://aclanthology.org/P18-2031
17. Subramanian, S., Lample, G., Smith, E.M., Denoyer, L., Ranzato, M., Boureau, Y.: Multiple-attribute text style transfer. CoRR abs/1811.00552 (2018). http://arxiv.org/abs/1811.00552
18. Sudhakar, A., Upadhyay, B., Maheswaran, A.: "Transforming" delete, retrieve, generate approach for controlled text style transfer. In: Proceedings of the 2019 Conference on Empirical Methods in Natural Language Processing and the 9th International Joint Conference on Natural Language Processing (EMNLP-IJCNLP), pp. 3269–3279. Association for Computational Linguistics, Hong Kong, China (2019). https://doi.org/10.18653/v1/D19-1322. https://aclanthology.org/D19-1322
19. Vaswani, A., et al.: Attention is all you need. In: Proceedings of the 31st International Conference on Neural Information Processing Systems, NIPS 2017, Curran Associates Inc., Red Hook, pp. 6000–6010 (2017)
20. Williams, R.J.: Simple statistical gradient-following algorithms for connectionist reinforcement learning. Mach. Learn. **8**(3–4), 229–256 (1992)
21. Xu, J., et al.: Unpaired sentiment-to-sentiment translation: a cycled reinforcement learning approach. In: Proceedings of the 56th Annual Meeting of the Association for Computational Linguistics (Volume 1: Long Papers), pp. 979–988. Association for Computational Linguistics, Melbourne, Australia (2018). https://doi.org/10.18653/v1/P18-1090. https://aclanthology.org/P18-1090
22. Zhang, T., Kishore, V., Wu, F., Weinberger, K.Q., Artzi, Y.: Bertscore: Evaluating text generation with BERT. CoRR abs/1904.09675 (2019). http://arxiv.org/abs/1904.09675
23. Zhou, C., et al.: Exploring contextual word-level style relevance for unsupervised style transfer. In: Proceedings of the 58th Annual Meeting of the Association for Computational Linguistics, pp. 7135–7144. Association for Computational Linguistics, Online (2020). https://doi.org/10.18653/v1/2020.acl-main.639. https://aclanthology.org/2020.acl-main.639

Exploiting Phrase Interrelations in Span-level Neural Approaches for Aspect Sentiment Triplet Extraction

Iwo Naglik and Mateusz Lango(✉)

Institute of Computing Science, Poznan University of Technology, Poznań, Poland
inaglik.put.poznan@gmail.com, mlango@cs.put.poznan.pl

Abstract. Aspect Sentiment Triplet Extraction (ASTE) is a challenging task in modern natural language processing concerning the automatic extraction of (aspect phrase, opinion phrase, sentiment polarity) triplets from a given text. Current state-of-the-art methods achieve relatively high results by analyzing all possible spans extracted from a text. Due to a high number of analyzed spans, span-level methods usually apply some kind of pruning operators that interrupt the gradient flow. They also do not analyze interrelations between spans while constructing model output, relying on independent, sequential predictions for candidate triplets. This paper presents a new span-level approach that applies a learnable extractor of spans and a differentiable span selector that enables end2end training. The approach relies on a fully connected pairwise CRF model to capture interrelations between spans while constructing the output. Conducted experiments demonstrated that the proposed approach achieves superior results in terms of F1-score in comparison to other, state-of-the-art ASTE methods.

Keywords: aspect sentiment triplet extraction · span-level approaches · sentiment analysis · natural language processing · deep learning

1 Introduction

Sentiment analysis (SA) is an important and challenging area within modern natural language processing, with applications ranging from interactive marketing [12] to preventing the breakdown of online conversation [6]. Among many problems considered in SA, Aspect Sentiment Triplet Extraction (ASTE) [11] is arguably one of the most complex ones. The goal of ASTE is to simultaneously extract aspect and opinion phrases from a given input text, link them into corresponding pairs, and consecutively assign them appropriate sentiment polarity. Therefore, the result of ASTE for a given sentence can be expressed by a set of (aspect phrase, opinion phrase, sentiment polarity) triples. For instance, in the sentence "The price is reasonable." there is one triplet (price, reasonable, Positive), where "price" is the aspect phrase, "reasonable" is the opinion phrase and the sentiment polarity is positive. See Fig. 1 for more examples. Note that one aspect phrase (AP) can be potentially used in several triplets with different

© The Author(s), under exclusive license to Springer Nature Switzerland AG 2023
H. Kashima et al. (Eds.): PAKDD 2023, LNAI 13938, pp. 222–233, 2023.
https://doi.org/10.1007/978-3-031-33383-5_18

opinion phrases and sentiment polarities, and conversely, an opinion phrase (OP) can be a part of several triples with different aspect phrases.

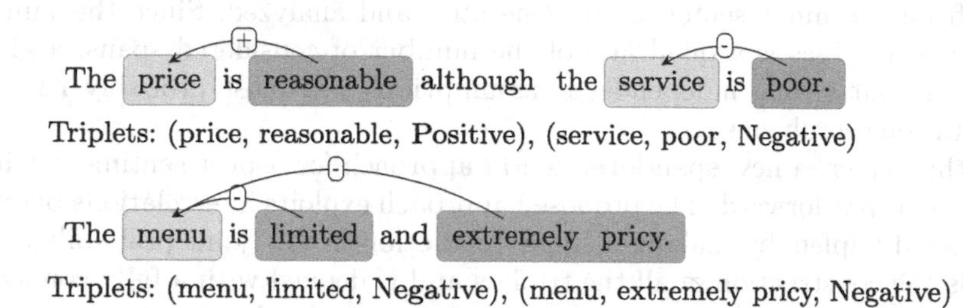
The price is reasonable although the service is poor.
Triplets: (price, reasonable, Positive), (service, poor, Negative)
The menu is limited and extremely pricy.
Triplets: (menu, limited, Negative), (menu, extremely pricy, Negative)

Fig. 1. Two examples of input sentences and ASTE triplets. The spans highlighted in blue are opinion phrases, whereas spans highlighted in orange are aspect phrases. The +/- sign denote positive/negative sentiment, respectively. (Color figure online)

In the last few years, the ASTE task received considerable research attention, which resulted in the development of various types of methods that include, among others, methods modeling word-to-word interactions [14], usage of joint tagging schemas [16] and generative models [17]. Currently, the so-called span-level approaches [2,15] obtain state-of-the-art predictive performance by generating all possible spans from a given sentence and then considering which spans should be linked together to constitute a triplet. Such methods generally construct some set of OP-AP candidate pairs and later apply a softmax classifier that decides which pairs are correct, additionally assigning a sentiment polarity to the valid pairs. Every prediction is done separately and is independent of the already constructed triplets and other triplet candidates. This prevents the models from using the information about OP-AP interrelations to improve the predictive performance.

One example of such dependencies between predictions for a given input sentence is the relation between all triplets containing the same opinion phrase. Even though the sentiment of an opinion phrase depends on the aspect being described, all triplets with the same opinion phrase usually have the same sentiment polarity within a given sentence [18]. Similarly, a phrase that carries sentiment polarity should be recognized as an opinion phrase only if in the considered sentence there is an aspect phrase to which it can be assigned. Furthermore, if the set of OP-AP candidate pairs contains elements with the overlapping aspect or opinion phrases (e.g. "extremely pricy" and "pricy"), the construction of a triplet with one of the phrases (e.g. "extremely pricy") should automatically invalidate the candidate triplets with the others (e.g. "pricy"). Also, even though the same OP/AP can be used in many triplets, one would expect that in most cases a given phrase would be assembled into one or two triplets only, with a decaying probability of multiple assignments [10].

One practical issue obstructing the usage of more global prediction models that take these dependencies into account is the high number of interrelated predictions to consider. In the current span-level ASTE approaches, all possible spans from the input sentence are generated and analyzed. Since the number of possible pairings is a quadratic of the number of considered spans, a global model without strong independence assumptions will most probably run into computational problems.

In this paper, a new span-level neural approach for aspect sentiment triplet extraction is put forward. The proposed approach exploits interrelations between constructed triplets by making a single prediction from a joint probability that controls the construction of all the triplets and is defined with a fully connected pairwise CRF model [7]. In contrast to previous span-level methods, the method splits the input sentences into non-overlapping pieces, thus not analyzing all possible spans of different lengths and enabling more effective global prediction. The paper also presents an experimental evaluation of the proposed architecture on standard ASTE datasets and compares the obtained results with the related works, demonstrating that considering interrelations between predictions leads to superior results in terms of F1-score.

2 Related Works

Aspect Sentiment Triplet Extraction (ASTE) was proposed by Peng et al. [11] as a task of aspect-level sentiment analysis that combines several other problems like opinion term extraction (OTE), aspect term extraction (ATE) and aspect-level sentiment classification (ASC). Several methods proposed to ASTE are briefly described below.

The authors of JET [16] aim to solve ASTE by converting it into a sequence labeling task. They extend the standard BIOES tagging with the information about triplet sentiment polarity, as well as with the offsets to another phrase in the triplet. The method have two variants: JET_t that labels aspect phrase with tags that include offsets to opinion phrases, and JET_o that, conversely, labels opinion phrases and aspect phrases are defined by the offsets. To label the sequence, JET uses a model based on linear-chain CRF. However, the method fails to capture all dependencies between predictions, for instance JET_t (JET_o) by design is not able to produce triplets with the same opinion (aspect) phrase.

Another approach that relies on sequence tagging is PBF [10], which consist of three sequential tagging predictors to construct triplets. First, aspect phrases are extracted by sequence tagging with BIO encoding. Then, opinion phrases for each extracted aspect are found by predicting a new sequence of BIO tags. Finally, the prediction of sentiment polarity is performed for each encountered OP-AP pair by softmax classifier.

Another approach was taken by Wu et al. [14] who proposed predicting a word-by-word matrix of tags G which encodes ASTE triplets with Grid Tagging Scheme (GTS). The diagonal of the predicted matrix contains information whether the i-th word is part of an opinion or aspect phrase ($G_{ii} = \{\mathtt{O}, \mathtt{A}\}$) or can be omitted during triplet construction ($G_{ii} = \texttt{I}$). Multiple words are merged into opinion/aspect phrases by predicting `O`/`A` tags on corresponding positions in the word-by-word matrix G, e.g. to connect i and j word into an aspect phrase G_{ij} should be equal to `A`. The whole triplets are also denoted in the same matrix G by predicting Positive/Negative/Neutral tag on the intersection of row and columns corresponding to words in aspect and opinion phrases. The method uses a softmax classifier to predict each element of G_{ij}, but also have an iterative heuristic procedure to make predictions more coherent. Another similar method is SAMBERT [1], which uses GTS for decoding the final triplets on top of the matrix representation built by extracting 2D attention maps from various layers of BERT and then processing them with a convolutional encoder-decoder to construct word-to-word matrix.

In contrast to GTS which relies on word-to-word interactions, Span-ASTE method [15] generates from the input sentence all possible text spans up to a certain length and later operates on span representations to perform predictions. First, a classifier assigns `O`/`A`/`I` tags to each span, filtering out the invalid spans and dividing the rest of spans into opinion/aspect phrases. To limit the number of considered spans, both sets of aspect/opinion phrases are punned by eliminating spans with the lowest classifier confidence. Next, all possible OP-AP pairs are analyzed by a softmax classifier that assigns sentiment to the correct pairs and eliminates incorrect ones. SBC [2] is also a span-level approach, which uses span representation constructed with a special separation loss and has bidirectional structure to generate OP-AP pairs.

FTOP [5] divides the input sentence into opinion/aspect phrases using sequence prediction on top of BERT with BIOES tagging. Next, the method considers all OP-AP pairings by concatenating their representations to the input sentence and again using BERT with restricted attention field to perform prediction. The prediction is made by a classifier applied to every possible pair.

Finally, the work [19] presents yet another approach called GAS, which uses prompting of the generative T5 language model to extract ASTE triplets.

3 Proposed Method

The proposed method is composed of four elements: 1) a pretrained masked language model which constructs the contextual embeddings of input tokens, 2) a span constructor which splits the text into spans from which triplets will later be formed, 3) a differentiable span selector which filters outs unnecessary or incorrect phrases and 4) triplet constructor which links the corresponding aspect and opinion phrases, additionally assigning them a proper sentiment. All above-mentioned elements constitute a single neural architecture which is trained by optimizing a loss function measuring the quality of constructed triplets, additionally enriched by loss functions of intermediary tasks. The proposed neural architecture is visualized on Fig. 2.

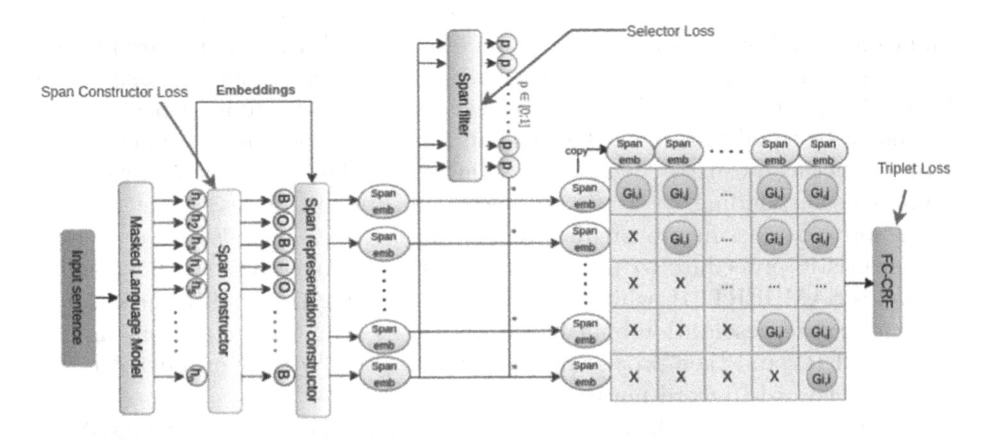

Fig. 2. The overall architecture of the proposed method.

3.1 Contextual Input Representation

In the first stage of the model's processing, the input text $\{w_1, w_2, \ldots, w_n\}$ is converted into a contextualized representation for each token $\{e_1, e_2, \ldots, e_n\}$ by a sentence encoder. To this end, we use the transformer architecture [13] initialized with pre-trained DeBERTa model [4], facilitating knowledge transfer about linguistic properties of words from the unsupervised masked language modeling task [3]. In comparison to other transformer-based language models, DeBERTa has a disentangled attention mechanism that uses two separate vectors for modelling word's content and position, which is particularly useful for further computation of text span representations.

3.2 Span Construction

The second element of our architecture is the span constructor, which divides the input text into a set of text spans which will be further considered as candidates for aspect and opinion phrases during the final prediction.

We employ BIO (beginning, inside, outside) encoding schema to tag each input token with corresponding information about its presence in an aspect/opinion phrase. Our tag set has five classes in total: `B-ASPECT`, `I-ASPECT` for denoting the first and following words in an aspect phrase; `B-OPINION`, `I-OPINION` for respective tokens in opinion phrases and `O` for all the other tokens. For example, for the input sentence $w =$ "The room was great" the corresponding tagging will be $t =$ `B-ASPECT I-ASPECT O B-OPINION`. Even though the goal of this phase is simply to extract candidate phrases and the distinction between aspect and opinion phrases is not later used by the model[1], we found it useful

[1] Our approach constructs triplets exploiting interrelations between phrases in the last layer of neural network. Splitting phrases into separate sets of aspect and opinion phrases at an earlier stage of processing would hinder the benefits of assigning phrase types while considering the entire triplet.

to include it in the tagging scheme since it provides an additional training signal and allows additional adjustment of token representations in aspect and opinion phrases respectively.

The sequence prediction model which is used to predict the sequence of BIO tags is a linear-chain conditional random field (CRF) layer [8] built on top of the contextual token representation. According to the CRF prediction, the corresponding phrases are extracted and passed to the next processing stage. The representation of each extracted span $s = [e_i; e_j; embedd(j - i + 1)]$ is computed as a concatenation of three vectors: the representation of the first word in the span e_i, the e_j vector of the span's last word and a learnable static embedding of span length $embedd(j - i + 1)$.

3.3 Span Filtering

Constructed span representations are later processed by a filtering mechanism, which goal is to detect ill formed or redundant spans and eliminate them from further consideration by zeroing their representations.

Span filtering mechanism has a binary classification head that basing on the span representation s returns the probability p of span being a correct phrase. The classification head is a fully connected neuron with sigmoid activation function $p = \sigma(w^T s + w_0)$ where w is the weight vector and w_0 is the bias term. Subsequently, an updated span representation is computed by multiplicating the original representation by the resulting probability $s' = s \cdot p$.

Note that if the classification head is certain that the span is a correct one ($p \approx 1$), the span representation remains unchanged $s' \approx s$. On the other hand, the detection of an unnecessary span will lead to zeroing its representation $s' \approx \vec{0}$. To strengthen this filtering effect, the logits in the classification head are multiplied by a constant $c > 1$, which makes the probabilities $p = \sigma(c(w^T s + w_0))$ closer to 0 or 1 during the model prediction phase. While training the model, $c = 1$ is used in order to avoid vanishing gradients.

3.4 Triplet Construction

The last stage of processing aims at constructing (aspect phrase, opinion phrase, sentiment polarity) triplets by pairing phrases selected in previous processing stages and additionally assigning them a sentiment polarity label.

Inspired by the GTS method [14], our approach considers all possible pairings of phrases by predicting a special $|S| \times |S|$ grid G, where $|S|$ is the number of considered spans. The diagonal of G introduces the distinction between spans constituting aspect and opinion phrases in a given triplet. For all spans s_i containing an aspect phrase $G_{ii} = \mathtt{A}$, whereas $G_{ii} = \mathtt{O}$ for spans with opinion phrases and $G_{ii} = \mathtt{I}$ for spans with invalid phrases. To each off-diagonal grid element, one out of four possible values: positive (`Pos`), negative (`Neg`), neutral (`Neu`), invalid (`I`) is assigned. If i-th span should constitute a triplet with the j-th span and positive sentiment polarity, the (i, j) element of the grid takes `Pos` value.

Similarly, when s_i and s_j are matching aspect and opinion phrases with negative or neutral sentiment, the value Neg or Neu is assigned to G_{ij}, respectively. In the case that spans s_i, s_j are not related, G_{ij} is set to I tag.

In summary, the method predicts a matrix of labels $G_{ij} \in \{\texttt{A}, \texttt{O}, \texttt{Pos}, \texttt{Neg}, \texttt{Neu}, \texttt{I}\}$ that encodes information about all triplets occurring in the given text. Clearly, there are dependencies between the tags in the predicted matrix. For instance, if a given i-th span is recognized as an aspect phrase ($G_{ii} = \texttt{A}$) then there should be a triplet containing it ($\exists_j G_{ij} \in \{\texttt{Pos}, \texttt{Neg}, \texttt{Neu}\}$). Therefore, instead of using independent predictions for every pair of spans, our approach uses fully connected pairwise conditional random fields (FC-CRF) [7] that models a global probability of the whole label matrix, taking into account label interrelations.

FC-CRF model is defined as follows:

$$P(G|X) = \frac{1}{Z(X)} \exp\left(-E(G|X)\right) \tag{1}$$

$$E(G|X) = \sum_{i,j} \phi_u(G_{ij}|X) + \sum_{i,j,k,l:i \neq k, j \neq l} \phi_p(G_{ij}, G_{kl}|X) \tag{2}$$

where $\phi_u()$ is the unary potential function, $\phi_p()$ is the pairwise potential function and $Z(X)$ is the partition function [8]. The unary potentials $\phi_u(G_{ij}|X)$ for every pair of text spans are computed by an MLP network containing several fully-connected layers, with the last one having the number of linear outputs equal to the number of possible tags. The input of the MLP network is the representation of a span pair constructed by the concatenation of the individual span representations, i.e. $[s_i; s_j]$. As proposed in [7], the pairwise potentials $\phi_p(G_{ij}, G_{kl}|X)$ are modelled by weighted Gaussian kernels applied on the feature vectors of span pairs, which are also computed by an MLP model. Since finding the most probable label assignment in FC-CRF model is intractable, we use the iterative approximation algorithm proposed in [20] that allows end2end training.

After the prediction of labels in grid G, the construction of triplets is quite straightforward. The decoding algorithm[2] iterates over the main diagonal of G looking for positions with A labels. For each found aspect phrase $G_{ii} = \texttt{A}$, the algorithm again iterates over the main diagonal of G looking for an element satisfying the condition $G_{jj} = \texttt{O} \wedge G_{ij} \in \{\texttt{Pos}, \texttt{Neg}, \texttt{Neu}\}$. If such element is found, the triplet with corresponding sentiment as well as aspect and opinion phrases is constructed.

Concluding the description of the proposed model, we would like to note that making the prediction using the grid G is computationally effective thanks to the span construction procedure used in our model. Most of the related span-level approaches generate all possible text spans up to a certain length and later consider them during prediction. The large number of generated spans makes considering all possible pairings computationally infeasible, therefore some form of pruning is typically used [2,15]. In contrast, our approach divides the input

[2] The pseudocode of the decoding procedure is available in the online appendix: https://www.cs.put.poznan.pl/mlango/publications/pakdd23.pdf.

text into non-overlapping pieces, processing an order of magnitude smaller set of spans. This keeps the grid size manageable and allows for faster FC-CRF calculations.

It is also worth noting, that although our method performs final prediction in a form of the matrix, there are some important differences between the Grid Tagging Scheme (GTS) [14] and our method. First, the proposed method is a span-level method which uses span representations to construct the matrix. Therefore, our method uses a smaller span-by-span matrix, whereas GTS predicts word-by-word matrices. Second, instead of independent softmax predictions with heuristic iterative procedure, the proposed method predicts the whole matrix of tags with a principled approach based on CRF model that allows the exploitation of dependencies between predictions.

3.5 Model Training

The proposed model is trained by minimizing the loss function, defined as a sum of the main loss function related to the final ASTE task and two loss functions related to intermediary tasks (constructing and selecting spans). More concretely, the loss function is defined as

$$\mathcal{L} = \mathcal{L}_{triplet} + \alpha \mathcal{L}_{SpanConstructor} + \beta \mathcal{L}_{SpanSelector} \quad (3)$$

where α, β are constant parameters that control the trade-off between optimizing particular loss elements during training. As $\mathcal{L}_{triplet}$ loss, we use standard negative log-likelihood loss defined over labels in the grid G which ultimately define the produced triplets. Similarly, $\mathcal{L}_{SpanConstructor}$ is also the negative log-likelihood but defined over the BIO tags that construct spans (see Sec. 3.2). For $\mathcal{L}_{SpanSelector}$ loss, we use binary Dice loss [9] since it is more suitable for class-imbalanced settings.

Due to the random initialization of neural network parameters, the spans generated at the beginning of the training are mostly inaccurate. Therefore, to speed up the training, we do not calculate the FC-CRF layer nor propagate gradient corresponding to $\mathcal{L}_{triplet}$ during the first few epochs of training.

4 Experimental Evaluation

4.1 Experimental Setup

We evaluated the predictive performance of the proposed approach in an experimental study performed on four ABSE datasets commonly used in the related works [2,11,15]: 14res, 14lap, 15res, 16res. Selected dataset statistics can be found in table 1. The models' performance is measured with three metrics: precision, recall and F1-score. The extraction of an aspect/opinion phrase is considered correct only when it exactly matches the gold standard. All reported metric values were computed on the corresponding test set and averaged over 5 independent training runs. The method's results were compared with the results of GTS [14], PBF [10], FTOP [5], GAS [19], JET_o [16], JET_t [16], Span-ASTE (Span) [15], SAMBERT (SAM.) [1] and SBC [2] – all were briefly described in Sec. 2.

The method was implemented in PyTorch and the code is publicly available[3]. The implementation of DeBERTa [4] from HuggingFace was used. The feature representation in Span constructor, Span filter and Triplet constructor are computed by fully connected layers with ReLU activation functions and dropout (0.1). The details of the architecture e.g. number of layers and neurons are presented in the appendix. The model was optimized by Adam algorithm for max. 120 epochs, but the training was automatically stopped when the performance measured by F1-score on the validation set didn't increase for 18 epochs. The values of α, β of the loss function (Eq. 3) were selected experimentally to $\alpha = \frac{2}{3}$, $\beta = \frac{1}{3}$.

Table 1. Selected datasets statistics

Dataset	14lap	14res	15res	16res
Number of sentences	328	492	322	326
Number of opinion phrases	470	845	456	470
Number of aspect phrases	463	848	432	452
Number of triplets	543	994	485	514
Average sentence length	15.76	16.34	15.62	14.69
Average length of opinion phrases	1.14	1.10	1.14	1.11
Average length of aspect phrases	1.40	1.27	1.29	1.27
Number of triplets with positive sentiment	364	773	317	407
Number of triplets with neutral sentiment	63	66	25	29
Number of triplets with negative sentiment	116	155	143	78
Number of one-to-many relations (opinions)	56	93	26	27
Number of one-to-many relations (aspects)	67	128	47	56
Number of single-word opinion phrases	405	759	393	418
Number of single-word aspects phrases	277	618	306	332
Number of multi-word opinion phrases	65	86	63	52
Number of multi-word aspect phrases	186	230	126	120

4.2 Results

The results of experimental evaluation are presented in Table 2. The global Friedman rank test performed on the results of F1-score indicated statistically significant differences between algorithms being compared. On tested datasets, the lowest (best) average rank equal to 1.5 was obtained by the proposed method. The second- and third-lowest rank was achieved by SBC and SAMBERT, respectively. The highest (worse) ranks were obtained by JET_o and JET_t methods.

[3] https://github.com/NaIwo/Span-ASTE.

Table 2. The experimental results of ASTE task. The best results according to F1-score are bolded, and second-best results are underlined. If the difference between the best result and other results is not statistically significant according to the T-test, these results are bolded as well.

	14lap			14res			15res			16res		
	Prec.	Rec.	F1	Prec.	Rec.	F1	Prec.	Rec.	F1	Prec.	Rec.	F1
GTS	58,54	50,65	54,30	68,71	67,67	68,17	60,69	60,54	60,61	67,39	66,73	67,06
PBF	56,60	55,10	55,80	69,30	69,00	69,20	55,80	61,50	58,50	61,20	72,70	66,50
FTOP	57,84	59,33	58,58	63,59	73,44	68,16	54,53	63,30	58,59	63,57	71,98	67,52
GAS	n/a	n/a	60,78	n/a	n/a	72,16	n/a	n/a	62,10	n/a	n/a	70,10
JET_t	53,53	43,28	47,86	63,44	54,12	58,41	68,20	42,89	52,66	65,28	51,95	57,85
JET_o	55,39	47,33	51,04	70,56	55,94	62,40	64,45	51,96	57,53	70,42	58,37	63,83
Span	63,44	55,84	59,38	72,89	70,89	71,85	62,18	64,45	63,27	69,45	71,17	70,26
SAM	62,26	59,15	60,66	70,29	74,92	72,53	65,12	63,51	**64,30**	68,01	75,44	71,53
SBC	63,64	61,80	62,71	77,09	70,99	**73,92**	63,00	64,95	63,96	75,20	71,40	**73,25**
Ours	66,98	60,55	**63,56**	75,29	72,56	**73,89**	66,44	64,74	**65,54**	71,12	72,45	71,77

On three out of four datasets, the results on F1-score obtained by our method are within the confidence interval constructed for the highest result. On the fourth dataset, our method achieved the second-highest F1-score result among ten methods being compared. Interestingly, despite considering a much smaller number of candidate phrases, the proposed method does not considerably suffer from low recall. For instance, on all datasets it achieves higher values of recall than Span-ASTE method that generates all possible spans up to a certain length (our method, on average, analyze less than 5% of spans considered by Span-ASTE).

Finally, the analysis of errors made by the proposed algorithm was performed in order to provide better insight into model operation and find problems that could be addressed in the future research. Due to the page limit, the detailed error statistics are presented in the online appendix and below we provide a summary of the most important observations.

According to the collected results, a large part of the wrong predictions was due to the incorrect assignment of sentiment to correctly generated opinion and aspect phrases (around 33 per test set). In addition, the model was much more likely to fail in correctly detecting the neutral class than in the case of stronger emotions (positive vs. negative). The reason for this behavior may be the class imbalance, i.e. the number of examples in the dataset containing neutral sentiment is much smaller than for the other sentiments (see Table 1). In general, the model assigned neutral class 42% less frequently than the true neutral class frequency. On the other hand, it is worth noting that errors resulting from incorrect pairing of correctly generated aspect/opinion phrases are rare (6 on average per dataset), which indicates that the model correctly handles the structure of the predicted matrix.

Another method's element that could cause a lower final score is the phrase construction stage - an error at this stage automatically propagate into an incorrect final prediction. An analysis of phrase construction error reveled that the model quite often failed to include additional consecutive words or constructed a too long text span. Some part of these errors were, in our opinion, insignificant. An example of an insignificant error is the inclusion or omission in the constructed phrase of punctuation marks, prepositions or parentheses - these errors do not change the meaning of the phrases and could be removed in the post-processing. However, the usage of metrics based on exact match makes a whole triplet with such phrases incorrect. This can indicate the need for less restrictive metrics in ASTE or, alternatively, the design of simple post-processing algorithms as an add-on to already existing methods. We would like to stress that despite abovementioned problems, the phrase construction stage achieves a pretty high F1 score for phrase extraction, on average equal 0.81.

5 Summary

The presented paper presents a new span-level method for Aspect Sentiment Triplet Extraction, that in contrast to earlier proposed methods does not perform independent predictions for each triplet candidate, but exploits the dependencies between predictions with a fully-connected CRF. The performed experimental evaluation proved the usefulness of the method, which achieves superior results of F1-score in comparison to other, state-of-the-art ASTE approaches.

Acknowledgement. This research has received funding from the National Center for Research and Development under the Infostrateg program (project: INFOSTRATEG-III/0003/2021-00 "Development of an IT system using AI to identify consumer opinions on product safety and quality" realized in a consortium of Poznan Institute of Technology and Poznan University of Technology).

References

1. Anonymous: sambert: improve aspect sentiment triplet extraction by segmenting the attention maps of bert, November 2021. https://openreview.net/forum?id=Z9vIuaFlIXx
2. Chen, Y., Chen, K., Sun, X., Zhang, Z.: Span-level bidirectional cross-attention framework for aspect sentiment triplet extraction. In: Proceedings of the 2022 Conference on Empirical Methods in Natural Language Processing (EMNLP). Association for Computational Linguistics (2022)
3. Devlin, J., Chang, M.W., Lee, K., Toutanova, K.: BERT: Pre-training of deep bidirectional transformers for language understanding. In: Proceedings of the 2019 Conference of the North American Chapter of the ACL: Human Language Technologies, Volume 1 (Long and Short Papers), pp. 4171–4186. Association for Computational Linguistics (2019)
4. He, P., Liu, X., Gao, J., Chen, W.: Deberta: decoding-enhanced bert with disentangled attention. In: International Conference on Learning Representations (2021)
5. Huang, L., et al.: First target and opinion then polarity: enhancing target-opinion correlation for aspect sentiment triplet extraction (2021)

6. Janiszewski, P., Lango, M., Stefanowski, J.: Time aspect in making an actionable prediction of a conversation breakdown. In: Dong, Y., Kourtellis, N., Hammer, B., Lozano, J.A. (eds.) Machine Learning and Knowledge Discovery in Databases. Applied Data Science Track, pp. 351–364 (2021)
7. Krähenbühl, P., Koltun, V.: Efficient inference in fully connected crfs with gaussian edge potentials. In: Shawe-Taylor, J., Zemel, R., Bartlett, P., Pereira, F., Weinberger, K. (eds.) Advances in Neural Information Processing Systems, vol. 24 (2011)
8. Lafferty, J.D., McCallum, A., Pereira, F.C.N.: Conditional random fields: Probabilistic models for segmenting and labeling sequence data. In: Proceedings of the Eighteenth International Conference on Machine Learning, pp. 282–289 (2001)
9. Li, X., Sun, X., Meng, Y., Liang, J., Wu, F., Li, J.: Dice loss for data-imbalanced NLP tasks. In: Proceedings of the 58th Annual Meeting of the ACL, pp. 465–476. Association for Computational Linguistics (2020)
10. Li, Y., Wang, F., Zhang, W., hua Zhong, S., Yin, C., He, Y.: A more fine-grained aspect-sentiment-opinion triplet extraction task (2021)
11. Peng, H., Xu, L., Bing, L., Huang, F., Lu, W., Si, L.: Knowing what, how and why: a near complete solution for aspect-based sentiment analysis. Proc. AAAI Conf. Artif. Intell. **34**(05), 8600–8607 (2020)
12. Rambocas, M., Pacheco, B.: Online sentiment analysis in marketing research: a review. J. Res. Interact. Mark. **12**(2), 146–163 (2018)
13. Vaswani, A., et al.: Attention is all you need. In: Guyon, I., Luxburg, U.V., Bengio, S., Wallach, H., Fergus, R., Vishwanathan, S., Garnett, R. (eds.) Advances in Neural Information Processing Systems, vol. 30 (2017)
14. Wu, Z., Ying, C., Zhao, F., Fan, Z., Dai, X., Xia, R.: Grid tagging scheme for aspect-oriented fine-grained opinion extraction. In: Findings of the Association for Computational Linguistics: EMNLP 2020, pp. 2576–2585. Association for Computational Linguistics (2020)
15. Xu, L., Chia, Y.K., Bing, L.: Learning span-level interactions for aspect sentiment triplet extraction. In: Proceedings of the 59th Annual Meeting of the ACL and the 11th IJCNLP (Volume 1: Long Papers), pp. 4755–4766. Association for Computational Linguistics (2021)
16. Xu, L., Li, H., Lu, W., Bing, L.: Position-aware tagging for aspect sentiment triplet extraction. In: Proceedings of the 2020 Conference on Empirical Methods in Natural Language Processing (EMNLP), pp. 2339–2349. Association for Computational Linguistics (2020)
17. Yan, H., Dai, J., Ji, T., Qiu, X., Zhang, Z.: A unified generative framework for aspect-based sentiment analysis. In: Proceedings of the 59th Annual Meeting of the ACL and the 11th IJCNLP (Volume 1: Long Papers), pp. 2416–2429. Association for Computational Linguistics (2021)
18. Yang, B., Cardie, C.: Extracting opinion expressions with semi-Markov conditional random fields. In: Proceedings of the 2012 Joint Conference on Empirical Methods in Natural Language Processing and Computational Natural Language Learning, pp. 1335–1345. Association for Computational Linguistics (2012)
19. Zhang, W., Li, X., Deng, Y., Bing, L., Lam, W.: Towards generative aspect-based sentiment analysis. In: Proceedings of the 59th Annual Meeting of the ACL and the 11th IJCNLP (Volume 2: Short Papers), pp. 504–510. Association for Computational Linguistics (2021)
20. Zheng, S., et al.: Conditional random fields as recurrent neural networks. In: International Conference on Computer Vision (ICCV) (2015)

What Boosts Fake News Dissemination on Social Media? A Causal Inference View

Yichuan Li[1(✉)], Kyumin Lee[1], Nima Kordzadeh[1], and Ruocheng Guo[2]

[1] Worcester Polytechnic Institute, Worcester, MA 01605, USA
{yli29,kmlee,nkordzadeh}@wpi.edu
[2] Bytedance Research, London, UK

Abstract. There has been an upward trend of fake news propagation on social media. To solve the fake news propagation problem, it is crucial to understand which media posts (e.g., tweets) cause fake news to disseminate widely, and further what lexicons inside a tweet play essential roles for the propagation. However, only modeling the correlation between social media posts and dissemination will find a spurious relationship between them, provide imprecise dissemination prediction, and incorrect important lexicons identification because it did not eliminate the effect of the confounder variable. Additionally, existing causal inference models cannot handle numerical and textual covariates simultaneously. Thus, we propose a novel causal inference model that combines the textual and numerical covariates through soft-prompt learning, and removes irrelevant information from the covariates by conditional treatment generation toward learning effective confounder representation. Then, the model identifies critical lexicons through a post-hoc explanation method. Our model achieves the best performance against baseline methods on two fake news benchmark datasets in terms of dissemination prediction and important lexicon identification related to the dissemination. The code is available at https://github.com/bigheiniu/CausalFakeNews.

Keywords: Causal inference on text · Fake news propagation

1 Introduction

People often create various news related posts on social media platforms (e.g., sports, politics and finance), and the posts are shared by their friends or influencers, and are re-shared by other users as illustrated in Fig. 1(a). Some posts start a viral "chain reaction" [11], which amplifies the influence of the news. Fake news get the same benefit, and some users intentionally or unintentionally rephrase and summarize the fake news content to encourage other users to share them via social networks.

To increase the dissemination of a news post/tweet (e.g., number of retweets), posters may create the posts more clickbait [18], and use fewer jargon words [32] in them. However, these insights are mainly based on observations or statistical correlations between social media posts and corresponding quantity of engagements, and sometimes these insights may fail [18]. For example, news topics affect the posters' writing style and tendency to share [1]. Posts created by posters, who have many followers, intrinsically

© The Author(s), under exclusive license to Springer Nature Switzerland AG 2023
H. Kashima et al. (Eds.): PAKDD 2023, LNAI 13938, pp. 234–246, 2023.
https://doi.org/10.1007/978-3-031-33383-5_19

receive more share than ones who have fewer followers [33]. Thus, merely capturing the correlation between observed properties and dissemination of a given news post is a less robust estimation, and limits finding meaningful patterns/true causes

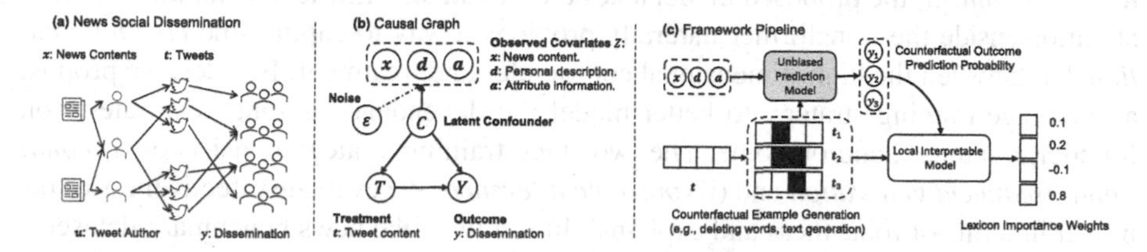

Fig. 1. (a) Overview of fake/real news dissemination on social media. A tweet author **u** imports news **x** from external websites and writes a tweet t to attract other Twitter users to disseminate **y**. (b) The causal graph contains hidden confounder C, noise ϵ, retweet status label **y**, and observed covariates Z, which consist of news content **x**, tweet author's profile information (**a**, **d**) and tweet content t. (c) The pipeline of identifying the important tweets' lexicon which causes fake/real news get disseminated.

To overcome the aforementioned limitations and understand news dissemination via social media (e.g. Twitter), we aim to answer the following research questions: **RQ1.** which news tweet[1] will receive more retweets? and **RQ2.** what textual features (lexicons) given a tweet corpus play decisive roles for the news dissemination via social networks? To answer **RQ1.**, we build a structural causal model (SCM) [19] as shown in Fig. 1(b) to model the causal relationship between news tweet and its dissemination (i.e., number of retweets). The SCM contains a hidden confounder C that influences both the probability of receiving treatment T (tweet) and outcome Y (a level of retweets). In particular, we go beyond correlation prediction $P(Y|T = \mathbf{t})$ and propose to use $P(Y|do(T = \mathbf{t}))$. By using *do*-operation, SCM can reduce the spurious correlations caused by confounder C (e.g. news topic) in news tweet dissemination prediction Y, leading to unbiased prediction. Since we do not have direct access to hidden confounder C, we adapt the idea of proximal variables [15,16]. We assume an approximate measurement can model C, given the observed covariates Z, including news content and tweet poster's personal information. To answer **RQ2.**, we follow the previous work [7] by utilizing a post-hoc explanation method to interpret our model's prediction. The whole procedure is illustrated in Fig. 1(c).

Existing works of causal inference on text still cannot completely answer **RQ1.**. They can only handle either numerical [24,25,29] or textual [7,23] covariates instead of both simultaneously. Other previous works [6,18] extracted latent properties (e.g. sentiment) from the text and treated these properties as the treatment. Usually, the latent properties are binary or continuous scalar values. This approximation would not only propagate the error from the property extraction model to the causal inference model, but also cannot answer which specific post/tweet causes the fake/real news dissemination on social media. Although existing works [7,24,25] can model high-dimensional textual treatment, they mainly relied on the multi-layer perception (MLP) for modeling the causal relationship. For both high-dimensional confounder and treatment, the MLP lacks the scalability, expressiveness and generalizability [5].

[1] A news tweet means a tweet mentions a certain news.

Thus, we propose a novel causal inference model based on the transformer model [34]. To represent *multimodal covariates*, our model adopts the soft prompt learning [14] to align the numerical and textual features. To represent *high-dimensional textual treatment*, the proposed model takes the treatment's raw text as input. The cross-attention inside the transformer naturally provides a way to capture the *complex relationship* between the high-dimensional confounder and treatment. Besides, we propose a two-stage training strategy to better model the dependency among covariates, confounder, treatment, and outcome. The two-stage training strategy consists of (i) *conditional treatment generation* and (ii) *outcome inference*. We evaluate effectiveness of our model in terms of robustness and explainability in two fake news benchmark datasets.

In short, this work has the following contributions: *i)* We propose a causal inference model that handles multimodal covariates and textual treatments, and estimates the outcome robustly; *ii)* We unbiasedly understand which lexicons inside tweets boost fake/real news dissemination on social media; *iii)* Our model achieves the best quantitative and qualitative results on the benchmark datasets. Experiments include the adjustment of different data distributions and interpreted lexicon explanation evaluation.

2 Problem Definition

Notation. Let boldface lowercase letter denote a vector or a sequence of words (e.g., $\mathbf{x}$), and boldface uppercase letter represent the matrix (e.g., $\mathbf{X}$), italic uppercase letter denote a causal inference variable (e.g., T), italic lowercase letter denote a word (e.g., w), and a calligraphic font represent a vocabulary set (e.g., $\mathcal{V}$). Let $\mathbf{X} = \{\mathbf{x}_i\}_{i=1}^{|\mathbf{X}|}$ denote a corpus of news contents. For each news content $\mathbf{x}_i$, there are $\mathbf{T}_i = \{\mathbf{t}_j^i\}_{j=1}^{|\mathbf{T}_i|}$ tweets that mention the news $\mathbf{x}_i$. An author of the tweet $\mathbf{t}_j^i$ is $\mathbf{u}_j^i$, and her profile consists of numerical attributes $\mathbf{a}_j^i$ and personal textual description $\mathbf{d}_j^i$. It should be noticed that each of $\mathbf{x}_i$, $\mathbf{t}_j^i$ and $\mathbf{d}_j^i$ is a sequence of words $\{w_k\}_{k=1}^{||\cdot||}$. Each tweet $\mathbf{t}_j^i$ is associated with a discrete y_j^i retweet status (i.e., assigning each tweet to a class/bin based on the number of retweets). To answer the aforementioned two research questions in Sect. 1, we conduct the following studies:

- **Predict tweets causing fake/real news dissemination**. We aim to learn the causal relation $P(Y|do(T = \mathbf{t}))$ as shown in Fig. 1(b), where Y is the outcome, T is treatment. For more information about *do*-operation, please check Sec. 3.1.
- **Understand syntax playing decisive roles.** Then, we identify K important words $\{w_k\}_{k=1}^{K} \subseteq \mathcal{V}^{\mathbf{t}}$ that most significantly influence news tweets $\mathbf{T}$ dissemination. Here, $\mathcal{V}^{\mathbf{t}}$ is a vocabulary set of tweets.

3 Our Framework

Our proposed framework is designed based on a well-known sequence-to-sequence (seq2seq) model, `BART` [13], as shown in Fig. 2. In the following subsections, we will firstly introduce the preliminary knowledge. We then propose a two-stage training strategy to better capture the causal dependency showed in Fig. 1(b). Lastly, we will discuss our approach for learning multimodal covariates embedding.

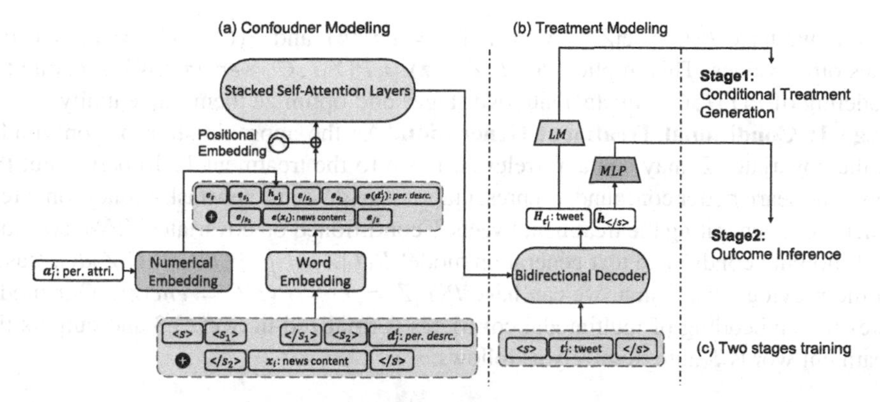

Fig. 2. The illustration of our model. The covariates (news content and the tweet author's profile) are input into the encoder and the treatment (tweet) is input into the decoder. The whole training process contains two stages. The first stage is to learn the hidden confounder's representation through conditional text generation. The second stage is to classify a level of news tweet dissemination by text classification.

3.1 Preliminary

***Do*-Notation** ($P(Y|do(T = \mathbf{t}))$) [19] is different from conditional correlation $P(Y|T = \mathbf{t})$, which is based on the sub-population of the dataset. The *Do*-Notation will change the data distribution by making intervention [9], even if the sub-population is unseen in the collected data. For example, in the collected data, if users whose followers are greater than 10 never posting $\mathbf{t}$, the $P(Y|T = \mathbf{t})$ cannot answer *what-if* this group of users have tweeted t, how many retweets they will receive. But $P(Y|do(T = \mathbf{t}))$ will model all groups of users including follower greater than 10 have posted $\mathbf{t}$. The observation/collected data often lacks this intervention. To estimate $P(Y|do(T = \mathbf{t}))$, it often requires the model to block all the incoming paths to the treatment ($C \rightarrow T$ as shown in Fig. 1(b)). This incoming path blocking can be modeled as conditioning on the confounder C: $P(Y|do(T = \mathbf{t})) = \int_C P(Y|T = \mathbf{t}, C = \mathbf{c})P(C = \mathbf{c})d\mathbf{c}$.
BART [13] is a denoising transformer-based language model. Both encoder and decoder of BART are stacked with the Transformer blocks. Each block contains a self-attention layer to interact with and aggregate the information from either the encoder or decoder: $Self-Attention(\mathbf{Q}, \mathbf{K}, \mathbf{V}) = softmax(\frac{\mathbf{QK}^{\top}}{\sqrt{d_k}}\mathbf{V})$. But the decoder has an additional cross attention layer, where $\mathbf{Q} = \mathbf{Q}_{dec}$, $\mathbf{K} = \mathbf{K}_{enc}$, $\mathbf{V} = \mathbf{V}_{enc}$. This difference allows us to use different inputs (see Sect. 3.3) for the encoder (multimodal covariates as input) and decoder (treatment as input) for better feature interaction modeling.

3.2 Causal Feature Representation Learning

We aim to learn the causal relation $P(Y|do(T = \mathbf{t}))$ to predict how dissemination (Y) would be changed if a tweet (T) had been written differently about the same news by the same author (Z). Based on the backdoor criterion [19] and the causal graph Fig. 1(b), we can predict each tweet's dissemination status by $P(Y|T = \mathbf{t}, Z) = \int_C P(Y|T, C)P(C|Z)dC$. As the encoder is a deterministic function, given covariates

$Z = \mathbf{z}$, we have $P(C = \mathbf{c}|Z = \mathbf{z}) = 1$ if $\mathbf{c} = enc(\mathbf{z})$, and $P(C = \mathbf{c}|Z = \mathbf{z}) = 0$ if $\mathbf{c}$ takes other values. This implies $P(Y|T, Z = \mathbf{z}) = P(Y|T, C = enc(\mathbf{z}))$. We separate the modeling of $P(Y|T, C = enc(\mathbf{z}))$ into two stages and optimize them sequentially.
Stage 1: Conditional Treatment Generation. As the approximation of confounder C, the covariates Z may contain irrelevant noise to the treatment T. To partial out the noise and learn better confounder representation, we design a new task - conditional text generation, generating the treatment tweets T conditioned on covariates Z. We consider the following conditional text generation model $P(T|Z) = \int_C P(T|C)P(C|Z)dC$. Based on the previous discussion, we can have $P(T|Z = \mathbf{z}) = P(T|C = enc(\mathbf{z}))$. Our model takes the embedding of multimodal covariates formalized in Sect. 3.3 and outputs the treatment words T autogressively as follows:

$$P(w_1, \cdots, w_{|\mathbf{t}_j^i|} \mid z = (\mathbf{x}, \mathbf{d}, \mathbf{a})) = \prod_{k=1}^{|\mathbf{t}_j^i|} P(w_k|C = enc(z), \mathbf{w}_{<k}) \tag{1}$$

where $w_{<k}$ is a sequence of previously generated $k - 1$ tokens.
Stage 2: Outcome Inference. Inspired by previous works [7,23,25,29], to avoid the underrepresentation of treatment, we separate the location of input of covariates and treatment. The previous works do the outcome inference by assigning confounder's hidden representation into different MLPs [23,29] or boosting the outcome prediction from confounder to treatment [7,25]. Instead, in this work, the encoder takes the embedding of covariates as input, while the decoder takes treatment/tweet embedding as input. These isolated inputs for the encoder and decoder bring well-discriminated representation learning for confounder and treatment [36].

After that, we add a MLP f to the seq2seq model to predict outcome based on the interacted representation of treatment and covariates. Specifically, we take the hidden representation of $\langle/s\rangle$ token, $\mathbf{h}_{\langle/s\rangle}$, from treatment to predict outcome (i.e., a level of tweet dissemination). Usually, the $\langle/s\rangle$ is appended to the end of treatment. Overall, the objective function for outcome inference is:

$$L_{outcome} = \mathbb{E}\left[CrossEntropy\left(y_j^i, f(\mathbf{h}_{\langle/s\rangle})\right)\right] \tag{2}$$

3.3 Multimodal Covariates Embedding

To better learn representation of the hidden confounder C, we consider as many observed covariates as possible. In particular, besides the news content, we also consider a tweet author's personal information. Previous researches [21,33] found that the number of followers and friends, status, lists, and user's verification status influenced the retweetability. Thus, our covariates consist of textual news content $\mathbf{x}$, personal description $\mathbf{d}$, and numerical attribute information $\mathbf{a}$ (i.e., *verified status*, and *number of tweets, followers count, friends count* and *number of lists*).

However, BART [13] is pre-trained on text only but not the tweet author's numerical data. To fill the gap between the pre-training and our downstream fine-tuning, in our framework, we utilize the soft-prompt [14] to align the modality. It wraps the input with unique tokens. These unique tokens do not need to be included in the vocabulary, and their embeddings are learned from scratch. During the training process, they will gradually capture the modality information. This setting will provide more flexibility and better coverage than actual tokens [12,26]. Specifically, the input of our encoder is:

$$\left[\mathbf{e}_{</s>}, \mathbf{e}_{</s_1>}, \mathbf{h}_{a^i_j}, \mathbf{e}_{</s_1>}, \mathbf{e}_{<s_2>}, \mathbf{e}(\mathbf{d}^i_j), \mathbf{e}_{</s_2>}, \mathbf{e}(\mathbf{x}_i), \mathbf{e}_{</s>}\right] \quad (3)$$

where $\{\mathbf{e}_.\}$ are the soft-prompts to wrap different inputs; $\mathbf{h}_{a^i_j}$ is the hidden vector of numerical features. $\mathbf{e}(\mathbf{d}^i_j)$ and $\mathbf{e}(\mathbf{x}_i)$ are the word embeddings for the tweet author's personal description, and news content, respectively.

4 Experiment

In this section, we answer the following research questions. **RQ 1.1.** How accurately can our model estimate a level of news tweet dissemination (i.e., predict the class/bin of the number of retweet given the news tweet)? **RQ 1.2.** What contribution does each observed covariate make for learning representation of the hidden confounder? **RQ 2.** What is the effectiveness of the identified words $\{w_k\}_{k=1}^{K}$ from the tweets $\mathbf{T}$ in terms of boosting the news dissemination?

4.1 Evaluation Datasets

We use two fake news benchmark datasets from FakeNewsNet [30]: (1) PolitiFact and (2) GossipCop. They contain news content, veracity labels (fake vs. real), and social context tweets (tweets mention the news). The statistical information of the tweets is listed in Table 1.

Since most of the tweets received zero or one retweet, we group the tweets based on their corresponding $\# \, of \, retweets$ into binary ($|reweets| = 0$ and $|reweets| > 0$), ternary ($|reweets| = 0$, $|reweets| = 1$ and $|reweets| > 1$), and quaternary ($|reweets| = 0$, $|reweets| = 1$, $1 < |reweets| <= 10$ and $|reweets| > 10$) classes as the ground truth. For example, in the ternary setting, class 0 is a tweet receiving 0 retweet, class 1 is a tweet receiving one retweet, and class 2, the most viral class, is a tweet receiving more than one retweet.

Table 1. Basic statistics of FakeNewsNet [30].

Dataset	Veracity	News	Tweets	Retweets
PolitiFact	Real	335	16,376	35,586
	Fake	247	11,975	27,627
GossipCop	Real	13,601	563,056	218,760
	Fake	4,111	101,910	225,447

These three different settings can help evaluate the robustness of the proposed model. Because fake news inherently received more retweets than real news [35], we train and evaluate two different models – one for fake news and the other one for real news. We split each dataset into training, validation, and test sets by a ratio of 70%:10%:20% based on the number of tweets. To evaluate the model's robustness, we follow the previous work [4] by creating biased data distributions for the training and validation sets, and unbiased distribution for test set. Because of transportability and omitted spurious dependency from the confounder [4,20] a causal model is expected to be more robust to distribution shift than non-causal model. We repeat the experiment for five times and report the averaged results.

4.2 Experiment Setting

Baseline Methods. Since there is no prior work that involves learning the hidden confounder from textual and numerical data and taking the textual data as the treatment, we mainly focus on four variations of our model and one causal inference work as baseline methods. To prove the effectiveness of the SCM mentioned in Fig 1(b), we discard covariates (`w/o` Z) and treatment (`w/o` T) of our model. Secondly, to understand the effectiveness of the treatment generation in the hidden confounder representation learning, we consider the following two variations without the first stage. The *Con. w/o seq2seq* concatenates the observed covariates and treatment as input for both encoder and decoder while *w/o seq2seq* isolates the inputs for encoder and decoder like our model. All the variation-based baseline methods will only minimize the objective function of the outcome inference which is the main task. Besides, we customize an existing *causal inference model*, `DeepResidual` (`DR`) [7,22], to handle the textual and numerical data. Different from our model, `DR` models the causal dependency through boosting. It firstly estimates the outcome relying on only the confounder and then takes the concatenation of the predicted probability and treatment's representation to conduct the outcome prediction. In Tables 2 and 3, each cell contains two performance numbers, for fake and real news data respectively. For example, 75.02/74.78 of our model under the binary setting in Table 2 mean the accuracy (ACC) for fake news and real news test sets on PolitiFact, respectively. The best performance is **bold** and the second-best is underlined.

Implementation Details. We utilize the `BART-base` [13] as our main module for feature representation learning and outcome inference. For a fair comparison, the DR takes the confounder as input for the encoder and treatment for the decoder like our model and `w/o seq2seq`. The learning rate for conditional text generation is set to $1e-5$, the learning rate for outcome prediction's seq2seq module is selected from $[1e-5, 1e-6, 0]$, and the learning rate for the prediction head is selected from $[1e-4, 1e-5, 1e-3]$ by the grid search. Because the size of the GossipCop dataset is larger than the PolitiFact, the number of training epochs for the first stage is set to 3 and 5 in GossipCop and PolitiFact, respectively. We report the test results based on the best validation results.

4.3 Main Results

RQ1.1 : Unbiased Dissemination Estimation. As the results showed on Table 2, we firstly observe that incomplete causal dependency-based baselines (i.e., `w/o` T and `w/o` Z) achieve worse performance than the complete causal dependency-based baselines (i.e., `Con. w/o seq2seq`, `w/o seq2seq` and `DR`). Secondly, our model after the conditional treatment generation shows better performance than `w/o seq2seq` and `Con. w/o seq2seq`. This indicates the importance of learning hidden confounder by capturing the relationship between the covariates and treatment. Thirdly, isolating the input for encoder and decoder only contributes when combined with the Treatment Generation. This is observed because `w/o seq2seq` shows competitive average rank compared with `Con w/o seq2seq`. Fourthly, overall our model shows the best performance compared with all the baselines across three experiment settings. This indicates our model can learn better confounder feature representation and provide robust outcome prediction by overcoming the challenge of data distribution shifting between the training and test sets.

Table 2. News tweet dissemination prediction on PolitiFact and GossipCop datasets with handling the distribution difference between the training and test sets.

Models	Binary		Ternary		Quaternary	
	ACC(%)	micro-AUC(%)	ACC(%)	micro-AUC(%)	ACC(%)	micro-AUC(%)
PolitiFact						
DR	70.59/71.15	82.99/80.78	57.62/65.71	73.69/**83.18**	60.98/60.99	**85.59**/84.96
w/o *T*	70.97/73.61	79.79/81.58	66.62/63.92	83.12/77.42	60.93/60.09	85.46/84.82
w/o *Z*	70.55/72.12	80.06/81.58	68.72/65.08	83.33/79.59	60.25/59.40	84.77/84.37
Con. w/o seq2seq	71.30/73.82	79.87/82.59	69.42/67.16	83.22/80.88	60.19/60.79	84.91/85.01
w/o seq2seq	74.46/70.38	82.37/80.29	69.76/63.59	84.13/79.37	60.59/60.35	84.86/84.91
Our model	**75.02/74.78**	**83.17/82.61**	**72.08/67.49**	**85.24**/82.18	**63.35/62.69**	85.52/**86.02**
GossipCop						
DR	75.89/89.28	74.88/94.10	**74.66**/86.12	85.89/**97.25**	74.60/84.38	91.91/96.41
w/o T	74.49/85.98	74.76/94.08	72.10/84.10	82.64/94.04	67.07/84.05	88.97/94.48
w/o C	73.80/84.83	76.88/92.76	71.70/84.20	84.32/96.59	71.03/84.59	90.35/96.07
Con. w/o seq2seq	74.65/87.56	73.95/93.23	73.10/83.99	83.82/94.57	72.14/85.34	90.86/95.64
w/o seq2seq	74.11/88.02	74.55/93.26	73.17/84.22	86.27/88.23	72.77/85.60	91.43/96.29
Our model	**76.01/92.44**	**79.22/94.83**	74.17/**87.73**	**86.30**/96.83	**74.83/86.46**	**92.06/96.94**

RQ1.2: Contribution of Different Covariates. This ablation study ablates several components of the tweet author's profile, such as personal text description (named as `w/o pers. descr.`), personal numerical attributes (`w/o pers. ATT`), and all the profile information (`w/o pers.`). As shown in Fig. 3, our model achieves the best performance in the average rank of accuracy and micro-AUC. This indicates that learning the hidden confounder's representation from the tweet author's profile has positive contributions. Besides, we can observe that the user's both personal textual description and numerical attributes have positive contributions to the social engagements prediction (Avg. Rank `w/o pers. descr.` and `w/o pers. descr.` < `w/o pers.`).

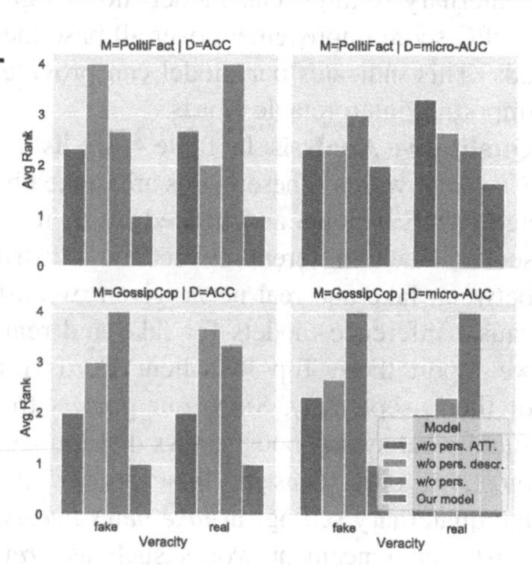

Fig. 3. Ablation study of covariates for fake and real news on PolitiFact and GossipCop datasets. A lower value ("rank") is better.

So far, we have discussed three different label settings. From now on, we will focus on the binary and quaternary settings due to limited space.

4.4 Lexicons Boosting Dissemination

To understand the key syntax of tweets causing news dissemination (**RQ2.**), we need to interpret the causal inference model. Since our model itself is not interpretable[2], we

[2] We leave the syntax self-interpreted causal inference model as our future work.

utilize the model-agnostic explanation method - LIME [27] to do the model interpretation. LIME will create many counterfactual tweets and fit a locally-linear model based on the pairs of counterfactual tweets and their unbiased retweet status prediction. It should be noticed that the unbiased prediction is from our model's output. In this experiment, we only generate counterfactual examples on the tweet content (treatment) and keep the news content and tweet author's profile (covariates) static. Since the baseline method w/o T did not utilize the treatment, it is impossible to make an intervention on the treatment tweet. Therefore, we discard w/o T in the following evaluations. To measure the effectiveness of our model with LIME in identifying words for the news dissemination, we provide quantitative and qualitative analysis. Due to the high computation costs for LIME, we follow the previous work [3], randomly sampling 500 instances from test sets.

Quantitative Evaluation. A well-trained model can capture the important information, and thus provide a more meaningful explanation compared with a poorly trained model under the same explanation method [28]. We use area over the perturbation (AOPC) score [17,28] as an evaluation metric. It will calculate the average prediction probability change when deleting the top-L words from LIME [3]. A higher AOPC score is preferred.

Table 3 shows the results under binary and quaternary settings. Our model shows significant AOPC score improvement over all baseline methods. This indicates our model can provide more important/interpretable words.

Table 3. AOPC [28] score(%■) for LIME in interpreting counterfactual outcome inference.

PolitiFact	Binary	Quaternary
DR	0.67/-0.60	0.94/1.21
w/o Z	0.26/0.10	2.11/1.48
Con. w/o seq2seq	0.13/0.99	3.19/2.28
w/o seq2seq	0.32/-0.60	6.57/6.11
Our model	**5.10/3.53**	**9.88/9.97**
GossipCop	Binary	Quaternary
DR	0.35/-0.60	0.98/0.50
Con. w/o seq2seq	-1.25/1.04	0.66/2.92
w/o seq2seq	0.37/1.84	0.43/0.01
w/o seq2seq	0.58/2.03	0.98/4.12
Our model	**1.04/2.03**	**1.45/8.47**

Qualitative Analysis. In Table 4, we list the Top-30 salient words. These words are ranked by average LIME weights and filtered by their TF-IDF scores. Due to different retweet count distribution between fake and real news, given two different causal inference models for fake and real news, we report these Top-30 salient words for both of them, separately. Since our goal is to understand what syntax boosts news dissemination the most, we only focus the $|reweets| > 10$ under the quaternary setting. In *fake news tweets*, there exists announcement words such as "*revealed*" and "*claim*". PolitiFact contains the sequential connectives like "*until*" and "*then*". It could indicate that the fake news tweets have similar writing style with real news to cause more dissemination [31]. Besides, we observe the emotional words like "*hating*", "*bullshit*" in PolitiFact and "*trouble*", "*death*", "*jealous*", "*happy*" in GossipCop. These words drive audiences' emotions and encourage them to engage [10]. In *real news tweets*, we observe referenced words (e.g., "*call*", "*posted*", "*approve*", "*info*") play important roles in both datasets. These words will evoke users' curiosity to share [10]. Besides, these real news tweets contain many time-relevant words such as "*year*" and "*week*". This finding indicates the use of more precise instruction strategy in real news tweets. Lastly, in GossipCop, there are many degree adverbs (e.g., words like "*very*", "*really*" and "*super*"). These degree words make the speakers' utterance expressive [8], which may receive more retweets.

Table 4. Top-30 salient words for the tweets' dissemination prediction.

	PolitiFact	GossipCop
Fake	revealed, emoji, hashtag, engages, doing, center, stands, goals, joining, flag, claims, until, loyalty, claim, really, then, crap, that, video, aside, betrayal, pleas, hating, normal, bullshit, protesters, honor, private, has, right	these, thanks, children, director, very, hired, paid, separate, year, contributed, scale, fifth, with, said, happy, confusing, between, picture, hashtag, announced, ice, jealous, until, detective, mystery, situation, trouble,complicated, death, bus
Real	emoji, station, attack, info, hashtag, programme, approve, lapse, posted, audio, space, advisor, climate, globe, hideaway, cancel, nite, cannabis, week, quot, dedicada, uncooperative, blaring, call, via, morphing, easily, here, soon, year	peaceful, design, why, reports, emoji, toughness, pic, already, hashtag, tribute, premiere, disagree, fight, knows, reluctantly, pieces, had, kill, myself, confirms, moment, are, been, star, victory, match, super, really, that, his

We also provide a fake news tweet[3] (shown in Table 5) to understand the context information of the salient word. The less retweeted tweet were very similar to the most retweeted tweet except the missing of identified salient word. This confirms that the importance of these salient words to cause fake/real news tweets get disseminated.

Table 5. Examples of the most and least retweeted tweets of fake news. The most retweeted tweets contain salient words identified via LIME from our model.

Retweet	Fake-PolitiFact, Keyword: "*claim*"	Fake-GossipCop, Keyword: "*reveal*"
Most	Dying 78 year old cia agent admits to killing marilyn monroe *claim*s he carried out 37	AQUAMAN Movie *reveal* First Look at Nicole Kidman as Atlanna
Less	Dying 78 year old cia agent admits to killing marilyn monroe	Nicole Kidman was pelted with rocks while filming new flick Aquaman

5 Related Work

User Engagements on News. Social media platforms provide a new way for news organizations to distribute content and receive feedback from the users through user engagement [1]. Many works try to predict the user engagements based on the news content and users' reactions [18]. Although they have achieved excellent performance in the prediction, they did not tackle the causal relationships between the social media posts and user engagements. The most similar work is [18], which estimated the causal effects of tweets' editing styles on boosting users' engagements. However, identifying the treatment (editing styles) is dependent on an out-of-domain style classifier, which will inevitably bring the measurement error for the downstream causal effect estimation. [35] revealed that fake news spreads faster, deeper, and more broadly than the truth. The authors found that fake news was more novel than the truth and inspired the replies'

[3] Due to space limit, we only report the fake news' tweets.

fear, disgust and surprise. [4] identified several tweet authors' attributes in causing the fake news spread on social media. Different from these works, we revealed which tweet caused news disseminated and what types of lexicons in these tweets played the decisive roles.

Causal Inference on Text. Text data provides a new perspective for researchers to understand the causal effects of the treatment's intervention. In this paper, the textual data appeared in both the proxy of confounder and treatment. To learn the hidden confounder from observed covariates, existing works map the covariates into a low-dimensional vector through methods like Latent Dirichlet Allocation (LDA) [2] and auto-encoder [15]. As for textual treatment, most works extract text properties from the text. They utilize a classifier to predict the text properties such as sentiment [23] and clickbait [18], which will inevitably generate measurement errors and require additional efforts to label the dataset for classifier training [6]. Others map the treatment text to latent vectors [7,24]. However, conventional methods cannot handle the multimodal covariates and cannot exploit the pre-trained language model in outcome inference. In addition, these methods ignore the dependency between the textual treatment and multimodal covariates. Thus, they cannot provide the correct outcome estimation.

6 Conclusion

In this paper, we propose a causal inference model to unbiasedly know which tweets cause fake/real news disseminated on social media and what lexicons play a critical role inside the tweet. our model successfully represents the multimodal covariates (news content and user personal attributes) and textual treatments (tweet). The comprehensive experiment results indicate the robustness and effectiveness of our model in resolving the confounding bias. In our qualitative analysis, we identify salient words from two fake news benchmark datasets. These salient words can not only be used as the fake news detection features, but also help further research in better understanding of behind mechanism of fake news dissemination on social media.

Acknowledgement. This work was supported in part by NSF grant CNS-1755536 and WPI TRIAD. Any opinions, findings and conclusions or recommendations expressed in this material are the author(s) and do not necessarily reflect those of the sponsors.

References

1. Aldous, K.K., An, J., Jansen, B.J.: View, like, comment, post: analyzing user engagement by topic at 4 levels across 5 social media platforms for 53 news organizations. In: ICWSM (2019)
2. Blei, D.M., Ng, A.Y., Jordan, M.I.: Latent dirichlet allocation. J. Mach. Learn. Res. **3**, 993–1022 (2003)
3. Chen, H., Ji, Y.: Learning variational word masks to improve the interpretability of neural text classifiers. In: EMNLP (2020)
4. Cheng, L., Guo, R., Shu, K., Liu, H.: Causal understanding of fake news dissemination on social media. In: KDD (2021)

5. Deng, Z., Zheng, X., Tian, H., Zeng, D.D.: Deep causal learning: representation, discovery and inference. arXiv preprint arXiv:2211.03374 (2022)
6. Egami, N., Fong, C.J., Grimmer, J., Roberts, M.E., Stewart, B.M.: How to make causal inferences using texts. CoRR abs/1802.02163 (2018)
7. Fytas, P., Rizos, G., Specia, L.: What makes a scientific paper be accepted for publication? (2021)
8. Indhiarti, T.R., Chaerunnisa, E.R.: A corpus-driven collocation analysis of degree adverb very, really, quite, and pretty (2020)
9. Keith, K.A., Jensen, D., O'Connor, B.: Text and causal inference: a review of using text to remove confounding from causal estimates. In: ACL (2020)
10. Kilgo, D.K., Sinta, V.: Six things you didn't know about headline writing: sensationalistic form in viral news content from traditional and digitally native news organizations. In: ISOJ, vol. 6, pp. 111–130 (2016)
11. Kushin, M.J., Yamamoto, M.: Did social media really matter? college students' use of online media and political decision making in the 2008 election. Mass Commun. Soc. **13**(5), 608–630 (2010)
12. Lester, B., Al-Rfou, R., Constant, N.: The power of scale for parameter-efficient prompt tuning. In: EMNLP (2021)
13. Lewis, M., et al.: BART: denoising sequence-to-sequence pre-training for natural language generation, translation, and comprehension. In: ACL (2020)
14. Liu, X., Zheng, Y., Du, Z., Ding, M., Qian, Y., Yang, Z., Tang, J.: Gpt understands, too (2021)
15. Louizos, C., Shalit, U., Mooij, J.M., Sontag, D., Zemel, R., Welling, M.: Causal effect inference with deep latent-variable models. In: NeurIPS (2017)
16. Miao, W., Geng, Z., Tchetgen Tchetgen, E.J.: Identifying causal effects with proxy variables of an unmeasured confounder. Biometrika **105**(4), 987–993 (2018)
17. Nguyen, D.: Comparing automatic and human evaluation of local explanations for text classification. In: NAACL HLT (2018)
18. Park, K., Kwak, H., An, J., Chawla, S.: How-to present news on social media: a causal analysis of editing news headlines for boosting user engagement. In: ICWSM (2021)
19. Pearl, J.: Causality. Cambridge University Press, Cambridge (2009)
20. Pearl, J., Bareinboim, E.: Transportability of causal and statistical relations: a formal approach. In: AAAI (2011)
21. Petrovic, S., Osborne, M., Lavrenko, V.: RT to win! predicting message propagation in twitter. In: ICWSM, The AAAI Press (2011)
22. Pryzant, R., Basu, S., Sone, K.: Interpretable neural architectures for attributing an ad's performance to its writing style. In: EMNLP Workshop BlackboxNLP (2018)
23. Pryzant, R., Card, D., Jurafsky, D., Veitch, V., Sridhar, D.: Causal effects of linguistic properties. In: NAACL HLT (2021)
24. Pryzant, R., joo Chung, Y., Jurafsky, D.: Predicting sales from the language of product descriptions. In: eCOM@SIGIR (2017)
25. Pryzant, R., Shen, K., Jurafsky, D., Wagner, S.: Deconfounded lexicon induction for interpretable social science (2018)
26. Qin, G., Eisner, J.: Learning how to ask: Querying LMS with mixtures of soft prompts. In: NAACL-HLT (2021)
27. Ribeiro, M.T., Singh, S., Guestrin, C.: "why should I trust you?": Explaining the predictions of any classifier. In: KDD (2016)
28. Samek, W., Binder, A., Montavon, G., Lapuschkin, S., Müller, K.: Evaluating the visualization of what a deep neural network has learned. IEEE Trans. Neural Netw. Learn. Syst **28**(11), 2660–2673 (2017)
29. Shi, C., Blei, D.M., Veitch, V.: Adapting neural networks for the estimation of treatment effects. In: NeurIPS (2019)

30. Shu, K., Mahudeswaran, D., Wang, S., Lee, D., Liu, H.: Fakenewsnet: a data repository with news content, social context and dynamic information for studying fake news on social media. arXiv preprint arXiv:1809.01286 (2018)
31. Shu, K., Sliva, A., Wang, S., Tang, J., Liu, H.: Fake news detection on social media: a data mining perspective. ACM SIGKDD Explor. Newsl. **19**(1), 22–36 (2017)
32. Strekalova, Y.A., Krieger, J.L.: Beyond words: amplification of cancer risk communication on social media. J. Health Commun. **22**(10), 849–857 (2017)
33. Suh, B., Hong, L., Pirolli, P., Chi, E.H.: Want to be retweeted? large scale analytics on factors impacting retweet in twitter network. In: SocialCom (2010)
34. Vaswani, A., et al.: Attention is all you need. In: NeurIPS (2017)
35. Vosoughi, S., Roy, D., Aral, S.: The spread of true and false news online. Science **359**(6380), 1146–1151 (2018)
36. Zhang, Y.F., Zhang, H., Lipton, Z.C., Li, L.E., Xing, E.P.: Can transformers be strong treatment effect estimators? arXiv preprint arXiv:2202.01336 (2022)

Topic-Selective Graph Network for Topic-Focused Summarization

Zesheng Shi[1] and Yucheng Zhou[2(✉)]

[1] Nankai University, Tianjin, China
2120210083@mail.nankai.edu.cn

[2] University of Technology Sydney, Ultimo, Australia
yucheng.zhou-1@student.uts.edu.au

Abstract. Due to the success of the pre-trained language model (PLM), existing PLM-based summarization models show their powerful generative capability. However, these models are trained on general-purpose summarization datasets, leading to generated summaries failing to satisfy the needs of different readers. To generate summaries with topics, many efforts have been made on topic-focused summarization. However, these works generate a summary only guided by a prompt comprising topic words. Despite their success, these methods still ignore the disturbance of sentences with non-relevant topics and only conduct cross-interaction between tokens by attention module. To address this issue, we propose a topic-arc recognition objective and topic-selective graph network. First, the topic-arc recognition objective is used to model training, which endows the capability to discriminate topics for the model. Moreover, the topic-selective graph network can conduct topic-guided cross-interaction on sentences based on the results of topic-arc recognition. In the experiments, we conduct extensive evaluations on NEWTS and COVIDET datasets. Results show that our methods achieve state-of-the-art performance.

Keywords: Text Summarization · Topic Model · Graph Neural Network

1 Introduction

Text summarization aims to compress a long article into a short and clear summary, which is a fundamental task in many NLP applications. With the success of sequence-to-sequence (seq2seq) language models, it is widely integrated into many real-world applications, e.g., document snippets generation in search engines [25], automatic news summaries [1] and legal document summarization [14]. In recent years, text summarization has been an essential area in academia and industry.

With advanced deep learning, summarization models are generally designed based on the seq2seq framework [5]. Recently, many pre-trained language models (PLM) are proposed by pre-training a Transformer [26] in a large-scale unlabeled corpus in a self-supervised manner. Since the PLM encapsulates large-scale language prior knowledge, it shows an excellent generative capability on many natural language generation tasks, e.g., caption generation [33], machine translation

© The Author(s), under exclusive license to Springer Nature Switzerland AG 2023
H. Kashima et al. (Eds.): PAKDD 2023, LNAI 13938, pp. 247–259, 2023.
https://doi.org/10.1007/978-3-031-33383-5_20

By . Associated Press . Baggage screeners at Chicago's O'Hare International Airport have discovered two World War I artillery shells in checked luggage that arrived on a flight from London. The Transportation Security Administration says the bags belonged to a 16-year-old and a 17-year-old who were returning from a school field trip to Europe. TSA spokesman Jim McKinney says a bomb disposal crew determined the shells were inert and no one was ever in danger. Baggage claim: Two World War I artillery shells discovered by baggage screeners in checked luggage that arrived on a flight from London at Chicago's O'Hare International Airport . Odd import: The teens told law enforcement they obtained the shells at a French World War I artillery range. It was not clear how . Suspect objects: FBI and Chicago Police officers evacuated the baggage room until the items could be clears as inert and interviewed the two minors . The teens told law enforcement they obtained the shells at a French World War I artillery range. It was not clear how. TSA explosives experts believe they are French 77 mm shells. They were seized Monday evening while the teens were transferring to a flight to Seattle. The teens were questioned then allowed to travel onward. They weren't charged. Historical artifacts: Transportation Security Administration officers at O,àöá ¨ ∏Hare Airport spotted two military-grade shells in the checked baggage of two minors, who were part of a field trip returning from London . Checking out: The TSA says the bags belonged to a 16-year-old and a 17-year-old who were returning from a school field trip to Europe. A bomb disposal crew determined the shells were inert and no one was ever in danger .

transport — war — no obvious or other topic

Fig. 1. Sentences with multiple topics in an article.

[19] and event generation [35]. Therefore, PLM-based models also have become a mainstream paradigm for text summarization. However, since the training summarization model by only finetuning is still somewhat insufficient, researchers propose many methods to improve the PLM-based summarization model, e.g., contrastive learning [23] and information retrieval [4].

Although it is very successful in text summarization based on PLM, these methods suffer from a non-focused problem. Since most existing summarization models are trained on general-purpose datasets, they focus on generating general-purpose summaries. A general-purpose summary fails to satisfy the needs of different readers and reflects the full range of content of the article. Recently, many text generation methods focus on controllable generation processes guided by sentiment polarity [24] and specific topic distributions [13]. However, these methods lack effective evaluation due to no topic-focused summarization dataset existing. Therefore, *Bahrainian et al.* [2] introduce a NEWs Topic-focused Summarization (NEWTS) corpus to close this gap and propose prompt-based methods to improve PLM-based summarization methods.

Despite the success of these works, they neglect topic-guided cross-interaction between sentences. As shown in Fig. 1, there are multiple sentences with different topics in an article. To generate a summary more relevant to a given topic, a topic-focused summarization model is required to distinguish topics in sentences and conducts cross-interaction between sentences with the given topic. In contrast, existing methods [22] generate a summary only guided by a prompt comprising topic words, which ignores the disturbance of sentences with non-relevant topics and only conducts cross-interaction between tokens by attention module.

Due to sentences with multiple topics in an article, we propose a topic-arc recognition objective to distinguish the topic of each sentence in the article. Since the summary has a definite topic, we leverage summaries and their topics on an article to train the model that can predict the topic of sentences in the article. Moreover, summaries selected from an article have a similar context, which pushes the model to focus on the topic instead of bias in content. In addition, we propose a topic-selective graph network, which can conduct topic-guided cross-interaction on sentences in the article. Specifically, we first construct a graph based on sentence and topic candidates sampled from prediction results of topic-arc recognition on the article. Then, the graph nodes are updated by relational graph convolution layers. Lastly, the updated node representations are delivered to the decoder for summary generation.

In the experiments, we conduct extensive evaluations on two datasets, i.e., NEWTS [2] and COVIDET [31]. Experimental results show that our method outperforms other strong competitors and achieves state-of-the-art performance on NEWTS and COVIDET. Moreover, we further analyze the effectiveness of our method by providing additional quantitative and qualitative results.

2 Related Work

2.1 PLM-based Summarization

With the rise of seq2seq models, researchers are increasingly interested in text summarization. The original topic-based model was the TOPIARY model proposed by *Li et al.* [11] in 2004, which combined language-driven compression techniques and unsupervised topic detection methods. With the advance of the pre-training technique [16], the development of PTMs in text summarization is also booming [10]. There are many sequence-to-sequence pre-trained language models that show their powerful capability for summarization, e.g., BART [9], T5 [18] and Prophetnet [15]. In addition, to be better compatible with discriminative and generation tasks, *Dong et al.* [7] propose UniLM that can be used in natural language understanding and generation tasks. The model also shows excellent summarization capability.

2.2 Topic-Guided Summarization

With the advance of text summarization, increasing researchers are interested in generating topic-specific summaries. Initially, the LDA model is used to guide the topic of summary [11]. For example, *Xing et al.* [29] propose a topic aware seq2seq model named Twitter LDA for a response, which introduces topic information through a joint attention mechanism and a bias generation probability. Another work, CATS [3], is a neural sequence-to-sequence model based on an attentional encoder-decoder architecture, which introduces a new attention mechanism named topic attention controlled by an unsupervised topic model. In PTM-based models, the Plug and Play Language Model (PPLM) [6] is based on GPT-2 [17]. In addition, the BART-FT-JOINT proposed in [31] can simultaneously use a sentiment inducer for sentiment-specific summary generation.

2.3 Graph Neural Network

Graph neural network [20] has been valued in the field of deep learning for its excellent processing ability on unstructured data and node-centric information aggregation mode. With the advance of graph neural networks, there are many graph networks with special structures, e.g., GCN [21], GAT [27], HAN [30] and r-GCN [21]. Moreover, GNN is often used for downstream tasks such as text classification, information extraction, and text generation. In text summarization, *Wang et al.* [28] propose a heterogeneous graph-based neural network for extracting summaries, which contains semantic nodes of different granularity levels except sentences. These extra nodes act as "intermediaries" between sentences and enrich cross-sentence relations. The introduction of document nodes allows the graph structure to be flexibly extended from a single document setup to multiple documents. Another work [8] proposes a multiplex graph summary (Multi-GraS) model based on multiplex graph convolutional networks that can be used to extract text summaries. This model not only considers Various types of inter-sentential relations (such as semantic similarity and natural connection), and intra-sentential relations (such as semantic and syntactic relations between words) are also modeled.

3 Method

This section starts with a base topic-focused summarization model, followed by our proposed methods, i.e., topic-arc recognition and topic-selective graph network. Lastly, we elaborate on the details of our model training.

3.1 Base Topic-Focused Summarization Model

Topic-focused summarization aims to generate a topic-relevant summary based on a long article. Since pre-trained language models (PLMs) based on Transformer show powerful capability for text generation, a recent trend is to finetune a PLM as a summarization model. In this work, we first introduce a base topic-focused summarization model based on PLM. Specifically, given an article $\mathcal{A}$ and topic words $\mathcal{W}$ corresponding to topic t, we first use topic words as a prefix prompt for the article and pass them into the Transformer encoder, i.e.,

$$\boldsymbol{H} = \text{Trans-Enc}([\text{CLS}]\, \text{Summarization by}\, \mathcal{W} : \mathcal{A}\, [\text{SEP}]), \tag{1}$$

where $\boldsymbol{H}$ denotes token representations generated by the Transformer encoder, and $\boldsymbol{H} \in \{\boldsymbol{h}_0, \boldsymbol{h}_1, \cdots, \boldsymbol{h}_l\}$. l is the length of the input.

Next, we deliver the token representations $\boldsymbol{H}$ and the gold summary $\mathcal{S}$ into the Transformer decoder, i.e.,

$$\boldsymbol{P} = \text{Trans-Dec}(\boldsymbol{H}, \mathcal{S}), \tag{2}$$

where $\boldsymbol{P} = \{\boldsymbol{p}_0, \boldsymbol{p}_1, \cdots, \boldsymbol{p}_n\}$ and $\boldsymbol{p}_i$ is a probability distribution over vocabulary $\mathcal{V}$. n is the length of the summary.

Lastly, we train the pre-trained Transformer by maximum likelihood estimation, and its loss function is defined as:

$$\mathcal{L}^{(ce)} = -\frac{1}{|n|} \sum_{n} \log \boldsymbol{p}_i[y = s_i] \tag{3}$$

where s_i is i-th word in the ground truth summary $\mathcal{S}$.

3.2 Topic-Arc Recognition

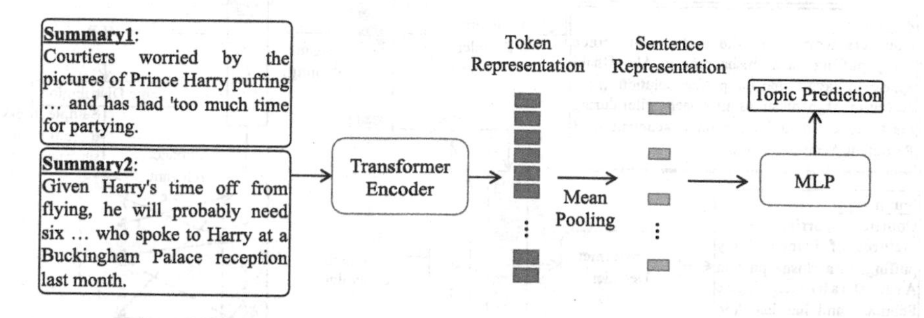

Fig. 2. Overview of topic-arc recognition.

Due to multiple sentences with different topics in an article, distinguishing topics of sentences in an article is essential to topic-focused summarization. To recognize the sentence's topic in articles, we propose a topic-arc recognition (TAR) objective, as shown in Fig. 2. In this objective, we concatenate two summaries ($\mathcal{S}_0$ and $\mathcal{S}_1$) with different topics (t_1 and t_2) from an article and pass them into the Transformer encoder, i.e.,

$$\boldsymbol{H} = \text{Trans-Enc}(\mathcal{S}_0, \mathcal{S}_1). \tag{4}$$

Next, we conduct mean pooling to token representations $\boldsymbol{H}$ to obtain sentence representations $\hat{\boldsymbol{H}} = \{\hat{\boldsymbol{h}}_1, \hat{\boldsymbol{h}}_2, \cdots, \hat{\boldsymbol{h}}_m\}, m = m_1 + m_2$, and m_1 and m_2 denote number of sentences in two summaries, respectively. Then, we deliver sentence representations $\hat{\boldsymbol{H}}$ to a multilayer perceptron (MLP) to predict the topic of each sentence, i.e.,

$$\boldsymbol{p}_i^t = \text{MLP}(\hat{\boldsymbol{h}}_i), i \in \{1, 2, \cdots, m\}, \tag{5}$$

where $\boldsymbol{p}_i^t$ denotes the topic probability distribution of i-th sentence on all topic categories. Lastly, we optimize the Transformer encoder and MLP by a cross-entropy loss, i.e.,

$$\mathcal{L}^{(tar)} = -\frac{1}{|m|} \sum_{m} \log \boldsymbol{p}_i^t[y = t_i], \tag{6}$$

where t_i is the ground truth topic category of i-th sentence.

3.3 Summarization with Topic-Selective Graph Network

As mentioned in Fig. 1, sentences with the same topic in an article are usually not connected together. To integrate semantic information on the same topic, we

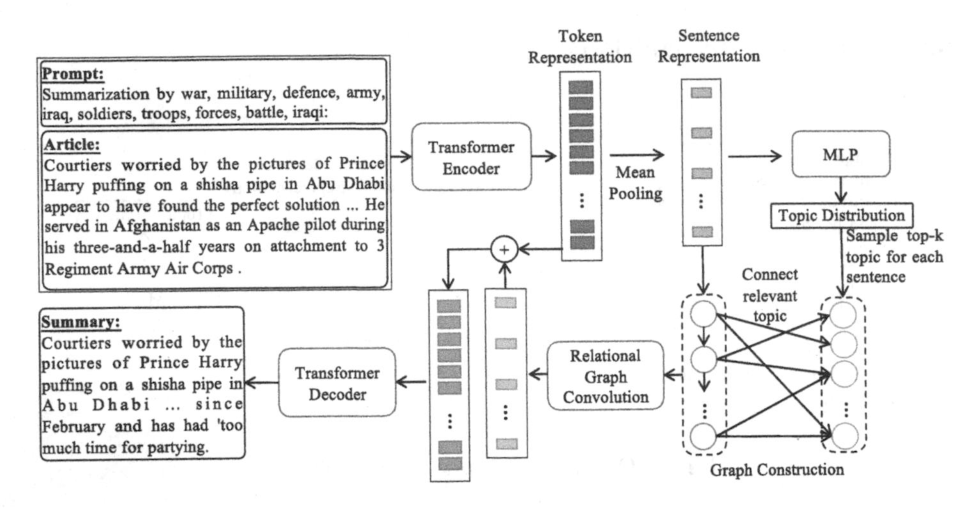

Fig. 3. Overview of summarization with topic-selective graph network.

propose a topic-selective graph network (TSGN) to conduct sentence-level cross-interaction through topic nodes as bridge. As shown in Fig. 3, we first extract token representations $\boldsymbol{H}$ in an article via Eq. 1. Next, we conduct a mean pooling operation to token representations $\boldsymbol{H}$ to obtain sentence representations $\bar{\boldsymbol{H}} = \{\bar{\boldsymbol{h}}_1, \bar{\boldsymbol{h}}_2, \cdots, \bar{\boldsymbol{h}}_m\}$. Then, we deliver sentence representations $\bar{\boldsymbol{H}}$ to the MLP in Eq. 5 to predict the topic probability distribution $\bar{\boldsymbol{p}}_i^t$ of each sentence. According to the ranking of probability $\bar{\boldsymbol{p}}_i^t$, we select top-k topics for each sentence as its topic candidates.

To conduct topic-guided cross-interaction, we construct a semantic graph, and sentence and topic nodes consist of sentence representations $\bar{\boldsymbol{H}}$ and topic representations $\boldsymbol{T} = \{\boldsymbol{t}_1, \boldsymbol{t}_2, \cdots, \boldsymbol{t}_v\}$ that are derived from an embedding layer; and v is number of topic categories. Due to heterogeneous nodes of sentence and topic, we connect these nodes through edges with different relations $\mathcal{E} \in \{\boldsymbol{e}_s, \boldsymbol{e}_t\}$. The semantic graph comprises two rules of connecting nodes: 1) We take the edge $\boldsymbol{e}_s$ to connect sentence nodes $\bar{\boldsymbol{h}}_i$ in sentence order. 2) Each sentence node $\bar{\boldsymbol{h}}_i$ is only connected to its topic candidate nodes $\boldsymbol{t}_u$ through edge $\boldsymbol{e}_t$, which can alleviate the disturbance of sentences with non-relevant topics. Moreover, we use relational graph convolution layer [21,34] to update sentence representations, i.e.,

$$\boldsymbol{w}_i^{l+1} = \text{ReLU}\Big(\sum_{e \in \mathcal{E}} \sum_{j \in \mathcal{N}_e(i)} \frac{1}{|\mathcal{N}_e(i)|} \boldsymbol{W}_e \cdot \boldsymbol{w}_j^l\Big), \boldsymbol{w}_j^0 \in \{\bar{\boldsymbol{H}}, \boldsymbol{T}\} \tag{7}$$

where $\mathcal{E}$ denote a set of all edges types, and $\mathcal{N}_e(i)$ is the neighborhood of node i under relation e. l is number of the relational graph convolution layer. Updated sentence representations are represented as $\widetilde{\boldsymbol{H}}$. Furthermore, we take $\widetilde{\boldsymbol{H}}$ plus into the corresponding token representations and pass them to Eq. 2 to predict probability distributions $\widetilde{\boldsymbol{P}} = \{\widetilde{\boldsymbol{p}}_0, \widetilde{\boldsymbol{p}}_1, \cdots, \widetilde{\boldsymbol{p}}_n\}$. n is length of the summary.

Lastly, we train our model by cross-entropy loss, which is defined as:

$$\mathcal{L}^{(tsgn)} = -\frac{1}{|n|}\sum_{n} \log \widetilde{\boldsymbol{p}}_i[y = s_i] \tag{8}$$

where s_i is i-th word in the ground truth summary $\mathcal{S}$.

3.4 Training

During training, we train our model by jointly topic-arc recognition and topic-selective graph network and minimize their loss functions:

$$L = \alpha \times \mathcal{L}^{(tar)} + \beta \times \mathcal{L}^{(tsgn)} \tag{9}$$

4 Experiments

4.1 Dataset and Evaluation Metrics

In the experiments, we train and evaluate our approach on two datasets, i.e., NEWTS [2] and COVIDET [31]. The NEWTS dataset is based on the famous CNN/Dailymail dataset and annotated by online crowd-sourcing. Every source article is paired with two summaries focusing on different topics and provides topic words to denote the topic. The dataset consists of 2,400 and 600 samples in training and test sets. The results of the test are reported by ROUGE (i.e., R-1, R-2, R-L) [12] and Topic Focus [2]. The COVIDET dataset is sourced from 1,883 English Reddit posts about the COVID-19 pandemic. Each post is annotated with 7 fine-grained emotion labels; for each emotion, annotators provided a concise, abstractive summary describing the triggers of the emotion. We follow the official data split with 2,234/1,526 samples in training/test sets. The evaluation metrics are ROUGE-L (R-L) and BERT Score [32].

4.2 Experimental Setting

The pre-trained transformer we used is BART-base. We use AdamW optimizer with learning rate of 5 $\times 10^{-5}$, and learning rate of the relational graph convolution layer is 1 $\times 10^{-4}$. The weight decay and the dropout rate are both 1.0. The maximum training epoch, sentence number and batch size are set to 3, 60 and 2. The length of input and output are set to 1024 and 128. We use NLTK to separate sentences. In Eq. 9, α and β are set to 1.0 and 0.8, respectively. Experiments are conducted on an NVIDIA RTX3080 GPU, and training time is around 3 h.

Table 1. Results on NEWTS test set.

Method	R-1	R-2	R-L	Topic Focus
BART [2]	16.48	0.75	11.71	0.0080
BART+T-W [2]	31.14	10.46	19.94	0.1375
BART+CNN-DM [2]	26.23	7.24	17.12	0.1338
T5+T-W [2]	31.78	10.83	20.54	0.1386
T5+CNN-DM [2]	27.87	8.55	18.41	0.1305
ProphetNet+T-W [2]	31.91	10.8	20.66	0.1362
ProphetNet+CNN-DM [2]	28.71	8.53	18.69	0.1295
PPLM [2]	29.63	9.08	18.76	0.1482
Ours	**34.24**	**12.65**	**23.08**	**0.1512**

Table 2. Results on COVIDET test set. "Ang", "Dis", "Fea", "Joy", "Sad", "Tru" and "Ant" denote anger, anticipation, joy, trust, fear, sadness and disgust, respectively.

Method	Ang	Dis	Fea	Joy	Sad	Tru	Ant	AVG.
Metric: R-L								
BART [31]	0.161	0.138	0.164	0.149	0.157	0.158	0.164	0.156
PEGASUS-FT [31]	0.185	0.155	0.199	0.158	0.173	0.164	0.193	0.175
BART-FT [31]	0.190	0.159	0.206	0.165	**0.177**	0.162	**0.198**	0.180
BART-FT-JOINT [31]	0.190	0.158	0.203	0.163	0.175	0.165	0.196	0.179
Ours	**0.202**	**0.177**	**0.223**	**0.208**	0.166	**0.201**	0.186	**0.195**
Metric: BERT Score								
BART [31]	0.587	0.558	0.529	0.551	0.559	0.571	0.558	0.559
PEGASUS-FT [31]	0.681	0.713	0.739	0.683	0.705	0.663	0.736	0.703
BART-FT [31]	0.705	0.695	0.748	0.699	0.718	0.653	0.749	0.710
BART-FT-JOINT [31]	0.701	0.706	0.729	0.694	0.713	0.659	0.746	0.707
Ours	**0.885**	**0.880**	**0.889**	**0.888**	**0.879**	**0.881**	**0.881**	**0.883**

4.3 Main Results

Comparison results of our methods and other strong competitors are shown in Table 1 and Table 2. From the tables, we find two observations. 1) Methods using topic words as a prompt outperform that of no prompt, which demonstrates that the topic words as prompt are significant in generating topic-focused summarization. 2) our approach is superior to others and achieves state-of-the-art, demonstrating the effectiveness of our method. Meanwhile, it shows that making a topic-guided sentence-level cross-interaction can improve topic-focused summarization by capturing the topic-relevant content in the article.

4.4 Ablation Study

As shown in Table 3, we conduct an ablation study on our method. First, we remove the topic-arc recognition (TAR) objective, and the results show performance drops. It demonstrates that the TAR objective can distinguish the topic

Table 3. Ablation study of our approach.

Method	R-1	R-2	R-L	Topic Foucs
Ours	**34.24**	**12.65**	**23.08**	**0.1512**
Ours w/o TSGN	34.12	12.08	22.61	0.1407
Ours w/o TAR	34.30	11.97	22.27	0.1453
Ours w/o TSGN, TAR	31.14	10.46	19.94	0.1375

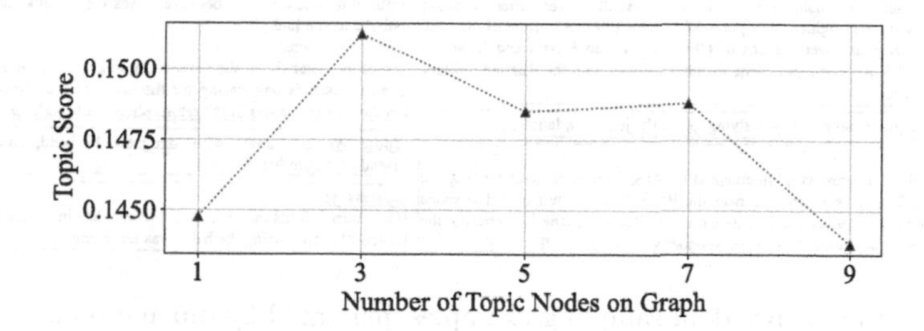

Fig. 4. Results on different number of topic nodes in topic-selective graph network.

category of the sentence, which improve the capability of topic awareness in the model. Moreover, we investigate the effectiveness of TSGN by removing it from our method. Results show that the performance drops, which demonstrates that TSGN can improve topic-focused summarization through cross-sentence interaction. Lastly, we remove TAR and TSGN to verify the effectiveness of our method. The results show a large drop, which further supports the effectiveness of our method.

4.5 Impact of Topic Node

It can be seen from Fig. 4 that the best performance is achieved when the number of topics selected in TSGN is 3. Moreover, the performance drops as the number of topic numbers increases or decreases. The reason is that the topic was sampled from the distribution with some noise. Introducing too few topic nodes (e.g., 1 node) into TSGN leads to missing the correct topic. In addition, to ensure that the correct topic node is introduced, the introduction of too many topic nodes (e.g., more than 3) leads to too much noise into the TSGN, which is also not conducive to the performance of the model.

4.6 Case Study

To conduct an extensive evaluation of our model, we take a qualitative comparison of our model. We randomly sample some generated summaries and their article and topic words, as are shown in Fig. 5 In articles, the green area denotes that the sentence's topic is recognized as the given topic. Moreover, we can

Article:
In a maneuver with ominous echoes of the Cold War, a Russian fighter jet 'aggressively' intercepted an American plane over Poland, the Pentagon claims. Filing an official complaint to Russia, the State Department alleges a U.S. RC-135U reconnaissance aircraft was flying near the Baltic Sea in international airspace when a Russian SU-27 Flanker cut into its path. **Pentagon officials have slammed the move as 'unprofessional' and 'unsafe'. A Russian jet 'aggressively' intercepted a U.S. RC-135U plane (pictured) over Poland, the Pentagon claims . Spokesman Mark Wright told CNN the Russian jet performed 'aggressive maneuvers' at high speed in close proximity to the plane. He added that the State Department will consequently be filing a complaint to Russia 'through diplomatic channels'. Russia has rubbished claims it was a deliberate and aggressive interception.** State news agency Sputnik reported that the jet was circling the plane to determine its tail number, as its transponder was switched off. **The Russian SU-27 Flanker 'aggressively' cut into the American plane's path, the spokesman stated . The Pentagon and the U.S. European Command have dismissed the account. The incident occurred almost exactly a year after Pentagon officials accused a Russian fighter jet of purposely flying 100 feet in front of the nose of an American spy plane over the Sea of Okhotsk between Russia and Japan. In August last year, U.S. Air Force spy plane avoided a run-in with the Russian military over Swedish airspace.**

Topic Words: air, plane, aircraft, flight, flying, pilot, fly, jet, crew, landing

Summary:
A Russian fighter jet 'aggressively' intercepted an American plane over Poland, the Pentagon claims. The plane was flying near the Baltic Sea in international airspace when a Russian SU-27 Flanker cut into its path. The Russian plane was circling the plane to determine its tail number, as its transponder was switched off.

Article:
An infant's parents have been charged after police say they couldn't agree who would take care of their three-month-old over the weekend and the father left her at a McDonald's in Ohio. Twenty-two-year-old Rachel Jaajaa and 29-year-old Aaron Tate's daughter was found on Friday night at a McDonald's in Elyria when a patron called 911 after seeing the baby was left alone. **They were both arrested Monday and charged with child endangering. Aaron Tate, 29, is accused to leaving his daughter alone on the floor of a McDonald's and then driving away .** Police say Rachel Jaajaa, 22, refused to take her daughter because it was the father's 'turn to have the baby' They were both charged with child endangering after a patron from the McDonald's in Elyria, Ohio, called 911 . **Police say Jaajaa was leaving McDonald's after her shift ended on Friday around 8pm and found Tate in the parking lot with the infant. Another employee told police Jaajaa refused to take the baby because it was the father's 'turn to have the baby' and she had plans,** Fox 8 reported. They say Tate then left the child in McDonald's and drove off, prompting the 911 call to police. **An officer later found the parents arguing in front of a Burger King. Jaajaa's sister is now caring for the baby. Tate declined to comment on the charges. Jaajaa didn't return phone calls seeking comment.**

Topic Words: family, wife, daughter, husband, couple, pictured, friends, left, brother, friend

Summary:
The father left his daughter at a McDonald's in Elyria when a patron called 911 after seeing the baby was left alone.

Fig. 5. Random sampling examples generated by our method.

Table 4. Human evaluation.

Type of evaluation	Win Percentage	
	BART+T-W	Ours
Topic Relevance	0.27	**0.73**
Content Consistency	0.36	**0.64**
Logic	0.41	**0.59**

observe that the recognized sentences and given topics are relevant, which verifies the effectiveness of the TAR objective. In addition, we can see that the generated summaries comprise the sentence content of the green area, which demonstrate the effectiveness of TSGN.

4.7 Human Evaluation

To comprehensively evaluate our method, d, we conducted a human evaluation to compare our model and BART+T-W [2]. We considered the topic relevance, content consistency and logic. Therefore, there are three types of human evaluation. We randomly sampled 150 samples from the test set, and each sample includes an article, topic words and generated summary. We displayed these samples to 3 recruited annotators. They need to distinguish which summaries are better quality based on the type of evaluation. As shown in Table 4, results show that the performance of our model is significantly better than BART+T-W.

5 Conclusion

In this work, we dive into the limitations of previous topic-guided summarization methods, i.e., these methods still ignore the disturbance of sentences with non-relevant topics and only conduct cross-interaction between tokens by attention module. To address the limitations and improve the summarization model, we propose a topic-arc recognition objective and topic-selective graph network. The topic-arc recognition objective aims to discriminate topics of sentences. Moreover, the topic-selective graph network conducts topic-guided cross-interaction on sentences based on the results of topic-arc recognition. Experimental results show that our methods achieve state-of-the-art performance on NEWTS and COVIDET datasets.

References

1. Ahuja, O., Xu, J., Gupta, A., Horecka, K., Durrett, G.: ASPECTNEWS: aspect-oriented summarization of news documents. In: ACL 2022, pp. 6494–6506. Association for Computational Linguistics (2022)
2. Bahrainian, S.A., Feucht, S., Eickhoff, C.: NEWTS: a corpus for news topic-focused summarization. In: Findings of the Association for Computational Linguistics: ACL 2022, pp. 493–503. Association for Computational Linguistics, Dublin, Ireland (2022)
3. Bahrainian, S.A., Zerveas, G., Crestani, F., Eickhoff, C.: Cats: Customizable abstractive topic-based summarization. ACM Trans. Inf. Syst. **40**(1), 1–24 (2021)
4. Bouras, C., Tsogkas, V.: Improving text summarization using noun retrieval techniques. In: Lovrek, I., Howlett, R.J., Jain, L.C. (eds.) KES 2008. LNCS (LNAI), vol. 5178, pp. 593–600. Springer, Heidelberg (2008). https://doi.org/10.1007/978-3-540-85565-1_73
5. Cho, K., et al.: Learning phrase representations using RNN encoder-decoder for statistical machine translation. arXiv preprint arXiv:1406.1078 (2014)
6. Dathathri, S., et al.: Plug and play language models: a simple approach to controlled text generation. arXiv preprint arXiv:1912.02164 (2019)
7. Dong, L., et al.: Unified language model pre-training for natural language understanding and generation. In: Advances in Neural Information Processing Systems, vol. 32 (2019)
8. Jing, B., You, Z., Yang, T., Fan, W., Tong, H.: Multiplex graph neural network for extractive text summarization. arXiv preprint arXiv:2108.12870 (2021)
9. Lewis, M., et al.: Bart: Denoising sequence-to-sequence pre-training for natural language generation, translation, and comprehension. arXiv preprint arXiv:1910.13461 (2019)
10. Li, J., Tang, T., Zhao, W.X., Wen, J.R.: Pretrained language models for text generation: a survey. arXiv preprint arXiv:2105.10311 (2021)
11. Li, Y., Hong, J.I., Landay, J.A.: Topiary: a tool for prototyping location-enhanced applications. In: Proceedings of the 17th Annual ACM Symposium on User Interface Software and Technology, pp. 217–226 (2004)
12. Lin, C.Y.: ROUGE: a package for automatic evaluation of summaries. In: Text Summarization Branches Out, pp. 74–81. Association for Computational Linguistics, Barcelona, Spain, July 2004

13. Liu, Z., Ng, A., Lee, S., Aw, A.T., Chen, N.F.: Topic-aware pointer-generator networks for summarizing spoken conversations. In: 2019 IEEE Automatic Speech Recognition and Understanding Workshop (ASRU), pp. 814–821. IEEE (2019)
14. Polsley, S., Jhunjhunwala, P., Huang, R.: Casesummarizer: a system for automated summarization of legal texts. In: COLING 2016, pp. 258–262. ACL (2016)
15. Qi, W., et al.: Prophetnet: predicting future n-gram for sequence-to-sequence pre-training. arXiv preprint arXiv:2001.04063 (2020)
16. Radford, A., Narasimhan, K., Salimans, T., Sutskever, I., et al.: Improving language understanding by generative pre-training (2018)
17. Radford, A., Wu, J., Child, R., Luan, D., Amodei, D., Sutskever, I., et al.: Language models are unsupervised multitask learners. OpenAI blog **1**(8), 9 (2019)
18. Raffel, C., Shazeer, N., Roberts, A., Lee, K., Narang, S., Matena, M., Zhou, Y., Li, W., Liu, P.J., et al.: Exploring the limits of transfer learning with a unified text-to-text transformer. J. Mach. Learn. Res. **21**(140), 1–67 (2020)
19. Ren, S., Zhou, L., Liu, S., Wei, F., Zhou, M., Ma, S.: Semface: pre-training encoder and decoder with a semantic interface for neural machine translation. In: ACL/IJCNLP 2021, pp. 4518–4527 (2021)
20. Scarselli, F., Tsoi, A.C., Gori, M., Hagenbuchner, M.: Graphical-based learning environments for pattern recognition. In: Fred, A., Caelli, T.M., Duin, R.P.W., Campilho, A.C., de Ridder, D. (eds.) SSPR /SPR 2004. LNCS, vol. 3138, pp. 42–56. Springer, Heidelberg (2004). https://doi.org/10.1007/978-3-540-27868-9_4
21. Schlichtkrull, M., Kipf, T.N., Bloem, P., van den Berg, R., Titov, I., Welling, M.: Modeling relational data with graph convolutional networks. In: Gangemi, A., Navigli, R., Vidal, M.-E., Hitzler, P., Troncy, R., Hollink, L., Tordai, A., Alam, M. (eds.) ESWC 2018. LNCS, vol. 10843, pp. 593–607. Springer, Cham (2018). https://doi.org/10.1007/978-3-319-93417-4_38
22. Shin, T., Razeghi, Y., Logan IV, R.L., Wallace, E., Singh, S.: Autoprompt: eliciting knowledge from language models with automatically generated prompts. arXiv preprint arXiv:2010.15980 (2020)
23. Su, Y., Lan, T., Wang, Y., Yogatama, D., Kong, L., Collier, N.: A contrastive framework for neural text generation. arXiv preprint arXiv:2202.06417 (2022)
24. Titov, I., McDonald, R.: A joint model of text and aspect ratings for sentiment summarization. In: Proceedings of ACL-08: HLT, pp. 308–316. Association for Computational Linguistics, Columbus, Ohio, June 2008
25. Turpin, A., Tsegay, Y., Hawking, D., Williams, H.E.: Fast generation of result snippets in web search. In: ACM SIGIR, 2007. pp. 127–134. ACM (2007)
26. Vaswani, A., et al.: Attention is all you need. In: Advances in Neural Information Processing Systems, vol. 30 (2017)
27. Velickovic, P., Cucurull, G., Casanova, A., Romero, A., Lio, P., Bengio, Y.: Graph attention networks. Stat **1050**, 20 (2017)
28. Wang, D., Liu, P., Zheng, Y., Qiu, X., Huang, X.: Heterogeneous graph neural networks for extractive document summarization. arXiv preprint arXiv:2004.12393 (2020)
29. Xing, C., et al.: Topic aware neural response generation. In: AAAI, vol. 31 (2017)
30. Yang, Z., Yang, D., Dyer, C., He, X., Smola, A., Hovy, E.: Hierarchical attention networks for document classification. In: NAACL, pp. 1480–1489 (2016)
31. Zhan, H., Sosea, T., Caragea, C., Li, J.J.: Why do you feel this way? summarizing triggers of emotions in social media posts. p. To appear (2022)
32. Zhang, T., Kishore, V., Wu, F., Weinberger, K.Q., Artzi, Y.: Bertscore: evaluating text generation with bert. arXiv preprint arXiv:1904.09675 (2019)

33. Zhou, Y.: Sketch storytelling. In: IEEE International Conference on Acoustics, Speech and Signal Processing, ICASSP 2022, Virtual and Singapore, 23–27 May 2022, pp. 4748–4752. IEEE (2022)
34. Zhou, Y., Long, G.: Multimodal event transformer for image-guided story ending generation. arXiv preprint arXiv:2301.11357 (2023)
35. Zhou, Y., Shen, T., Geng, X., Long, G., Jiang, D.: Claret: Pre-training a correlation-aware context-to-event transformer for event-centric generation and classification. In: ACL 2022, pp. 2559–2575. Association for Computational Linguistics (2022)

Time-Series and Streaming Data

RiskContra: A Contrastive Approach to Forecast Traffic Risks with Multi-Kernel Networks

Changlu Chen[1], Yanbin Liu[2], Ling Chen[1](✉), and Chengqi Zhang[1]

[1] University of Technology Sydney, Sydney, Australia
Changlu.Chen@student.uts.edu.au, {Ling.Chen,Chengqi.Zhang}@uts.edu.au
[2] Australian National University, Canberra, Australia

Abstract. Traffic accident forecasting is of vital importance to the intelligent transportation and public safety. Spatial-temporal learning is the mainstream approach to exploring complex evolving patterns. However, two intrinsic challenges lie in traffic accident forecasting, preventing the straightforward adoption of spatial-temporal learning. First, the temporal observations of traffic accidents exhibit ultra-rareness due to the inherent properties of accident occurrences (Fig. 1(a)), which leads to the severe scarcity of risk samples in learning accident patterns. Second, the spatial distribution of accidents is severely imbalanced from region to region (Fig. 1(b)), which poses a serious challenge to forecast the spatially diversified risks. To tackle the above challenges, we propose ***RiskContra***, a ***Contrastive*** learning approach with multi-kernel networks, to forecast the ***Risk*** of traffic accidents. Specifically, to address the first challenge (i.e. temporal rareness), we design a novel contrastive learning approach, which leverages the periodic patterns to derive a tailored mixup strategy for risk sample augmentation. This way, the contrastively learned features can better represent the risk samples, thus capturing higher-quality accident patterns for forecasting. To address the second challenge (i.e. spatial imbalance), we design the multi-kernel networks to capture the hierarchical correlations from multiple spatial granularities. This way, disparate regions can utilize the multi-granularity correlations to enhance the forecasting performance across regions. Extensive experiments corroborate the effectiveness of each devised component in RiskContra.

Keywords: Traffic accident · Contrastive learning · Mixup

1 Introduction

Traffic accidents have posed a serious threat to public safety due to the severe destructiveness and negative impact. The casualty from traffic accidents all over the world has amounted to over 1.3 million according to the World Health Organization (WHO) [12]. Thus, it is of vital importance to establish more effective traffic accident forecasting systems, which could significantly protect lives and properties from potential risks.

© The Author(s), under exclusive license to Springer Nature Switzerland AG 2023
H. Kashima et al. (Eds.): PAKDD 2023, LNAI 13938, pp. 263–275, 2023.
https://doi.org/10.1007/978-3-031-33383-5_21

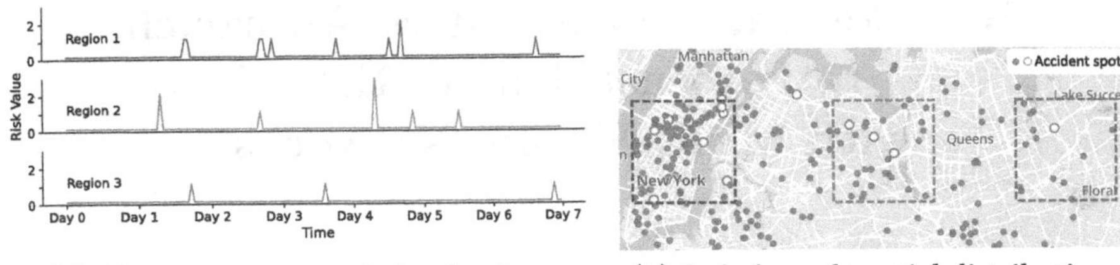

(a) Ultra-rare temporal distribution (b) Imbalanced spatial distribution

Fig. 1. Two intrinsic challenges of traffic accident forecasting.

Traffic accident forecasting aims to predict the risk value of the latent traffic accident that may occur in a target region (*spatial*) at future time steps (*temporal*) given the historical traffic accident observations. Early works employ the statistical or linear machine learning models such as SVM or ARIMA [20] to explore the traffic accident patterns. However, they can hardly capture the complex and non-linear spatial-temporal correlations in the traffic accident data. Thus, recent works [11,17,18,25] resort to deep learning models to promote the forecasting performance by taking advantage of the hierarchically deep and non-linear architectures, which have achieved significant progress. Despite the effectiveness of these deep forecasting methods, two challenges have greatly impeded the further improvement of existing traffic accident forecasting models.

The first challenge is the ultra-rareness of the accidents distributed across the temporal domain, i.e. the majority of the accidental risk value are zero due to the infrequency of an accident (Fig. 1(a)). This hinders the effective training of deep neural networks, which requires sufficient examples to explore the data distribution and fit the model parameters. Moreover, since the number of non-risk samples is much larger than that of risk samples, the forecasting of traffic accidents will be easily biased to non-risks rather than potential risks. This temporal rareness issue, however, has not been well solved in previous works. To tackle this challenge, we devise a novel contrastive learning approach, which customizes the mixup [23] strategy to generate sufficient augmented risk samples for traffic accident forecasting. Our customized mixup strategy leverages the periodic patterns inherent in accident data. Specifically, we generate data augmentations by mixing two temporal samples with the same daily stamp from a fixed interval (e.g. one week). If either temporal sample is a risk sample, the generated sample will be labelled as risk with a mixing score. With these augmented risk samples, the ratio gap between the non-risk and risk samples is significantly reduced. Then, we utilize the generated risk/non-risk samples to construct the positive and negative pairs, on which the supervised contrastive learning [10] is conducted. Equipped with contrastive learning, the feature embeddings of risk samples will be more distinguishable from those of non-risk samples, thus facilitating the accident forecasting performance.

The second challenge lies in the imbalanced spatial distribution of traffic accidents, which exhibits diverse patterns of multiple granularities from region to region, as shown in Fig. 1(b). For example, an urban area with crowded traffic flows may bear a higher risk compared to a rural counterpart with fewer vehicles, and a district with underdeveloped traffic management systems may experience more accidents than those equipped with advanced traffic monitoring systems. This challenge is not well addressed by existing works. GSNet [17] used the Convolution Neural Networks (CNNs) with a fixed kernel size to model the spatial correlation, which only captured the local and single-granularity features due to the limited receptive field. HeteroLSTM [22] modeled each region with a different ConvLSTM and adopt an ensemble strategy to generate the final results, which is infexible and parameter in-efficient. Different from these attempts, we design a delicate Multi-kernel CNNs architecture to capture the multi-granularity spatial correlations for traffic accident forecasting. In this architecture, the local, global, and point-wise convolution layers are elegantly arranged in a unified model (Fig. 2). The local convolution layer targets at regions with densely-distributed accidents, while the global convolution layer focuses on regions with sparsely-distributed accidents. Moreover, the point-wise convolution layer is utilized to reduce the computation burden caused by the feature aggregation, thus guaranteeing the efficiency of the multi-kernel networks.

To sum up, the contributions of our work are as follows:

- We propose ***RiskContra***, the first attempt to apply contrastive learning to the traffic accident forecasting problem.
- By utilizing intrinsic periodic patterns of accident data, we integrate the mixup algorithm into the contrastive framework seamlessly to address the temporal rareness challenge.
- We devise a delicate Multi-kernel CNNs structure to explicitly capture the multi-granularity spatial correlations, thus addressing the spatial imbalance challenge.
- The state-of-the-art performance is achieved on two real-world benchmark datasets, showing the superiorlty of the proposed RiskContra.

2 Related Work

Traffic Accident Forecasting. The emerging deep learning techniques provide a promising alternative to the traffic accident forecasting, due to its capability to extract discriminative features and capture complicated dynamic correlations. Therefore, an assortment of deep network architectures have been applied to traffic accident forecasting. For example, [22] designs different ConvLSTM modules for different regions to simulate the spatial heterogeneity of traffic accidents. [8] leverages the heterogeneous contextual knowledge and utilizes attention mechanism to capture the temporal correlations. [25] adopts a multi-task learning scheme which employs the traffic volume prediction as an auxiliary task for minute-level traffic accident forecasting. [17] designs a CNN-GRU module to capture the geographical spatial-temporal correlations and a GCN-GRU module to explore the semantic spatial-temporal correlations. [18] designs a channel-wise CNN and multi-view GCN to capture both the local geographic and global semantic dependencies for traffic accident forecasting. Despite of the recent progress of these deep learning based forecasting models, the issues of imbalanced spatial distribution and ultra-rare temporal distribution of accident risks have not been comprehensively studied in a unified framework.

Contrastive Learning. Contrastive learning has shown remarkable successes in various domains [3,4,7]. It aims to distinguish the semantically similar samples (positive pairs) from the semantically dissimilar samples (negative pairs) in the latent space. Typically, SimCLR [3] employs data augmentations such as cropping, color distortion and blurring to generate different views of image as positives. MoCO [7] maintains the negative samples in a queue and employs the momentum encoder to ensure the consistency of the queue. SimSiam [4] explores a stop-gradient strategy to prevent contrastive learning from collapsing. Mixup [23] has proved to be an effective data augmentation strategy in supervised learning, which generates new samples by the convex interpolation of two different samples in both the data and label space. Albeit simple, mixup has been adopted as an effective regularization to mitigate the overfitting and improve the robustness of deep neural networks [9,15,16]. However, this effective strategy has not been studied in the context of traffic accident forecasting, which greatly suffers from the data lacking and overfitting issues. Motivated by this, we devise a customized mixup strategy to serve as an effective augmentation function in the proposed contrastive learning approach.

3 Methodology

3.1 Problem Definition

In the traffic accident forecasting task, we are given the historical traffic accident features $\{X_1, X_2, \cdots, X_T\}$, where $X_t \in \mathbb{R}^{W \times H \times D}$ represents D-dimensional traffic accident related features including accident risk, temporal features, POIs information, taxi order, and weather information at time step t in all regions (which are partitioned from grid map of the city under study with width W and height H). In addition, temporal features E_{T+1} (hour in a day, day in a week, and holiday) of the future time step $T+1$ is given. The aim is to predict the future accidental risk values for all regions at time step $T+1$:

$$Y_{T+1} = \mathcal{F}_\theta(\{X_1, X_2, \cdots, X_T\}, E_{T+1}), \tag{1}$$

where θ is the model parameter, and $Y_{T+1} \in \mathbb{R}^{W \times H}$ denotes the predicted risk values for all regions.

3.2 Spatial-temporal Accident Forecasting

Multi-kernel Networks. Existing works employ CNNs, GCNs or Transformer to capture the spatial correlations in spatial-temporal tasks [6,8,17,19,24]. However, CNNs and GCN can only handle the local spatial proximity with a limited receptive filed, and Transformer is inefficient to model large number of elements. Taking both the model capacity and computation efficiency into account, we

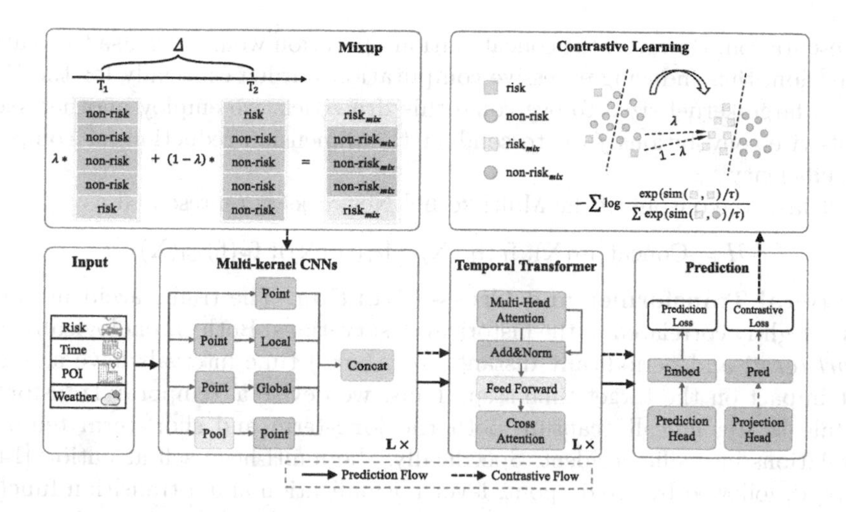

Fig. 2. The framework of RiskContra.

thus propose the Multi-kernel networks that can capture the hierarchical spatial correlations to address the spatial imbalance issue.

We define a single-layer convolution module as follows:

$$f(X; W, b) = \mathrm{ReLU}(\mathrm{W} * \mathrm{X} + \mathrm{b}), \tag{2}$$

where $*$ is the convolution operation, ReLU is the activation function, and W and b are the parameters. Our Multi-kernel networks are composed of diverse convolution modules targeting at regions with different spatial granularities (Fig. 2). We omit the layer index $l \in \{1, 2, \ldots, L\}$ for simplicity.

First, we adopt the ***Point-wise CNN*** $f_{\mathrm{P}}(\cdot)$ to capture the intra-region spatial representation. Besides, for the areas where risk is densely distributed across regions, the geographical proximity is a contributing factor of traffic accidents. For example, two spatially close roads in the busy commercial center tend to gain higher risks of traffic accidents. To capture local correlations among the adjacent regions such as road intersections and crowded streets, we design the ***Local CNN*** module $f_{\mathrm{L}}(\cdot)$ with *a small kernel size* to explore their *spatial proximity.*

Moreover, geographically faraway regions may share similar accident patterns due to similar POI distributions and temporal conditions. These semantic correlations are critical for analyzing accidents caused by external factors, which can significantly promote the forecasting performance. Thus, for these areas with sparsely distributed risks, a ***Global CNN*** module $f_{\mathrm{G}}(\cdot)$ with *a large kernel size* is devised in parallel to capture the spatial correlations among *distant regions.*

Except for the above parametric convolution modules, we also propose a ***non-parametric Pooling-based module*** $f_{\mathrm{pool}}(\cdot)$, which has the potential to avoid the overfitting issue of the parameterized modules.

To handle diverse regions, the concatenation operation Concat[·] is adopted to fuse the output from different CNN modules to obtain a comprehensive feature representation. However, the concatenation operation would increase the feature dimension, thus inducing excessive computation burden especially for the CNN with a large kernel size. To overcome this drawback, we employ another set of point-wise convolutions $f_P(\cdot)$ to conduct the dimension reduction for computation efficiency.

At last, the output of the Multi-kernel Networks is represented as:

$$H = \text{Concat}\left[\text{f}_\text{P}(\text{X});\ \text{f}_\text{L}(\text{f}_\text{P}(\text{X}));\ \text{f}_\text{G}(\text{f}_\text{P}(\text{X}));\ \text{f}_\text{P}(\text{f}_\text{pool}(\text{X}))\right]. \tag{3}$$

Temporal Transformer and Cross-Attention. The traffic accident condition is tightly correlated to the historical observations. Both chronologically near (*short-term*) and periodically distant (*long-term*) time intervals have a significant impact on the target time step. Thus, we devise a Temporal Transformer module to dynamically capture both the long-term and short-term temporal correlations in traffic accident. Specifically, the multi-head self-attention [14] is adopted, followed by the dropout, layer normalization and a transition function (2-layer feed-forward network) to obtain a more powerful representation.

To predict the risk of a future time step based on the historical spatial-temporal features, we devise a ***cross-attention module*** to establish the correlation between future and historical sequences. Specifically, we employ temporal features E_{T+1} (time of a day, day of a week, and holiday) of the target time step $T+1$ as the query to explore its correlation with the spatial-temporal features H from Eq. 3 at each historical time step. The attention score between $T+1$ and each historical time step $t \in \{1, 2, \ldots, T\}$ is computed:

$$\text{att} = \text{Softmax}\left(\text{ReLU}\left(W_H H + W_E E_{T+1} + b\right)\right), \tag{4}$$

where $W_H \in \mathbb{R}^{1 \times D_H}, W_E \in \mathbb{R}^{T \times D_T}$, and $b \in \mathbb{R}^T$ are weight and bias parameters. Then the future representation aggregates the most relevant historical features according to the attention score:

$$H_{T+1} = \sum_{t=1}^{T} \text{att}_t \cdot H_t. \tag{5}$$

Prediction. For risk prediction, H_{T+1} is fed into a two-layer prediction head (PredHead) to generate the final forecasting results: $\hat{Y}_{T+1} = \text{PredHead}(H_{T+1})$.

3.3 Contrastive Learning with Mixup

Mixup as the Augmentation Function. To mitigate the ultra-rareness issue of the risk samples, we ***design a customized mixup strategy*** as the augmentation function to generate sufficient augmented risk samples. Inspired by the periodic temporal pattern that exists in traffic accident data, we generate the augmented samples through the *convex combination* of temporal samples $\mathbf{x}_i$ and $\mathbf{x}_j$ (with the same daily time stamp) from a fixed time interval Δ for both the features and labels as follows:

$$\mathbf{X}_{mix} = \{\lambda \mathbf{x}_i + (1-\lambda)\mathbf{x}_j \mid t_i - t_j = \Delta\}, \tag{6}$$

$$\mathbf{Y}_{mix} = \{\lambda \mathbf{y}_i + (1-\lambda)\mathbf{y}_j \mid t_i - t_j = \Delta\}, \tag{7}$$

where λ is the mixing coefficient to control the impact of two data sources. Δ reflects the periodic patterns, e.g. $\Delta = 1$ denotes daily pattern and $\Delta = 7$ denotes weekly pattern. If either $\mathbf{x}_i$ or $\mathbf{x}_j$ is a risk sample, then the generated sample will be labelled as a risk sample. The mixed data could: (1) inherit the specific pattern from the original accident data, with the mixed label simulating diverse risk values of the future occurrence of potential accidents; (2) maintain the temporal schemes such as hour of day and day of week in the original dataset, which is an important factor in the traffic forecasting task; (3) significantly bridge the ratio gap between the risk and non-risk samples to facilitate the risk representation in the following contrastive learning.

RiskContra. After setting up the effective mixup augmentation function, we devise a novel RiskContra approach that employs the augmented samples to tackle the temporal rareness challenge.

Firstly, we group the generated risk/non-risk samples into two disjoint positive and negative sets according to their mixed risk values. Specifically, the samples with the risk value larger than 0 are combined into the positive set, while those with the risk value equal to 0 are combined into the negative set.

Then, both the positive and negative sets are fed into an encoder $f(\cdot)$ composed of the Multi-kernel networks and the Temporal Transformer, and followed by a projection head $g(\cdot)$ to obtain the latent embedding. Formally, the embeddings of the positive and negative sets are produced as follows:

$$\mathbf{Z}_{pos} = \{g(f(\mathbf{x}_i)) \,|\, \mathbf{x}_i \in \mathbf{X}_{mix}, \mathbf{y}_i > 0\}, \tag{8}$$

$$\mathbf{Z}_{neg} = \{g(f(\mathbf{x}_i)) \,|\, \mathbf{x}_i \in \mathbf{X}_{mix}, \mathbf{y}_i = 0\}. \tag{9}$$

Finally, supervised contrastive learning is conducted in the latent embedding space to encourage the similarity of all positive pairs from risk embeddings while to discourage the similarity of all negative pairs:

$$\mathcal{L}_C = -\sum_{z_i \in \mathbf{Z}_{pos}} \log \frac{\sum_{\mathbf{z}_j \in \mathbf{Z}_{pos}} \exp(\text{sim}(\mathbf{z}_i, \mathbf{z}_j)/\tau)}{\sum_{z_k \in \mathbf{Z}_{pos} \cup \mathbf{Z}_{neg}} \exp(\text{sim}(\mathbf{z}_i, \mathbf{z}_k)/\tau)}, \tag{10}$$

where sim$(\cdot)$ is instantiated by the cosine similarity, and τ is the temperature.

Benefiting from the mixup strategy, more augmented risk samples could be utilized to prevent the contrastive learning from trivially overfitting the non-risk samples, which could significantly facilitate the accident forecasting performance by learning better risk representation.

3.4 Loss Function

At training time, the overall loss includes the accident prediction loss $\mathcal{L}_F$ and the contrastive loss $\mathcal{L}_C$. Specifically, $\mathcal{L}_F$ is computed as the mean square error between prediction results and the ground truth:

$$\mathcal{L}_F = \frac{1}{W}\frac{1}{H}\sum_{w=1}^{W}\sum_{h=1}^{H}\left(Y[w,h] - \hat{Y}[w,h]\right)^2, \tag{11}$$

where Y and $\hat{Y}$ respectively denote the true traffic accident risk and the predicted accident risk accident at target time $T+1$, and W and H are the width and height of the grid map covering all regions. The final loss is a combination of the two losses with a regularization parameter α to control the impact of contrastive learning: $\mathcal{L} = \mathcal{L}_F + \alpha\mathcal{L}_C$.

4 Experiments

Datasets and Evaluation Metrics. We employ two widely-used datasets for traffic accident forecasting, i.e. New York City (NYC) and Chicago [17]. We evaluate the proposed RiskContra by employing three commonly used metrics: Root Mean Square Error (***RMSE***), ***Recall*** and Mean Average Precision (***MAP***).
Baselines. To verify the effectiveness of the proposed RiskContra, we compare with 2 statistical methods: Historical Average (HA) and XGBoost [2], and 7 deep learning methods: MLP, GRU [5], SDCAE [1], ConvLSTM [13], Hetero-ConvLSTM [22], Graph WaveNet [21] and GSNet [17].
Implementation Details: For the Multi-kernel CNNs, the kernel-sizes (k) for point-wise, local, and global convolution module are set to 1×1, 3×3, and 5×5 respectively, which capture spatial correlations of multiple granularities to solve the spatial imbalance issue. The number of network layers L is set to 2.

In the mixup strategy, we set the interval $\Delta = 7$ to represent weekly patterns in accident occurance, weight $\lambda = 0.1$ to generate the augmented risk samples. As to the contrastive learning module, we set the temperature τ to be 0.4 for NYC and 0.8 for Chicago respectively, which can reflect the different accident conditions of the dataset. The weight α of the contrastive loss is set to $1e^{-5}$ and $1e^{-3}$ for NYC and Chicago respectively. The model and hyper-parameters are chosen according to the best performance on the validation set.

4.1 Comparison with the State-of-the-Art

The results are reported on two settings: (1) "**All Hours**" that measures the risk w.r.t all time intervals ranging from 0:00 to 24:00; (2) "**Rush Hours**" that

Table 1. Comparison with the state-of-the-art methods.

Model	NYC						Chicago					
	All Hours			Rush Hours			All Hours			Rush Hours		
	RMSE	Recall (%)	MAP	RMSE	Recall (%)	MAP	RMSE	Recall (%)	MAP	RMSE	Recall (%)	MAP
HA	10.3243	24.42	0.1049	9.4994	26.94	0.1258	14.9581	13.80	0.0572	10.2564	15.89	0.0644
XGBoost	11.0165	23.14	0.1008	10.173	25.22	0.1119	15.6946	12.58	0.0545	10.3685	15.22	0.0614
MLP	8.4289	27.28	0.1196	7.6379	29.51	0.1338	12.5116	17.53	0.0631	8.9500	18.93	0.0748
GRU	8.3375	28.09	0.1228	7.3546	30.76	0.1301	12.6482	17.83	0.0664	9.0421	18.66	0.0758
SDCAE	7.9774	30.81	0.1594	7.2806	31.22	0.1536	11.3382	18.78	0.0753	8.7543	20.58	0.1002
ConvLSTM	7.9505	30.99	0.1526	7.2554	32.61	0.1557	11.1309	18.84	0.0789	8.5254	20.30	0.0925
Hetero-ConvLSTM	7.9731	30.42	0.1454	7.275	31.43	0.1498	11.3033	18.43	0.0716	8.5437	18.93	0.0770
GraphWaveNet	7.7358	31.78	0.1623	7.0958	33.04	0.1647	11.0835	18.95	0.0805	8.4484	20.42	0.0933
GSNet[a]	7.6722	33.41	0.1856	6.8406	34.50	0.1797	10.8229	20.27	0.0976	8.2822	21.26	0.1204
RiskContra	**7.3994**	**34.48**	**0.1974**	**6.7103**	**35.54**	**0.1924**	**10.3784**	**22.30**	**0.1079**	**7.8663**	**23.87**	**0.1373**

[a]Result obtained by running the official code with default parameter setting.

considers the time intervals in 7:00-9:00 and 16:00-19:00, which are supposed to be the high-risk hours in a day.

The overall comparison is presented in Table 1. The proposed RiskContra outperforms all existing state-of-the-art models in terms of all metrics and datasets, demonstrating the better accident forecasting capability. Among all three metrics, RiskContra shows especially superior performance in terms of Recall and MAP. For example, the relative MAP improvement over the second best (i.e. GSNet) is 6.8% for "Rush Hours" on NYC. The relative Recall improvement for "All Hours" and "Rush Hours" on Chicago is 10.0% and 12.3%.

Notably, for all settings, the performance of "Rush Hours" is more superior than that of the "All Hours", which indicates that the higher ratio of risk samples in "Rush Hours" can substantially facilitate the model training. This is consistent with our motivation to augment the risk samples for better contrastive learning.

4.2 Ablation Study and Visualization

Table 2. Ablation study for Multi-kernel CNNs and Contrastive learning.

Model	All Hours			Rush Hours		
	RMSE	Recall (%)	MAP	RMSE	Recall (%)	MAP
CNN+TT	7.6044	33.16	0.1864	6.7569	33.62	0.1704
MKCNN+TT	7.5279	33.76	0.1912	6.7559	34.53	0.1861
CNN+TT+Contra	7.5424	34.02	0.1899	6.7201	34.08	0.1769
RiskContra	**7.3994**	**34.48**	**0.1974**	**6.7103**	**35.54**	**0.1924**

Ablation Study. To evaluate the effectiveness of our devised modules, we first construct a **baseline** by removing both the Multi-kernel CNNs and mixup contrastive modules from RiskContra, dubbed as "CNN+TT" (kernel size $k = 3$, TT stands for Temporal Transformer). Then, we add the Multi-kernel CNNs to the baseline and obtain the "MKCNN+TT" model. Similarly, "CNN+TT+Contra" stands for adding the mixup contrastive module to the baseline. RiskContra is our final model with both modules.

Ablation experiments are conducted on NYC dataset (Table 2). "MKCNN+TT" achieves significant improvements over the baseline "CNN+TT", corroborating the effectiveness of the Multi-kernel networks in capturing the multi-granular spatial correlations. "CNN+TT+Contra" also outperforms the baseline in all settings, which indicates that the augmented risk samples can facilitate the contrastive feature learning to improve accident forecasting performance. By combining all modules, the final model (RiskContra) achieves the best performance, showing the complementary effect of the two devised modules.

Model Variants. To further investigate each component in our RiskContra respectively, we implement various variants to analyse the specific module design.

To study the impact of kernel size in convolution layers, we construct several model variants each with a different fixed kernel size (from $\{1 \times 1, 3 \times 3, 5 \times 5, 7 \times 7\}$), namely $CNN_{1\times1}, \ldots, CNN_{7\times7}$. From Table 3, we can conclude that

Table 3. The impact of kernel size in the convolution layer.

Model	All Hours			Rush Hours		
	RMSE	Recall(%)	MAP	RMSE	Recall(%)	MAP
$CNN_{1\times1}$	7.8586	33.55	0.1866	7.0871	34.67	0.1856
$CNN_{3\times3}$	7.5424	34.02	0.1899	6.7201	34.08	0.1769
$CNN_{5\times5}$	7.5790	33.17	0.1775	6.7975	34.39	0.1736
$CNN_{7\times7}$	7.7378	31.38	0.1668	6.8653	31.95	0.1545
RiskContra	**7.3994**	**34.48**	**0.1974**	**6.7103**	**35.54**	**0.1924**

Table 4. Comparisons of various mixup constractive variants.

Model	Mixup	Contra	Supervise	All Hours			Rush Hours		
				RMSE	Recall(%)	MAP	RMSE	Recall(%)	MAP
A	✗	✓	✓	7.5837	34.24	0.1885	6.8012	34.64	0.1777
B	✓	✗	✓	7.4852	32.87	0.1854	6.7366	33.90	0.1754
C	✓	✓	✗	7.6113	34.00	0.1879	6.7679	34.78	0.1836
$RiskContra_{\Delta 1}$	✓	✓	✓	7.4082	33.87	0.1909	**6.6443**	34.39	0.1809
$RiskContra_{\Delta 7}$	✓	✓	✓	**7.3994**	**34.48**	**0.1974**	6.7103	**35.54**	**0.1924**

no single kernel size can obtain better performance than Multi-kernel CNNs. This verifies the necessity of Multi-kernel CNNs in tackling the spatial imbalance.

To show the rationale of our mixup contrastive framework, we construct several mixup and contrastive variants in Table 4. Model **A** applies contrastive learning on original data from two consecutive weeks without the mixup augmentation. The inferior performance compared with RiskContra implies that the contrastive learning alone can hardly improve the accident forecasting, which is due to the lack of risk samples in contrastive pair construction. Model **B** utilizes the mixup data to compute a prediction loss instead of contrastive loss. The worse results show the advantage of contrastive mechanism over the redundant prediction task. Model **C** applies contrastive learning on the mixed data and the original data in an unsupervised manner without using the mixed label. The degraded performance verifies the importance of the virtual labels created by the mixup strategy to provide prior cues about risk/non-risk samples. Finally, ***RiskContra*** utilizes both the Mixup and Supervised Contrastive modules. To further verify the influence of the periodicity when performing the mixup strategy, we report the settings of both $\Delta = 1$ and $\Delta = 7$, representing daily and weekly periodicity, respectively. The results show that weekly mixup ($\Delta = 7$) achieves better performance compared to the daily counterpart in risk forecasting, which is consistent with the risk occurrence pattern in real life.

Parameter Analysis. We examine the impact of two important hyper-parameters, i.e. the loss weight α and the mixup weight λ. It can be observed from Fig. 3 that our model exhibits the robust performances with parameters varying in a reasonable range.

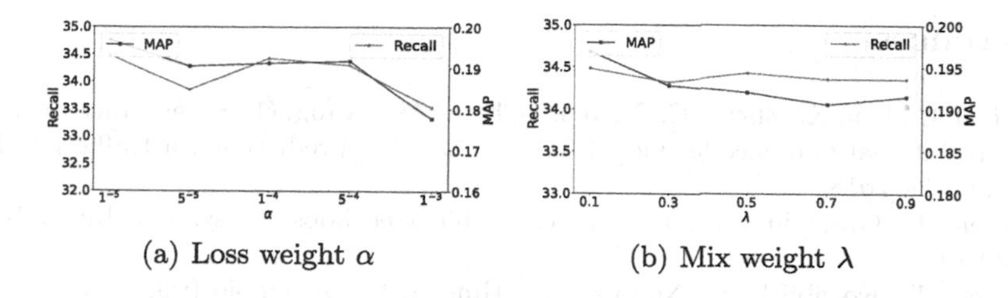

(a) Loss weight α (b) Mix weight λ

Fig. 3. The influence of two important parameters w.r.t two metrics.

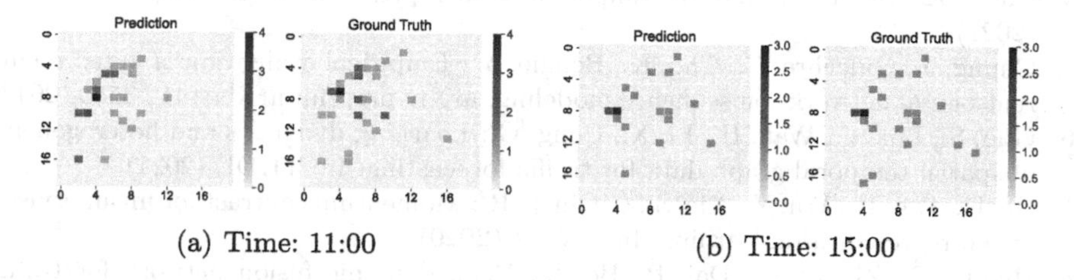

(a) Time: 11:00 (b) Time: 15:00

Fig. 4. The visualization of prediction results compared to the ground truth.

Visualization. To understand the effectiveness of RiskContra from a more intuitive perspective, we conduct a visualization of one-hour prediction results on the NYC dataset at different time. The heatmaps in Fig. 4 demonstrate the risk values at different regions with different colors, which are masked by the high risk region map to reflect more significant prediction results. It can be observed that the visualization results are generally consistent between the predicted traffic accident risks and the ground truth, which indicates the capability of our model to accurately forecast the future traffic risks.

5 Conclusion

This paper proposes to tackle two intrinsic challenges in the traffic accident forecasting task. To deal with the spatial imbalance, we design the Multi-kernel Networks to capture the hierarchical spatial correlations among disparate regions. To address the temporal rareness, we devise a customized mixup strategy to generate sufficient augmented risk samples for effective contrastive learning. All proposed modules are incorporated into a novel contrastive learning approach with extensive experiments corroborating its effectiveness.

Acknowledgement. This work is supported by the Australian Research Council under Grant DP210101347.

References

1. Chen, C., Fan, X., Zheng, C., Xiao, L., Cheng, M., Wang, C.: Sdcae: stack denoising convolutional autoencoder model for accident risk prediction via traffic big data. In: CBD (2018)
2. Chen, T., Guestrin, C.: Xgboost: a scalable tree boosting system. In: SIGKDD (2016)
3. Chen, T., Kornblith, S., Norouzi, M., Hinton, G.: A simple framework for contrastive learning of visual representations. In: ICML (2020)
4. Chen, X., He, K.: Exploring simple siamese representation learning. In: CVPR (2021)
5. Chung, J., Gulcehre, C., Cho, K., Bengio, Y.: Empirical evaluation of gated recurrent neural networks on sequence modeling. arXiv preprint arXiv:1412.3555 (2014)
6. Guo, S., Lin, Y., Wan, H., Li, X., Cong, G.: Learning dynamics and heterogeneity of spatial-temporal graph data for traffic forecasting. In: TKDE (2021)
7. He, K., Fan, H., Wu, Y., Xie, S., Girshick, R.: Momentum contrast for unsupervised visual representation learning. In: CVPR (2020)
8. Huang, C., Zhang, C., Dai, P., Bo, L.: Deep dynamic fusion network for traffic accident forecasting. In: CIKM (2019)
9. Kalantidis, Y., Sariyildiz, M.B., Pion, N., Weinzaepfel, P., Larlus, D.: Hard negative mixing for contrastive learning. In: NIPS (2020)
10. Khosla, P., et al.: Supervised contrastive learning. In: NIPS (2020)
11. Moosavi, S., Samavatian, M.H., Parthasarathy, S., Teodorescu, R., Ramnath, R.: Accident risk prediction based on heterogeneous sparse data: new dataset and insights. In: SIGSPATIAL (2019)
12. Pal, C., Hirayama, S., Narahari, S., Jeyabharath, M., Prakash, G., Kulothungan, V.: An insight of world health organization (who) accident database by cluster analysis with self-organizing map (som). Traffic injury prevention (2018)
13. Shi, X., Chen, Z., Wang, H., Yeung, D.Y., Wong, W.K., Woo, W.C.: Convolutional LSTM network: a machine learning approach for precipitation nowcasting. In: NIPS (2015)
14. Vaswani, A., et al.: Attention is all you need. In: NIPS (2017)
15. Verma, V., et al.: Manifold mixup: better representations by interpolating hidden states. In: ICML (2019)
16. Verma, V., Luong, T., Kawaguchi, K., Pham, H., Le, Q.: Towards domain-agnostic contrastive learning. In: ICML (2021)
17. Wang, B., Lin, Y., Guo, S., Wan, H.: Gsnet: learning spatial-temporal correlations from geographical and semantic aspects for traffic accident risk forecasting. In: AAAI (2021)
18. Wang, S., Zhang, J., Li, J., Miao, H., Cao, J.: Traffic accident risk prediction via multi-view multi-task spatio-temporal networks. In: TKDE (2021)
19. Wang, S., Zhang, M., Miao, H., Peng, Z., Yu, P.S.: Multivariate correlation-aware spatio-temporal graph convolutional networks for multi-scale traffic prediction. In: TIST (2022)
20. Williams, B.M., Hoel, L.A.: Modeling and forecasting vehicular traffic flow as a seasonal arima process: Theoretical basis and empirical results. J. Transp. Eng. **129**(6), 664–672 (2003)
21. Wu, Z., Pan, S., Long, G., Jiang, J., Zhang, C.: Graph wavenet for deep spatial-temporal graph modeling. In: IJCAI (2019)

22. Yuan, Z., Zhou, X., Yang, T.: Hetero-convlstm: a deep learning approach to traffic accident prediction on heterogeneous spatio-temporal data. In: SIGKDD (2018)
23. Zhang, H., Cisse, M., Dauphin, Y.N., Lopez-Paz, D.: mixup: beyond empirical risk minimization. In: ICLR (2018)
24. Zhou, Z.: Attention based stack resnet for citywide traffic accident prediction. In: MDM. IEEE (2019)
25. Zhou, Z., Wang, Y., Xie, X., Chen, L., Liu, H.: Riskoracle: a minute-level citywide traffic accident forecasting framework. In: AAAI (2020)

Petrel: Personalized Trend Line Estimation with Limited Labels from One Individual

Tong-Yi Kuo and Hung-Hsuan Chen(✉)

National Central University, Taoyuan, Taiwan
hhchen@acm.org

Abstract. This study proposes a framework for generating customized trend lines that consider user preferences and input time series shapes. The existing trend estimators fail to capture individual needs and application domain requirements. The proposed framework obtains users' preferred trends by asking users to draw trend lines on sample datasets. The experiments and case studies demonstrate the effectiveness of the model. Code and dataset are available at https://github.com/Anthony860810/Generating-Personalized-Trend-Line-Based-on-Few-Labelings-from-One-Individual.

Keywords: Time series analysis · Trend estimation

1 Introduction

Given a time series vector $\boldsymbol{x} = [x_1, \ldots, x_T]$, a trend line estimation algorithm returns a slowly varying time series $\hat{\boldsymbol{y}} = [\hat{y}_1, \ldots, \hat{y}_T]$ that aims to represent the global pattern of the original time series $\boldsymbol{x}$. However, a trend line has no precise definition: "slowly varying" and "global pattern" are both vague descriptions. As a result, even though many trend estimation algorithms are available [3,5, 10,20,23], none seems to have a dominant advantage over the others.

We believe the ambiguous definition of a trend is inevitable because a proper trend should depend on not only the input time series but also the nature of an application and a user's needs. In other words, even given the same time series, different users or applications may prefer different trends. To demonstrate this, we asked different users to draw trend lines given a fixed time series. The results indeed show that different users illustrate trends with distinct patterns. An example is shown in Fig. 1: given a time series, the left trend is sensitive to local turbulence, and the right trend is smoother. A simple survey shows that 30% of users preferred the left trend and 70% preferred the right one. This result motivates our study, which aims to design a framework to estimate a personalized trend given both a time series and a user's personal requirements or preferences.

Unfortunately, it is extremely challenging for a user to concretely illustrate the characteristic of a trend that meets her/his requirements or preferences. Ultimately, we decided to leverage machine learning algorithms to capture a user's needs based on the user's plotted trend lines on limited time series samples.

© The Author(s), under exclusive license to Springer Nature Switzerland AG 2023
H. Kashima et al. (Eds.): PAKDD 2023, LNAI 13938, pp. 276–288, 2023.
https://doi.org/10.1007/978-3-031-33383-5_22

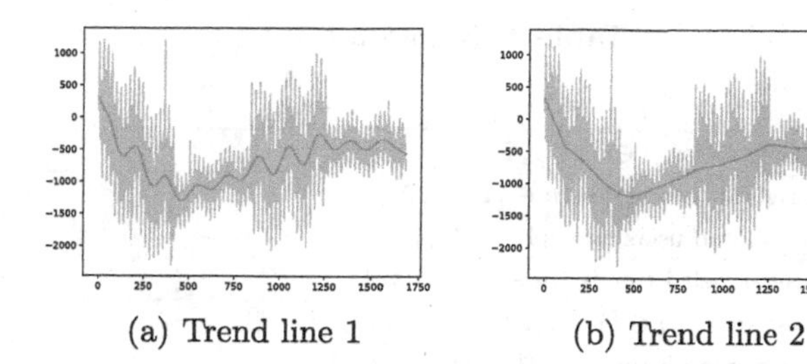

(a) Trend line 1 (b) Trend line 2

Fig. 1. Two distinct trend lines on the same time series dataset. A simple survey shows that 30% of users preferred the left one and 70% of them preferred the right. The results show that different users indeed prefer different types of trend lines.

We hope that once the machine learning algorithms determine a user's needs by observing these examples, the algorithms can automatically generate customized trend lines for a large number of time series that have the same requirements.

We present a scenario that is well-suited for the application of our personalized trend line estimator. Let us consider a situation where a user needs to produce trend lines for a large set of time series such as stock market prices, website log reports, or Internet traffic reports [1,7,13,15]. However, the currently popular trend estimators such as ℓ_1 trend filtering or Hodrick-Prescott filtering are not able to generate satisfactory trends that align with the user's specific requirements as these estimators are not personalized. Using our work, a user only needs to draw trends on a small number of time series samples, and our algorithm will identify the characteristics of the user's preferred trend lines and generate customized trend lines for all other time series.

As we will show later in the experiments, directly learning the relationship between an input time series and the trend line tends toward overfitting given limited training samples labeled by a user. As a result, the main technical challenge of our work becomes designing a framework to learn a user's preferred shapes of trend lines from limited samples.

The main contributions of this paper include the following. First, we propose a new research problem – estimating a customized trend from a time series with personalized or application-specific requirements. We explain why this task is challenging. Second, we propose a personalized trend line estimator named Petrel to address this problem. Petrel, by design, learns a user's requirements even when the user's labeled trend line samples are limited. Third, we conducted thorough experiments to compare our proposed method with supervised algorithms and few-shot learning algorithms, both of which learn to map a time series to a personalized trend line. Additionally, we conducted case studies to demonstrate that Petrel indeed identifies a user's preferences. Finally, we release both the source code and the experimental dataset for reproducibility. The dataset includes the trends plotted by the real users we recruited. The dataset alone could be an invaluable resource for researchers studying trend line estimation.

Table 1. Notation list

Indices:	
T	Length of a time series ($t \in \{1, \dots, T\}$)
N	Number of original time series ($n \in \{1, \dots, N\}$)
U	Number of simulated users ($u \in \{1, \dots, U\}$)
M	Number of user-labeled trends ($m \in \{1, \dots, M\}$)
J	Number of test series ($j \in \{1, \dots, J\}$)
K	Number of base trend estimators ($k \in \{1, \dots, K\}$)
Variables for simulated trend generation and model training:	
b_k	a non-customized base trend generating function that takes a time series as the input argument
$\boldsymbol{w}_u$	the affine coefficients for a simulated user; $\boldsymbol{w}_u = [w_{u,1}, \dots, w_{u,K}]$
$\boldsymbol{x}_n^{orig}$	a collected time series used during training; $\boldsymbol{x}_n^{orig} = [x_{n,1}^{orig}, \dots, x_{n,T}^{orig}]$
$\hat{\boldsymbol{y}}_{n,u}^{sim}$	a simulated trend line for simulated user u on $\boldsymbol{x}_n^{orig}$; $\hat{\boldsymbol{y}}_{n,u}^{sim} = \sum_{k=1}^{K} w_{u,k} b_k(\boldsymbol{x}_n^{orig})$
Variables for inference and verification:	
$\boldsymbol{x}_j^{te}$	a test series; $\boldsymbol{x}_j^{te} = [x_{j,1}^{te}, \dots, x_{j,T}^{te}]$
$\hat{\boldsymbol{y}}_j^{te}$	the estimated personalized trend for $\boldsymbol{x}_j^{te}$; $\hat{\boldsymbol{y}}_j^{te} = [\hat{y}_{j,1}^{te}, \dots, \hat{y}_{j,T}^{te}]$
$\boldsymbol{y}_j^{te}$	the ground truth of the personalized trend for $\boldsymbol{x}_j^{te}$; $\boldsymbol{y}_j^{te} = [y_{j,1}^{te}, \dots, y_{j,T}^{te}]$
$\boldsymbol{x}_m^{lab}$	a time series sample for a user to label with a trend line sample; $\boldsymbol{x}_m^{lab} = [x_{m,1}^{lab}, \dots, x_{m,T}^{lab}]$
$\hat{\boldsymbol{y}}_m^{lab}$	the estimated trend line for $\boldsymbol{x}_m^{lab}$; $\hat{\boldsymbol{y}}_m^{lab} = [\hat{y}_{m,1}^{lab}, \dots, \hat{y}_{m,T}^{lab}]$
$\boldsymbol{y}_m^{lab}$	the trend line sample for $\boldsymbol{x}_m^{lab}$ labeled by the user; $\boldsymbol{y}_m^{lab} = [y_{m,1}^{lab}, \dots, y_{m,T}^{lab}]$

2 Personalized Trend Line Estimation

2.1 Task Overview

We follow the notation list shown in Table 1 in this paper.

A standard trend line estimator outputs $\hat{\boldsymbol{y}} = [\hat{y}_1, \dots, \hat{y}_T]$ given a time series vector $\boldsymbol{x} = [x_1, \dots, x_T]$. In this paper, we call a standard trend line estimator b_k a "non-customized base trend generator" because b_k generates $\hat{\boldsymbol{y}}$ without considering any personalized factors. Famous examples of non-customized estimators include Hodrick-Prescott filtering (HP filtering) [6,12], ℓ_1 trend filtering [10], and seasonal-trend decomposition using LOESS (STL) [4].

In contrast, a personalized trend line estimator takes both $\boldsymbol{x}$ and the user's requirements as the input to generate $\hat{\boldsymbol{y}}$. Since it could be complicated for a user to specify the requirements directly, we ask the user to draw trend lines (called "trend line samples" below) on a small number of time series (called "time series samples" below) and let our model determine the user's preferred or required characteristics of the trend line. Specifically, let $\boldsymbol{y}_m^{lab} = [y_{m,1}^{lab}, \dots, y_{m,T}^{lab}]$ be the trend line sample labeled by a user on a time series sample $\boldsymbol{x}_m^{lab} = [x_{m,1}^{lab}, \dots, x_{m,T}^{lab}]$; we want to learn a customized trend line estimator f that transforms $\boldsymbol{x}_m^{lab}$ into $\boldsymbol{y}_m^{lab}$. Once f is obtained, we can estimate $\hat{\boldsymbol{y}}_j^{te}$ as a personalized trend for the time series $\boldsymbol{x}_j^{te}$ using $f(\boldsymbol{x}_j^{te})$.

2.2 Challenge of the Task

An obvious approach to performing the aforementioned task is to train a supervised learner to map a time series sample $\boldsymbol{x}_m^{lab}$ to a trend line sample $\boldsymbol{y}_m^{lab}$. However, such an approach requires a large number of user-labeled trends. In our scenario, since a user only draws trend line samples for few time series samples, a supervised learning algorithm tends to overfit the training data.

A possible strategy to address the issue of small training data is the pretraining and fine-tuning strategy used in few-shot learning [18,22]. However, as we will show later in the experiments, although few-shot learning performs moderately better than supervised learners, the estimated personalized trends are still unsatisfactory. In summary, it is challenging to learn a personalized trend estimator based on a limited number of $\boldsymbol{x}_m^{lab}$ and $\boldsymbol{y}_m^{lab}$ labeled by a single user.

2.3 Petrel Model – Training

We propose the Petrel model to estimate personalized trends. Instead of directly learning to map $\boldsymbol{x}_m^{lab}$ onto $\boldsymbol{y}_m^{lab}$, Petrel consists of a two-stage training process.

Collecting a large number of time series is simple, but labeling the personalized trends for them is laborious. In stage 1 of the training, we generate simulated personalized trends from a large collection of data series. For a collected time series $\boldsymbol{x}_n^{orig}$, we use K non-customized base trend generators to generate K different trends: $b_1(\boldsymbol{x}_n^{orig}), \ldots, b_K(\boldsymbol{x}_n^{orig})$. Next, we generate a simulated personalized trend line $\hat{\boldsymbol{y}}_{n,u}^{sim}$ for a simulated user u by assuming that $\hat{\boldsymbol{y}}_{n,u}^{sim}$ is composed of an affine combination of $b_k(\boldsymbol{x}_n^{orig})$, as shown by the following equation.

$$\hat{\boldsymbol{y}}_{n,u}^{sim} = \sum_{k=1}^{K} w_{u,k} b_k(\boldsymbol{x}_n^{orig}), \tag{1}$$

where the $w_{u,k}$s are the affine coefficients representing the characteristics of user u's preferred trend. We vary the values of $[w_{u,1}, \ldots, w_{u,K}]$ to simulate different user us so that different simulated personalized trends $\hat{\boldsymbol{y}}_{n,u}^{sim}$ are generated even when $\boldsymbol{x}_n^{orig}$ is fixed. For the non-customized base trend generating functions, we selected three trend estimators: b_1 is ℓ_1 trend filtering [10], b_2 is HP filtering [6], and b_3 is STL estimation [4]. It is straightforward to include other base trend generators, e.g., the ARIMA model [9] or a local regression model.

In stage 2, we train an affine coefficient estimator f_{coef}. Given a time series $\boldsymbol{x}_n^{orig}$ and a simulated personalized trend $\hat{\boldsymbol{y}}_{n,u}^{sim}$ that was generated in stage 1 as the input features, we want the affine coefficient estimator f_{coef} to return the affine coefficients $[w_{u,1}, \ldots, w_{u,K}]$ after training. In our experiment, f_{coef} is composed of 4 layers of 1D convolutions (each with ReLU as the activation function) followed by a fully connected layer (with softmax as the activation function to ensure that the sum of the outputs is 1). The input includes 2 channels – one for the time series $\boldsymbol{x}_n^{orig}$ and the other for the simulated personalized trend $\hat{\boldsymbol{y}}_{n,u}^{sim}$.

The first stage of this training strategy can generate a large number of training instances for the second stage to train the affine coefficient estimator f_{coef}, which plays a crucial role during inference, as described below.

2.4 Petrel Model – Inference

The Petrel inference method also involves two stages. In stage 1, the user is asked to plot M trend line samples (i.e., $\boldsymbol{y}_1^{lab}, \ldots, \boldsymbol{y}_M^{lab}$) on M time series samples (i.e., $\boldsymbol{x}_1^{lab}, \ldots, \boldsymbol{x}_M^{lab}$) for a small value of M. For each pair $\left(\boldsymbol{x}_m^{lab}, \boldsymbol{y}_m^{lab}\right)$, we estimate the affine coefficients to generate the trend $\boldsymbol{y}_m^{lab}$ using Eq. 2.

$$\boldsymbol{w}_m = [w_{m,1}, \ldots, w_{m,K}] = f_{\text{coef}}\left(\boldsymbol{x}_m^{lab}, \boldsymbol{y}_m^{lab}\right) \tag{2}$$

In stage 2, we estimate the personal affine coefficients ($[w_1^*, \ldots, w_K^*]$) for a user and generate the personalized trend $\hat{\boldsymbol{y}}_j^{te}$ for a test series $\boldsymbol{x}_j^{te}$. We use two methods of estimating the personal affine coefficients. The first is a simple average of $\boldsymbol{w}_1$ to $\boldsymbol{w}_M$, as shown in Eq. 3.

$$w_k^* = \frac{1}{M}\sum_{m=1}^{M} w_{m,k} \tag{3}$$

The second way to compute w_k^* is to take a weighted sum of $\boldsymbol{w}_1$ to $\boldsymbol{w}_M$. The weights should be inversely correlated with the distance between $\boldsymbol{y}_m^{lab}$ and $\hat{\boldsymbol{y}}_m^{lab}$ (the estimated trend line for $\boldsymbol{x}_m^{lab}$). Ultimately, we define the distance by the symmetric mean absolute percentage error (SMAPE) (defined in Eq. 7). The personal affine coefficient (estimated by weighted sum) is shown in Eq. 4.

$$w_k^* = \sum_{m=1}^{M} \alpha_m w_{m,k}, \tag{4}$$

where α_m is defined in Eq. 5.

$$\alpha_m = \frac{1 / \text{SMAPE}\left(\boldsymbol{y}_m^{lab}, \hat{\boldsymbol{y}}_m^{lab}\right)}{\sum_{i=1}^{M} 1 / \text{SMAPE}\left(\boldsymbol{y}_i^{lab}, \hat{\boldsymbol{y}}_i^{lab}\right)} \tag{5}$$

Once w_k^* is obtained by either a simple average (Eq. 3) or weighted sum (Eq. 4), we estimate the personalized trend for a test series $\boldsymbol{x}_j^{te}$ by

$$\hat{\boldsymbol{y}}_j^{te} = \sum_{k=1}^{K} w_k^* b_k\left(\boldsymbol{x}_j^{te}\right). \tag{6}$$

3 Experiments

3.1 Experimental Datasets

We used two datasets for the experiments. The first dataset is the Yahoo! S5 time series dataset [17]. We asked users to plot trends on some of these time series. The second dataset includes real users' plotted trends on manually created time series. Some of these user-plotted trends will be regarded as trend line samples (y_m^{lab}), and others will be regarded as the ground truth of the personalized trend lines that will be used for evaluation.

Table 2. A comparison of trend estimation algorithms on the Yahoo! S5 series

Type	Algorithm	SMAPE	MSE
Our method	Petrel (averaged)	**0.44**	5264.34
	Petrel (weighted)	**0.44**	**5258.34**
DNN models	ConvNet	0.83	176593.87
	LSTM	1.02	497312.33
	Transformer	1.08	579188.89
DNN with pretraining and fine-tuning	P&F ConvNet	**0.44**	5425.77
	P&F LSTM	0.52	7394.09
	P&F Transformer	0.47	9311.75
	P&F MLP	0.68	31934.92

Table 3. A comparison of trend estimation algorithms on the manually created series

Type	Algorithm	SMAPE	MSE
Our method	Petrel (averaged)	0.33	6164.38
	Petrel (weighted)	**0.32**	**6002.32**
DNN models	ConvNet	0.94	166951.8
	LSTM	1.11	323712.95
	Transformer	1.20	637955.96
DNN with pretraining and fine-tuning	P&F ConvNet	1.45	241890.91
	P&F LSTM	1.23	1292454.44
	P&F Transformer	0.81	1357013.58
	P&F MLP	1.18	242234.14

3.2 Experimental Scenario

To generate personalized trends for the participants, we asked each participant to draw trend lines on 10 time series samples. These time series samples, along with the trend line samples plotted by users, were used by the algorithms to learn or infer a user's preferences regarding the shapes of the trends.

Next, Petrel and each of the compared baselines generated personalized trends for 5 time series in Yahoo! S5 ($\boldsymbol{x}_n^{orig}$) and another 5 manually created time series ($\boldsymbol{x}_j^{te}$). We asked the users to plot the trends on these 10 time series without showing the machine-generated trends. We compare the distances between a user's plotted trends and the machine-estimated personalized trends.

Although the time series in Yahoo! S5 (i.e., $\boldsymbol{x}_n^{orig}$) are available in the beginning, the ground truths of the corresponding personalized trends are not given. In particular, Petrel uses the simulated personalized trend $\hat{\boldsymbol{y}}_{n,u}^{sim}$ as part of the input feature; the pretraining and fine-tuning models (the baseline models that will be introduced in the next section) use $\hat{\boldsymbol{y}}_{n,u}^{sim}$ as the target during the pretraining step, but the $\hat{\boldsymbol{y}}_{n,u}^{sim}$ values are not the ground truths of the personalized trends. To compare the generalizability of the Petrel model and the baseline models, we also tested each model on the manually generated time series $\boldsymbol{x}_j^{te}$ that differed from any time series in $\boldsymbol{x}_n^{orig}$ or $\boldsymbol{x}_m^{lab}$.

3.3 Baseline Methods

We compare Petrel with deep neural network (DNN) models and DNN models with pretraining and fine-tuning strategies. We do not include traditional trend estimators (e.g., ℓ_1 trend filter or HP filter) because they are non-personalized.

The first type of baseline model (DNNs) includes the convolutional neural network (ConvNet), long short-term memory (LSTM), and Transformer. Each model learns to use each of the 10 time series samples as the input features to predict the corresponding trend line sample drawn by a user. After training, each model estimates the personalized trend line for each of the test series.

The second type of baseline model applies the pretraining and fine-tuning strategy to the ConvNet (denoted as P&F ConvNet), LSTM (denoted as P&F LSTM), Transformer (denoted as P&F Transformer), and multilayer perceptron (denoted as P&F MLP). The network structure of each model is the same as that used in the first type. However, during the pretraining step, we use $\boldsymbol{x}_n^{orig}$, the collected time series, and $\hat{\boldsymbol{y}}_{n,u}^{sim}$, the simulated trend line, as the training feature and target. In the fine-tuning step, we use $\boldsymbol{x}_m^{sam}$, the time series samples, and $\boldsymbol{y}_m^{sam}$, the user-plotted trend line samples, as the features and targets to fine-tune the parameters in the last layer.

3.4 The Quality of the Estimated Personalized Trends

We quantify the quality of a trend line generating algorithm by comparing its generated trends with users' plotted trends based on the symmetric mean absolute percentage error (SMAPE) and mean squared error (MSE). Let $\hat{\boldsymbol{y}} = [\hat{y}_1, \ldots, \hat{y}_T]$ and $\boldsymbol{y} = [y_1, \ldots, y_T]$ be the algorithm-generated and user-plotted trend lines for a time series with T time steps. The SMAPE and MSE between $\hat{\boldsymbol{y}}$ and $\boldsymbol{y}$ are defined by Eq. 7 and Eq. 8, respectively.

$$\mathrm{SMAPE}(\hat{\boldsymbol{y}}, \boldsymbol{y}) = \frac{1}{T}\sum_{t=1}^{T} 2\frac{|\hat{y}_t - y_t|}{|\hat{y}_t| + |y_t|} \tag{7}$$

$$\mathrm{MSE}(\hat{\boldsymbol{y}}, \boldsymbol{y}) = \frac{1}{T}\sum_{t=1}^{T} (\hat{y}_t - y_t)^2 \tag{8}$$

While the MSE is the most widely used metric to measure the difference between the predictions and observed targets when the target variables are numeric, the MSE is scale dependent, i.e., the score depends on the scale of the time series. On the other hand, the SMAPE scales the value by considering the magnitude of each predicted $\hat{y}_t$ and the observed y_t, so the SMAPE score is always between 0 and 2 (a smaller SMAPE means that the prediction is more accurate). Unfortunately, given two estimated trends $\hat{\boldsymbol{y}}_1$ and $\hat{\boldsymbol{y}}_2$,

$\mathrm{SMAPE}(\hat{\boldsymbol{y}}_1, \boldsymbol{y}) > \mathrm{SMAPE}(\hat{\boldsymbol{y}}_2, \boldsymbol{y})$ does not imply $\mathrm{MSE}(\hat{\boldsymbol{y}}_1, \boldsymbol{y}) > \mathrm{MSE}(\hat{\boldsymbol{y}}_2, \boldsymbol{y})$ (and vice versa). For a fair comparison, we report both the SMAPE and MSE.

Table 2 gives the SMAPEs and MSEs of various methods in predicting personalized trends on the Yahoo! S5 dataset. These time series appear in the training step of Petrel and the pretraining steps of all the pretraining and fine-tuning models. However, the ground-truth labels of the personalized trends are not given. Petrel (averaged) and Petrel (weighted) refer to the strategies used to determine the personal affine coefficients w_k^*: either using Eq. 3 (averaged) or Eq. 4 (weighted). The results show that Petrel outperforms all the baseline methods in terms of both SMAPE and MSE on Yahoo! S5. The DNN models perform poorly, likely because of the limited number of training instances from the time series samples and trend line samples. For the pretraining and fine-tuning strategies, although the trends in the pretraining step are synthesized by Eq. 1, these synthetic trends are still helpful in alleviating overfitting.

Table 3 shows the quality of the predicted personalized trends for Petrel and the baseline models on the manually created time series that do not appear during all training (or pretraining) steps. The Petrel model shows a more obvious advantage. Both types of baseline models (DNNs and DNNs with pretraining and fine-tuning) generate trends that are very different from a user's plotted trends. When comparing the results with the results shown in Table 2, the performance of the pretraining and fine-tuning strategies becomes much worse. This is likely because the pretraining and fine-tuning strategy works only when the test series appear in the pretraining step but fails if the test series differ greatly from those appearing in the pretraining step.

3.5 Case Study

This section presents a case study of two users to demonstrate that Petrel indeed identifies the users' preferred trend shapes. We only show the averaged version of Petrel here because the weighted version gives similar results.

Figure 2 shows three trend samples plotted by users A and B: user A prefers the trends to be sensitive to local turbulence, but user B prefers smooth trends.

Figure 3 shows the personalized trends generated by Petrel on three series presented in the Yahoo! S5 dataset (each column represents the same series). The personalized trend for user A appears sensitive to local turbulence, while user B's trend is smoother. This matches our assessment of their preferences.

Figure 4 illustrates the personalized trends generated by Petrel on three manually crated series, i.e., series that do not appear in any training phase. Again, when given the same time series, the estimated personalized trend for user B is smoother than the one generated for user A.

4 Related Work

4.1 Trend Estimation

Trend line estimation is a fundamental problem in time series analysis. It aims to find a slowly varying curve that represents a given sequence well. It has been shown that estimating trend lines improves the performance of time series analysis tasks. For example, removing trends from a time series as a preprocessing step may improve the prediction quality of various target tasks [11,16,24].

Trend estimating algorithms can be parametric or nonparametric. Parametric models aim to find a function that transforms the observed time series into a trend. The simplest model of this form is probably the linear regression model – given a time series $\boldsymbol{x} = [x_1, \ldots, x_T]$, a linear regression model assumes a linear relationship between each x_j $(j \leq t)$ and $\hat{y}_t$ and looks for parameters that minimize a given objective. This concept can be easily extended to high-order polynomial regressions or even more complex functions. However, to ensure the "slowly varying" property that is usually required for a trend line, extremely high-order polynomial regressions or overcomplex functions are rarely used in practice for trend estimation. The famous autoregressive integrated moving average (ARIMA) model [8,9] also falls into the family of parametric models. ARIMA assumes that the difference between neighboring $\hat{y}_t$s is linearly correlated with both their lagged values and the previous error terms.

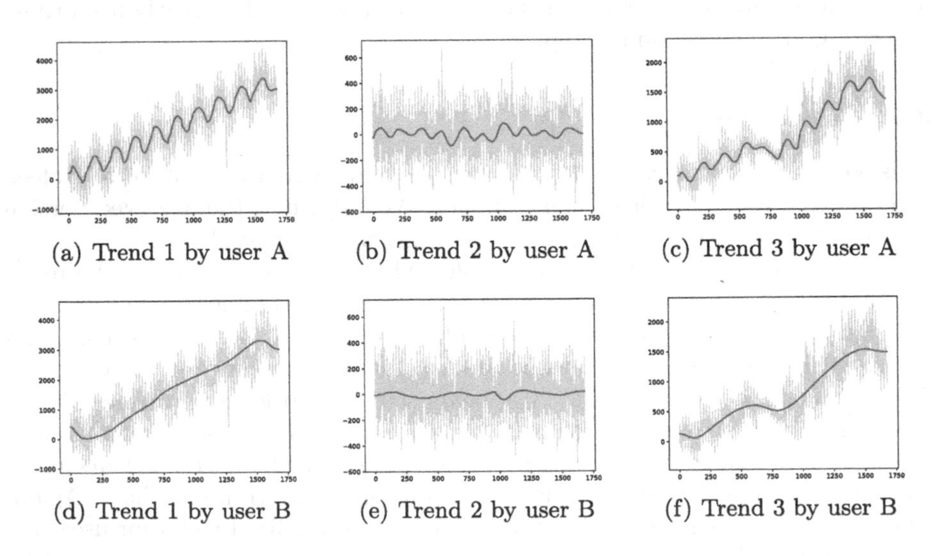

(a) Trend 1 by user A (b) Trend 2 by user A (c) Trend 3 by user A

(d) Trend 1 by user B (e) Trend 2 by user B (f) Trend 3 by user B

Fig. 2. Three trend line samples plotted by user A and user B. It appears that user A prefers the trends to be sensitive to local turbulence, but user B prefers smooth trends.

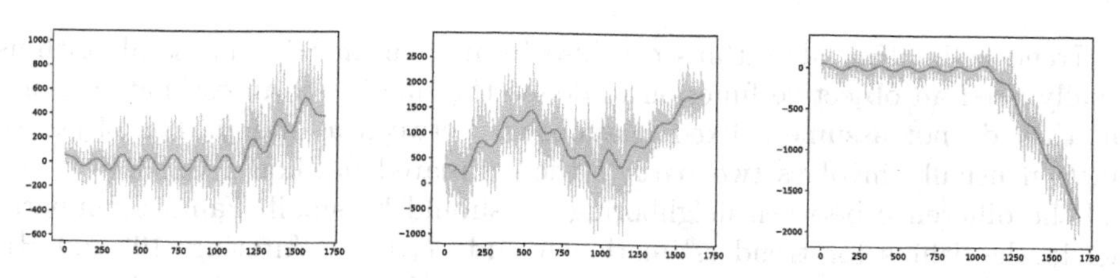

(a) Personalized trend 1 estimated by Petrel for user A (b) Personalized trend 2 estimated by Petrel for user A (c) Personalized trend 3 estimated by Petrel for user A

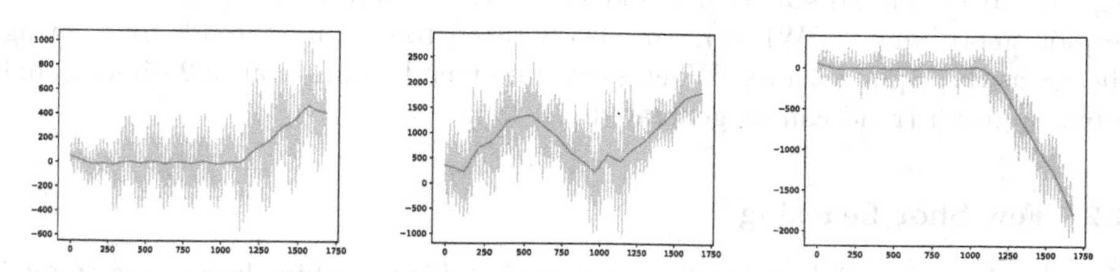

(d) Personalized trend 1 estimated by Petrel for user B (e) Personalized trend 2 estimated by Petrel for user B (f) Personalized trend 3 estimated by Petrel for user B

Fig. 3. The personalized trends for user A and user B generated by Petrel (averaged) on three Yahoo! S5 time series (the time series used to generate the simulated trends).

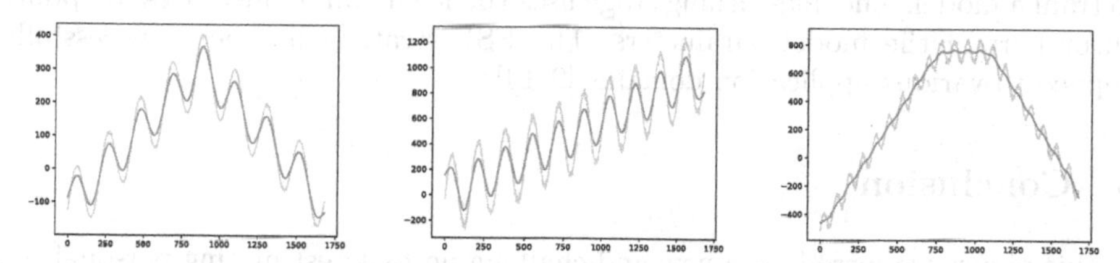

(a) Personalized trend 1 for unseen time series estimated by Petrel for user A (b) Personalized trend 2 for unseen time series estimated by Petrel for user A (c) Personalized trend 3 for unseen time series estimated by Petrel for user A

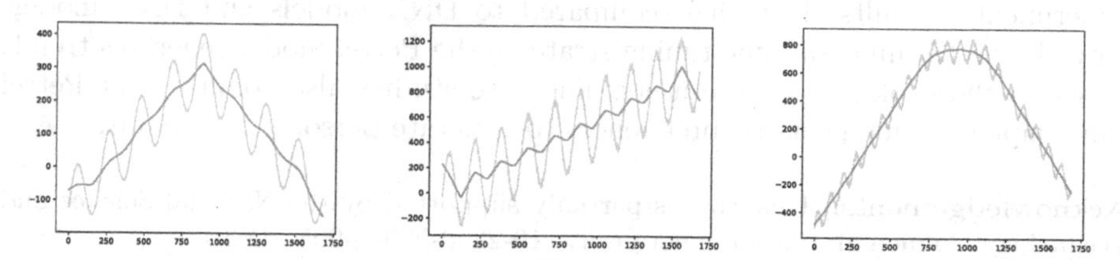

(d) Personalized trend 1 for unseen time series estimated by Petrel for user B (e) Personalized trend 2 for unseen time series estimated by Petrel for user B (f) Personalized trend 3 for unseen time series estimated by Petrel for user B

Fig. 4. The personalized trends for user A and user B generated by Petrel (averaged) on three manually created series (which did not appear during training).

Trend estimating algorithms can also be nonparametric. These algorithms usually need an objective function to define the quality of an estimated trend, but they do not assume a fixed relationship between $\boldsymbol{x}$ and $\hat{\boldsymbol{y}}$. The objective function usually involves two parts – the estimated $\hat{y}_t$ should be close to x_t, and the difference between neighboring $\hat{y}_t$s should be small. Famous nonparametric algorithms for trend estimation include ℓ_1-trend filtering [10] and HP filtering [6]. The main difference between ℓ_1-trend filtering and HP filtering is their definitions of closeness between neighboring $\hat{y}_t$s. Local regression methods, e.g., locally estimated scatterplot smoothing (LOESS) and locally weighted scatterplot smoothing (LOWESS), can also be used to estimate trends by treating the x_ts as the input features. When extending trend estimation to 2-dimensional input, a graph trend can be generated [21,23].

4.2 Few-Shot Learning

Few-shot learning (FSL) refers to a scenario in which machine learning is applied with a limited number of training instances [19,22]. In many cases, FSL tries to embed prior knowledge into the model and update the model based on the few training instances. A typical way to embed prior knowledge is to use a pretraining strategy, which usually leverages a relevant dataset with many training instances to train a model. The fine-turning stage uses the few training instances to update all or part of the model parameters. The FSL strategy has been successfully applied to various application domains [2,14].

5 Conclusion

In this paper, we introduce a new and challenging task: estimating personalized trends for a large number of time series based on limited samples labeled by a user. We propose a novel algorithm named Petrel to perform this task. We recruited users to plot personalized trends on synthetic and open datasets. The experimental results show that, compared to DNN models and DNN models with the pretraining and fine-tuning strategy, the Petrel model generates trends closer to those plotted by the users. The case studies also confirm that Petrel can adapt to different users' preferences to generate personalized trends.

Acknowledgements. This work is partially supported by the National Science and Technology Council of Taiwan under grant 110-2222-E-008-005-MY3.

References

1. Bai, G.J., Lien, C.Y., Chen, H.H.: Co-learning multiple browsing tendencies of a user by matrix factorization-based multitask learning. In: IEEE/WIC/ACM International Conference on Web Intelligence, pp. 253–257 (2019)

2. Bansal, T., Jha, R., McCallum, A.: Learning to few-shot learn across diverse natural language classification tasks. In: Proceedings of the 28th International Conference on Computational Linguistics, pp. 5108–5123 (2020)
3. Bianchi, M., Boyle, M., Hollingsworth, D.: A comparison of methods for trend estimation. Appl. Econ. Lett. **6**(2), 103–109 (1999)
4. Cleveland, R.B., Cleveland, W.S., McRae, J.E., Terpenning, I.: STL: a seasonal-trend decomposition. J. Off. Stat. **6**(1), 3–73 (1990)
5. Gray, K.L.: Comparison of Trend Detection Methods. University of Montana, Missoula (2007)
6. Hodrick, R.J., Prescott, E.C.: Postwar us business cycles: an empirical investigation. J. Money Credit Bank. 1–16 (1997)
7. Hsu, C.Y., Chen, T.R., Chen, H.H.: Experience: analyzing missing web page visits and unintentional web page visits from the client-side web logs. ACM J. Data Inf. Qual. (JDIQ) **14**(2), 1–17 (2022)
8. Hsu, C.J., Chen, H.H.: Taxi demand prediction based on LSTM with residuals and multi-head attention. In: VEHITS, pp. 268–275 (2020)
9. Hyndman, R.J., Athanasopoulos, G.: Forecasting: Principles and Practice. OTexts, Melbourne (2018)
10. Kim, S.J., Koh, K., Boyd, S., Gorinevsky, D.: ℓ_1 trend filtering. SIAM Rev. **51**(2), 339–360 (2009)
11. Laptev, N., Yosinski, J., Li, L.E., Smyl, S.: Time-series extreme event forecasting with neural networks at Uber. In: International Conference on Machine Learning. vol. 34, pp. 1–5 (2017)
12. Leser, C.E.V.: A simple method of trend construction. J. R. Stat. Soc.: Ser. B (Methodological) **23**(1), 91–107 (1961)
13. Lien, C.Y., Bai, G.J., Chen, H.H.: Visited websites may reveal users' demographic information and personality. In: IEEE/WIC/ACM International Conference on Web Intelligence, pp. 248–252 (2019)
14. Lifchitz, Y., Avrithis, Y., Picard, S., Bursuc, A.: Dense classification and implanting for few-shot learning. In: Proceedings of the IEEE/CVF Conference on Computer Vision and Pattern Recognition, pp. 9258–9267 (2019)
15. Lin, T.H., Zhang, X.R., Chen, C.P., Chen, J.H., Chen, H.H.: Learning to identify malfunctioning sensors in a large-scale sensor network. IEEE Sens. J. **22**(3), 2582–2590 (2021)
16. Liu, Z., Hauskrecht, M.: Learning adaptive forecasting models from irregularly sampled multivariate clinical data. In: Proceedings of the Thirtieth AAAI Conference on Artificial Intelligence, pp. 1273–1279. AAAI 2016. AAAI Press (2016)
17. Marlin, B.M., Zemel, R.S.: Collaborative prediction and ranking with non-random missing data. In: Proceedings of the Third ACM Conference on Recommender Systems, pp. 5–12 (2009)
18. Shen, Z., Liu, Z., Qin, J., Savvides, M., Cheng, K.: Partial is better than all: revisiting fine-tuning strategy for few-shot learning. In: Thirty-Fifth AAAI Conference on Artificial Intelligence, AAAI 2021, pp. 9594–9602. AAAI Press (2021)
19. Tao, R., Zhang, H., Zheng, Y., Savvides, M.: Powering finetuning in few-shot learning: domain-agnostic bias reduction with selected sampling. In: Thirty-Sixth Conference on Artificial Intelligence, AAAI 2022, pp. 8467–8475. AAAI Press (2022)
20. Tibshirani, R.J.: Adaptive piecewise polynomial estimation via trend filtering. Ann. Stat. **42**(1), 285–323 (2014)
21. Varma, R., Lee, H., Kovačević, J., Chi, Y.: Vector-valued graph trend filtering with non-convex penalties. IEEE Trans. Signal Inf. Process. Over Netw. **6**, 48–62 (2019)

22. Wang, Y., Yao, Q., Kwok, J.T., Ni, L.M.: Generalizing from a few examples: a survey on few-shot learning. ACM Comput. Surv. **53**(3), 1–34 (2020)
23. Wang, Y.X., Sharpnack, J., Smola, A., Tibshirani, R.: Trend filtering on graphs. In: Artificial Intelligence and Statistics, pp. 1042–1050. PMLR (2015)
24. Zhang, G.P., Qi, M.: Neural network forecasting for seasonal and trend time series. Eur. J. Oper. Res. **160**(2), 501–514 (2005)

A Global View-Guided Autoregressive Residual Network for Irregular Time Series Classification

Jianping Zhu[1], Haocheng Tang[2], Liang Zhang[3], Bo Jin[1], Yi Xu[2(✉)], and Xiaopeng Wei[1]

[1] Dalian University of Technology, Dalian, China
zhujp@mail.dlut.edu.cn, {jinbo,xpwei}@dlut.edu.cn

[2] Institute of Automation, Chinese Academy of Sciences, Beijing, China
{haocheng.tang,yi.xu}@ia.ac.cn

[3] Dongbei University of Finance and Economics, Dalian, China
liang.zhang@dufe.edu.cn

Abstract. Irregularly sampled multivariate time series classification tasks become prevalent due to widespread application of sensors. However, different collection frequencies or sensor failures presents nontrivial challenges since mainstream methods generally assume aligned measurements across sensors (variables). Besides, most existing studies fail to account for the relationship between misaligned patterns and classification tasks. To this end, we propose a Global view-guided Autoregressive Residual Network (GARNet), which mainly adopts a generation-and-sampling strategy to deal with the partially observed data at each timestamp. Specifically, we first leverage a Structure-augmented Global Information Extractor (SGIE) to capture the global semantic information in the whole conditioning window. Then, a Global view-guided Autoregressive Recurrent Neural Network (GARNN) is developed to capture the local temporal dynamics hidden in latent factors. Finally, a Masked Temporal Information Aggregator (MTIA) is proposed to attentively aggregate the extracted latent factors at each timestamp for the classification task. Experimental results on two real-world datasets show that GARNet outperforms state-of-the-art methods.

Keywords: Irregularly Sampled Multivariate Time Series · Graph Neural Network · Autoregressive Residual Network

1 Introduction

Irregularly sampled multivariate time series (MTS) are ubiquitous in various information systems and sensor-rich applications such as healthcare management, biological monitoring, etc. It is characterized by non-uniform time intervals between successive measurements [11]. For example, for a dialysis patient, venous pressure, glucose level, and cardiothoracic ratio (CTR) are collected at different frequencies. The misaligned data poses a great challenge for subsequent data mining tasks [19] since most existing MTS models typically assume well-aligned and fixed-size inputs.

© The Author(s), under exclusive license to Springer Nature Switzerland AG 2023
H. Kashima et al. (Eds.): PAKDD 2023, LNAI 13938, pp. 289–300, 2023.
https://doi.org/10.1007/978-3-031-33383-5_23

Current approaches generally adopt a missing value perspective by first aligning the data with missing values (via interpolation or imputation) and then building prediction models [11]. Traditional methods include interpolation [9], spectral analysis [13], and kernel methods [14]. Modern techniques, such as Recurrent Neural Networks [4], and Transformers [19], have also been widely used. As mentioned in [4], the separate two-step (imputation and prediction) learning paradigm overlooks the relation between missing patterns and prediction tasks, and can easily introduce errors when dealing with severe missing data, leading to suboptimal results. Hence, end-to-end learning is becoming the trend [3,4]. However, on the one hand, it is difficult to model temporal dynamics due to the lack of data at some consecutive timestamps; on the other hand, complex inter-variable relations [29] and sample-level relational structures [25] increase the learning difficulties.

To address the above issues, we propose a Global view-guided Autoregressive Residual Network (GARNet), which is composed of a Structure-augmented Global Information Extractor (SGIE), an Autoregressive Recurrent Neural Network (GARNN), and a Masked Temporal Information Aggregator (MTIA). First, inspired by the way how global information guides the generation process of multivariate time series data [8,22,24] and how the across-sample knowledge is accumulated [20], SGIE is devised to automatically learn the inter-variable relation structure, for global information extraction. Second, the global representation is utilized in GARNN, which helps capture the local temporal dynamics via a recurrent generation-and-sampling strategy. In each autoregressive recurrent neural network layer, GARNN adds a global information-aware forgetting gate, generating an informative latent factor at each timestamp via interacting with the global information. Finally, MTIA is designed to learn the mapping between latent factors and the classification task.

To summarize, our work has the following contributions:

- We propose an end-to-end classification method GARNet for irregularly sampled multivariate time series. It mainly adopts a generation-and-sampling strategy to deal with the misaligned values at each timestamp.
- We design an autoregressive residual network, which updates the latent feature vectors via an interaction with a sample-specific global vector at each timestamp while effectively avoiding the vanishing gradient problem caused by the stack of recurrent neural network layers.
- We demonstrate the effectiveness of our model by comparing with several state-of-the-art methods on real-world datasets. GARNet outperforms all baselines. With the increase of missing data or variables, GARNet still achieves the best results and is relatively stable.

2 Related Work

The main characteristic of irregularly sampled time series is the non-uniform time intervals between adjacent observations [6,21,28]. A more common approach to handling such problems is imputing missing values, which includes statistical methods and deep learning methods [11]. Some statistical models include

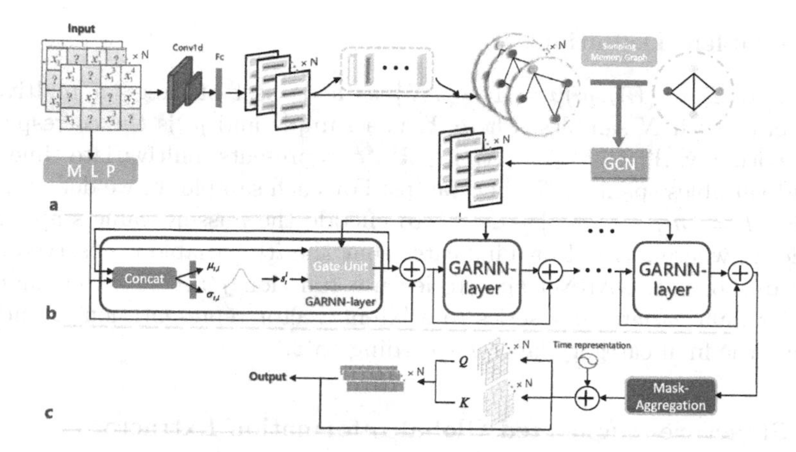

Fig. 1. General overview of the GARNet model framework.

ARIMA [2], ARFIMA [7], and SARIMA [7], which eliminate the non-stationary part of the time series by filling in missing locations with either the mean or last observed value [1]. In addition, some deep learning models were used to impute missing values, including recurrent neural networks (RNN) [4] and deep generative models [12]. Among GAN-based generative models, TSGAN [12] was a two-stage approach that optimized the generator's input vectors by learning the time series distribution. However, as mentioned in [4], two-step learning methods can often distort the underlying distribution of time series data, leading to suboptimal performance.

Hence, state-of-the-art approaches often adopt an end-to-end modeling strategy [5]. For example, Che et al. [4] proposed a variant of the gated recursive unit (GRU-D), which utilized the decay of a binary mask to capture past features. mTAND [19] used a multi-temporal attention mechanism to learn event correlations from irregularly sampled time series to obtain continuous embeddings. However, Wu et al. [25] pointed out that existing methods failed to model the across-sample relational structures, which can help improve the prediction performance from a global view. In contrast, GARNet can simultaneously construct complex variable relationships and correlations between samples. Specifically, we first adopt the ideas from GTS [17] which learns a probabilistic graph model by parameterizing its distribution using neural networks, combined with reparameterization for differential sampling, and finally learns an inter-variable graph structure for each sample. Then, related graph neural networks [10,26,30] can be employed to learn a global representation for the whole conditioning window. In addition, recent works in leveraging relational structures [20] motivates us to utilize a memory graph to capture the shared patterns in the classification task.

3 Methodology

This section introduces the problem definition and describes our proposed GARNet in detail. The overall structure of GARNet is shown in Fig. 1.

3.1 Problem Definition

We denote $\mathcal{D} = \{(\mathcal{X}_i, y_i)|i = 1, \cdots, N\}$ as a dataset of irregular multivariate time series with N samples, where $\mathcal{X}_i$ is a sample and y_i is the corresponding label. Formally, $\mathcal{X} = \{X_1, \cdots, X_C\} \in \mathbb{R}^{C\times T}$ represents multivariate time series with C variables spanning T timestamps. For each sample $\mathcal{X}$, we define a mask matrix $M = \{m_1, \cdots, m_C\} \in \mathbb{R}^{C\times T}$ to encode the missing value states of the sample $\mathcal{X}$, where m_{ij} is 1 or 0 indicates that the ith variable is observed at jth timestamp or not. GARNet aims to learn a function $f : \mathcal{X} \to z$ that maps the irregular multivariate time series to a latent vector representation z, and then predicts the final category label $\hat{y}$ according to z.

3.2 Structure-Augmented Global Information Extractor

Inspired from previous works [8,22,24] that global information can guide the generation process of complex temporal dynamics, in this paper, we propose a Structure-augmented Global Information Extractor (SGIE), which first extracts the complex structural relationships between variables within a sample, and then employs a graph neural network to learn global representations.

To efficiently extract complex inter-variable relations and across-sample relational structures, we first learn the adjacency matrix $A \in \mathbb{R}^{(C+K)\times(C+K)}$ via GTS method [17]. Specifically, we use a feature extractor to convolve along the time dimension, add a linear layer to reduce dimensionality, and perform initial feature extraction on the time series data X^c for each variable c, obtaining ε^c:

$$\varepsilon^c = W_c(vec(Conv(X^c))) + b_c, \tag{1}$$

where W_c and b_c are trainable parameters. $e = \{\varepsilon, \rho\} \in \mathbb{R}^{(C+K)\times Fe}$ denotes the set of vectors ε and the across-sample shared memory vectors $\rho \in \mathbb{R}^{K\times Fe}$. We connect the feature vectors e^i and e^j, and input them into a link predictor:

$$\theta_{ij} = Sigmoid(W_\theta[e^i||e^j] + b_\theta), \tag{2}$$

where W_θ and b_θ are trainable parameters, and $\theta_{ij} \in [0,1]$ denotes the probability that node i is connected to node j. The adjacency matrix is finally generated as:

$$A_{ij} = Sigmoid((\log(\theta_{ij}/(1-\theta_{ij})) + (\eta^1_{ij} - \eta^2_{ij}))/\tau), \tag{3}$$

where $\eta^1_{ij}, \eta^2_{ij} \sim Gumbel(0,1)$ for all i, j. When the temperature τ approaches 0, θ_{ij} is equivalent to the probability of $A_{ij} = 1$, $1 - \theta_{ij}$ is the probability of $A_{ij} = 0$, and τ gradually becomes close to 0 during the training process.

Finally, we use a graph neural network to update the global information. It is enriched by aggregating the information of other variables connected with the memory vector. The information propagation is performed based on the adjacency matrix A learned via the aforementioned neural network, which also allows each global vector to retain a certain amount of spatial information.

3.3 Global View-Guided Autoregressive Recurrent Neural Network

After obtaining structure-augmented global semantic vectors, we further propose a Global view-guided Autoregressive Recurrent Neural Network (GARNN). It uses the global vectors to help capture local temporal dynamics via a recursive generation and sampling process, and leverages a residual layer to transfer information between layers, thus improving the informativeness of latent factors.

In the case of missing data, traditional GRU model requires the hidden output vectors of the previous timestamp and the current timestamp as input. However, it is impossible to access to the data at each timestamp in temporal data with a large amount of missing information. Here we propose GARNN to construct features for each timestamp via recurrent autoregressions. Compared to traditional GRU, in the GARNN gating unit, we add the global information-aware forgetting gate, which helps learn an informative latent factor at each timestamp via interaction with the global vector. To directly model irregularly sampled time series, we introduce cascaded fusion sampling to generate the current timestamp's input. Specifically, we predict the possible distribution of the present timestamp data x_t based on the hidden output vector f_{t-1}^l from previous timestamp and current timestamp feature vector β_t^{l-1} in the last layer, from which the vector s_t^l is sampled as the input for the lth layer at tth timestamp:

$$\begin{aligned}
\mu_{t,l} &= {w_\mu^l}^\top [f_{t-1}^l || \beta_t^{l-1}] + b_\mu^l, \\
\sigma_{t,l} &= \log(1 + \exp({w_\sigma^l}^\top [f_{t-1}^l || \beta_t^{l-1}] + b_\sigma^l)), \\
s_t^l &\sim N(\mu_{t,l}, \sigma_{t,l}^2),
\end{aligned} \tag{4}$$

where w_μ^l, w_σ^l, b_μ^l, and b_σ^l are trainable parameters, the mean value μ is given by the affine function of the network output, and the standard deviation σ is obtained by applying the affine transformation and soft additive activation to ensure that $\sigma > 0$. The vector s_t^l is obtained by multiple sampling of the distribution, which ensures the diversity of information while minimizing the possible bias due to sampling.

The sampled s_t^l is taken as input along with the hidden output vector f_{t-1}^l at the last timestamp and the global vector g of this variable. Here the global vector g contains not only the variable's features but also implies the structured features of that sample. The global information-aware forget gate is used to introduce the global vector with structural features as a guide for the feature vector output at each timestamp:

$$\begin{aligned}
r_t^l &= Sigmoid(W_r^l \cdot [f_{t-1}^l, g] + b_r^l), \\
o_t^l &= Sigmoid(W_o^l \cdot [f_{t-1}^l, s_t^l] + b_o^l), \\
v_t^l &= Sigmoid(W_v^l \cdot [f_{t-1}^l, s_t^l] + b_v^l), \\
\tilde{f}_t^l &= tanh(W^l \cdot [f_{t-1}^l, r_t^l * g, v_t^l * s_t^l] + b^l), \\
f_t^l &= (1 - o_t^l) * f_{t-1}^l + o_t^l * \tilde{f}_t^l,
\end{aligned} \tag{5}$$

where W_r^l, W_o^l, W_v^l, W^l, b_r^l, b_o^l, b_v^l, and b^l are trainable parameters. r_t^l is used as a global information-aware forget gate to control how much global information is retained at t timestamps. The reset gate v_t^l and the forgetting gate o_t^l function similarly to the traditional GRU. Finally, we connect each GARNN layer through a residual structure:

$$\beta_t^l = \beta_t^{l-1} + f_t^l. \tag{6}$$

We use original data x_t as the initial input β_t^0 after a multilayer perceptron operation, i.e., $\beta_t^0 = MLP(x_t)$. The vector β is updated iteratively through multiple layers to make it more effective in representing the feature vectors of the missing timestamps. Here we introduce a Gaussian likelihood to further constrain the correlation between x_t with true values and its predictive distribution $N(\mu_{t,l}, \sigma_{t,l}^2)$. Since the generation of the predictive distribution possesses temporal continuity, the overall performance is also optimal if we add supervision to x_t with known true values. Thus, the sampled s_t^l can be constrained to be closer to the true x_t while ensuring the information diversity. The model can be learned by maximizing the log-likelihood:

$$\begin{aligned} &s_t^l \sim N(\mu_{t,l}, \sigma_{t,l}^2), x_t \sim N(\mu_{t,l}, \sigma_{t,l}^2), \\ &p(s_t^l|\mu_{t,l}, \sigma_{t,l}) = (2\pi\sigma_{t,l}^2)^{-\frac{1}{2}} \exp(\frac{-(x_t - \mu_{t,l})^2}{2\sigma_{t,l}^2}), \\ &\mathcal{L}_{GL} = -\sum_{i=1}^{N}\sum_{l=1}^{L}\sum_{t=1}^{T} m_{i,t} * \log p(s_{i,t}^l|\mu_{i,t,l}, \sigma_{i,t,l}), \end{aligned} \tag{7}$$

where $m_{i,t}$ determines whether the ith sample at tth timestamp has true value $x_{i,t}$, and we sum only the Gaussian likelihood with the true value.

3.4 Masked Temporal Information Aggregator

After learning the hidden vectors at each timestamp, we further propose a Masked Temporal Information Aggregator (MTIA). It uses a self-attention layer with the masked matrix to selectively aggregate the latent factors of each timestamp to obtain the final prediction vector z.

Based on the masked matrix, we simultaneously aggregate variables with true observations and normalize them as:

$$h_t = \frac{\sum_{c=1}^{C} m_t^c * \beta_{t,c}}{\sum_{c=1}^{C} m_t^c}. \tag{8}$$

It is worth noting that we preprocess the masked matrix so that if no observations are available for all variables at tth timestamp, we let $h_t = \frac{1}{C}\sum_{c=1}^{C} \beta_{t,c}$. Since the contribution of different timestamps to the final prediction results is different, we leverage an attention mechanism. The vectors $H = \{h_1, \cdots, h_T\}$ of every timestamp is concatenated with the position codes $P = \{p_1, \cdots, p_T\}$, and mapped into query vectors, key vectors, and value vectors respectively after linear layers, i.e., $K = [H||P]W_K$, $Q = [H||P]W_Q$, $V = [H||P]W_V$. Matrices Q and K are used to calculate the attention weights of vectors at each timestamp, selectively aggregating information from matrix V:

$$Z = softmax(\frac{QK^\top}{\sqrt{d}})V, \tag{9}$$

where d is the scaling factor. Finally, the aggregation of the time vectors is realized by using an aggregation function:

$$z = Agg(z_t|t = 1, \cdots, T), \tag{10}$$

where Average pooling is used for aggregation in this paper. Ultimately, the learned representation z will be utilized to perform downstream tasks.

3.5 Loss Function

The final loss function can be expressed as $\mathcal{L} = \mathcal{L}_{CE} + \lambda\mathcal{L}_{GL}$, where $\mathcal{L}_{CE}$ is cross entropy loss:

$$\mathcal{L}_{CE} = -\sum_{i=1}^{N} y_i \cdot \log(\hat{y}_i) + (1 - y_i) \cdot \log(1 - \hat{y}_i), \tag{11}$$

where $\hat{y}_i$ is the prediction result of the ith sample. $\mathcal{L}_{GL}$ constrains the predictive distribution of each layer and λ is a hyperparameter to control the importance of $\mathcal{L}_{GL}$.

4 Experiments

4.1 Datasets

Here we briefly describe the healthcare and human activity datasets used in this paper: (1) P19 [16] includes 38803 patients, who were monitored by 34 irregularly sampled sensors. The dataset presents a dichotomous task representing whether sepsis occurs in the next 6 h. (2) PAM [15] contains 5333 segments (samples) of sensory signals from 9 subjects. Each sample was measured by 17 sensors and contained 600 consecutive observations. Sixty percent of the observations were randomly removed to ensure the irregularity of the data. The dataset classifies human activity into eight categories.

4.2 Baselines

We compare GARNet with eight state-of-the-art baselines: Transformer [23], Trans-mean [29], GRU-D [4], SeFT [3], mTAND [19], IP-Net [18], DGM^2-O [25], and MTGNN [27]. Among them, Trans-mean is a typical separate fill-and-predict method that first utilizes average interpolation followed by the combination of the Transformer model, all the others are end-to-end learning methods.

4.3 Implementation Details

Based on testing performance on validation dataset, we summarize final configuration of hyperparameters. We use ADAM optimizer with a learning rate of 0.0001 and batch size is set to 64. 50 shared memory vectors are utilized to ensure the richness of across-sample shared information. In addition, the number of sampling operation of s_t is set to 20, and the number of GARNN-layers is set to 3. Finally, considering the extra static information of the P19 dataset, we exploited various information fusion methods, such as gated fusion, concatenation, and summation. Experimental results show little difference. The parameter λ used to control $\mathcal{L}_{GL}$ is set to 0.1 in our experiments.

4.4 Main Results

We split the data into 80%, 10%, 10% for training, validation and testing. The final results are shown in Table 1. GARNet achieves the best performance on both P19 and PAM datasets. The indices of these splits are fixed across all methods. In the binary classification task of the P19 dataset, GARNet outperforms the strongest baseline by 1.7% on AUROC and 14.3% on AUPRC. In the 8-way classification task for the PAM dataset, our model achieves at least 2.7% improvement on Precision value and 1.8% improvement on $F1$ score compared to the strongest baseline.

Table 1. Irregularly sampled time series classification results.

Methods	P19		PAM			
	AUROC	AUPRC	Accuracy	Precision	Recall	F1 score
Transformer	83.2 ± 1.3	47.6 ± 3.8	83.5 ± 1.5	84.8 ± 1.5	86.0 ± 1.2	85.0 ± 1.3
Trans-mean	84.1 ± 1.7	47.4 ± 1.4	83.7 ± 2.3	84.9 ± 2.6	86.4 ± 2.1	85.1 ± 2.4
GRU-D	83.9 ±1.7	46.9 ± 2.1	83.3 ± 1.6	84.6 ± 1.2	85.2 ± 1.6	84.8 ± 1.2
SeFT	78.7 ± 2.4	31.1 ± 2.8	67.1 ± 2.2	70.0 ± 2.4	68.2 ± 1.5	68.5 ± 1.8
mTAND	80.4 ± 1.3	32.4 ± 1.8	74.6 ± 4.3	74.3 ± 4.0	79.5 ± 2.8	76.8 ± 3.4
IP-Net	84.6 ± 1.3	38.1 ± 3.7	74.3 ± 3.8	75.6 ± 2.1	77.9 ± 2.2	76.6 ± 2.8
DGM^2-O	86.7 ± 3.4	44.7 ± 11.7	82.4 ± 2.3	85.2 ± 1.2	83.9 ± 2.3	84.3 ± 1.8
MTGNN	81.9 ± 6.2	39.9 ± 8.9	83.4 ± 1.9	85.2 ± 1.7	86.1 ± 1.9	85.9 ± 2.4
GARNet	**88.2 ± 1.9**	**54.4 ± 4.2**	**84.8 ± 2.1**	**87.2 ± 1.7**	**86.5 ± 2.2**	**86.5 ± 1.8**

Prediction Distribution Accuracy Verification. In irregularly sampled temporal data, each autoregressive recurrent neural network layer predicts the distribution and then conducts sampling from the distribution to obtain the input of the gating unit in GARNN-layers at each timestamp. To verify the fit of the predicted distributions to the original data, we predicted the distribution of the test set data via using the well-trained model with three GARNN-layers, and then we visualized the predicted distribution and true values. Figure 2 shows that GARNet can fit the original data well by predicting the distribution, and the fit increases gradually with successive residual concatenations.

Ablation Studies. To further evaluate the effectiveness of the components in the model, we conducted ablation studies on the PAM dataset, and the results are shown in Fig. 3. We separately removed the memory vectors to test the effect of using only the variables in the adjacency relationship. Removing the SGIE and changing the gating unit in the GARNN-layers to a normal GRU verify the necessity of global guidance. The Gaussian likelihood constraint on the predictive distribution is removed to prove the importance of the constraints. As the results show, all model components are necessary.

Table 2. Classification performance of samples with random omitted sensors on the PAM dataset.

Missing sensor ratio	Methods	PAM (leave-random-sensors-out)			
		Accuracy	Precision	Recall	F1 score
10%	Transformer	60.9 ± 12.8	58.4 ± 18.4	59.1 ± 16.2	56.9 ± 18.9
	Trans-mean	62.4 ± 3.5	59.6 ± 7.2	63.7 ± 8.1	62.7 ± 6.4
	GRU-D	68.4 ± 3.7	74.2 ± 3.0	70.8 ± 4.2	72.0 ± 3.7
	SeFT	40.0 ± 1.9	40.8 ± 3.2	41.0 ± 0.7	39.9 ± 1.5
	mTAND	53.4 ± 2.0	54.8 ± 2.7	57.0 ± 1.9	55.9 ± 2.2
	GARNet	**79.0 ± 1.7**	**80.2 ± 1.4**	**80.5 ± 2.0**	**79.8 ± 2.4**
20%	Transformer	62.3 ± 11.5	65.9 ± 12.7	61.4 ± 13.9	61.8 ± 15.6
	Trans-mean	56.8 ± 4.1	59.4 ± 3.4	53.2 ± 3.9	55.3 ± 3.5
	GRU-D	64.8 ± 0.4	69.8 ± 0.8	65.8 ± 0.5	67.2 ± 0.0
	SeFT	34.2 ± 2.8	34.9 ± 5.2	34.6 ± 2.1	33.3 ± 2.7
	mTAND	45.6 ± 1.6	49.2 ± 2.1	49.0 ± 1.6	49.0 ± 1.0
	GARNet	**72.8 ± 3.8**	**73.8 ± 1.8**	**72.9 ± 2.7**	**72.4 ± 3.8**
30%	Transformer	52.0 ± 11.9	55.2 ± 15.3	50.1 ± 13.3	48.4 ± 18.2
	Trans-mean	**65.1 ± 1.9**	63.8 ± 1.2	**67.9 ± 1.8**	64.9 ± 1.7
	GRU-D	58.0 ± 2.0	63.2 ± 1.7	58.2 ± 3.1	59.3 ± 3.5
	SeFT	31.7 ± 1.5	31.0 ± 2.7	32.0 ± 1.2	28.0 ± 1.6
	mTAND	34.7 ± 5.5	43.4 ± 4.0	36.3 ± 4.7	39.5 ± 4.4
	GARNet	63.0 ± 3.9	**65.3 ± 3.7**	64.5 ± 4.7	**65.9 ± 4.2**
40%	Transformer	43.8 ± 14.0	44.6 ± 23.0	40.5 ± 15.9	40.2 ± 20.1
	Trans-mean	48.7 ± 2.7	55.8 ± 2.6	54.2 ± 3.0	55.1 ± 2.9
	GRU-D	47.7 ± 1.4	**63.4 ± 1.6**	44.5 ± 0.5	47.5 ± 0.0
	SeFT	26.8 ± 2.6	24.1 ± 3.4	28.0 ± 1.2	23.3 ± 3.0
	mTAND	23.7 ± 1.0	33.9 ± 6.5	26.4 ± 1.6	29.3 ± 1.9
	GARNet	**59.3 ± 3.1**	62.0 ± 2.6	**57.2 ± 1.3**	**59.3 ± 3.5**
50%	Transformer	43.2 ± 2.5	52.0 ± 2.5	36.9 ± 3.1	41.9 ± 3.2
	Trans-mean	46.4 ± 1.4	59.1 ± 3.2	43.1 ± 2.2	46.5 ± 3.1
	GRU-D	49.7 ± 1.2	52.4 ± 0.3	42.5 ± 1.7	47.5 ± 1.2
	SeFT	26.4 ± 1.4	23.0 ± 2.9	27.5 ± 0.4	23.5 ± 1.8
	mTAND	20.9 ± 3.1	35.1 ± 6.1	23.0 ± 3.2	27.7 ± 3.9
	GARNet	**57.3 ± 2.9**	**60.2 ± 1.7**	**56.1 ± 3.5**	**57.0 ± 2.7**

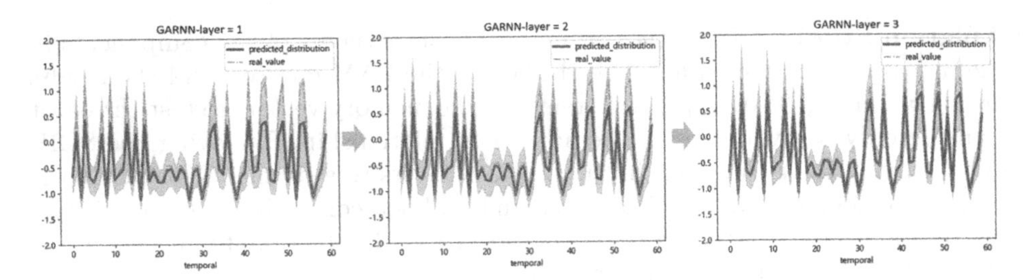

Fig. 2. Predicted distribution with its corresponding true values in three consecutive GARNN-layers. The experiment is conducted on a test set of PAM data using the trained model. The predicted distribution (mean and variance) was obtained for each layer based on the residual concatenations of the previous layer and the output of the previous timestamp. It shows that GARNet can fit the original data well by predicting the distribution, and the fit increases gradually with successive residual concatenations.

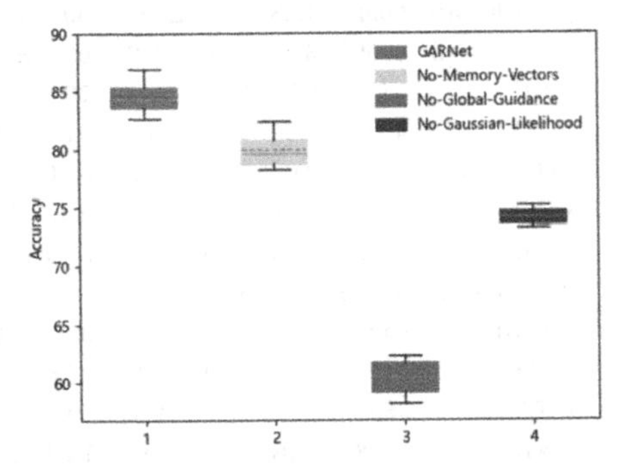

Fig. 3. Accuracy performance analysis of each component in the model.

Missing Sensors. We conducted experiments on the PAM dataset under conditions of sensor failures or increasing amounts of missing data to validate the model's performance. Specifically, we experimentally validated the missing random sensor condition. The results are shown in Table 2. Our model can achieve more stable results as the number of missing sensors keeps increasing, i.e., the percentage of missing sensors varies from 10% to 50%. In addition, our model achieves the best performance in 17 of the 20 settings. GARNet outperforms baselines by up to 12.3% in accuracy, 3.2% in precision, 11.0% in recall, and 9.5% in F1 score.

5 Conclusion

We introduce GARNet for irregularly sampled multivariate time series classification tasks. GARNet consists of three important components: SGIE, GARNN, and MTIA. SGIE automatically extracts complex inter-variable relations and leverages sample-level relational structures from which a structure-augmented global feature vector can be obtained. Further, guided by the global vector, the

GARNN captures latent factors at each timestamp. It uses the residual structure to complete the iterative update between GARNN layers. Finally, we use MTIA with selective aggregation to learn the final mapping for the classification task. Experimental results show that our model can predict missing values of multivariate temporal data with irregular sampling very well and achieves state-of-the-art performance in the classification tasks.

Acknowledgements. This work was financially supported by China National Key R & D (or Research and Development) Program (No. 2020AAA0105000 and 2020AAA0105003) and National Natural Science Foundation of China (No. 61877008, 62172074).

References

1. Amiri, M., Jensen, R.: Missing data imputation using fuzzy-rough methods. Neurocomputing **205**, 152–164 (2016)
2. Bartholomew, D.J.: Time series analysis forecasting and control (1971)
3. Horn, M.: Set functions for time series. In: International Conference on Machine Learning (2020)
4. Che, Z., Purushotham, S., Cho, K., Sontag, D., Liu, Y.: Recurrent neural networks for multivariate time series with missing values. Sci. Rep. **8**(1), 1–12 (2018)
5. Chen, R.T., Rubanova, Y., Bettencourt, J., Duvenaud, D.K.: Neural ordinary differential equations. In: Advances in Neural Information Processing Systems, vol. 31 (2018)
6. Chen, Z., Jiaze, E., Zhang, X., Sheng, H., Cheng, X.: Multi-task time series forecasting with shared attention. In: 2020 International Conference on Data Mining Workshops (ICDMW), pp. 917–925. IEEE (2020)
7. Hamzaçebi, C.: Improving artificial neural networks' performance in seasonal time series forecasting. Inf. Sci. **178**(23), 4550–4559 (2008)
8. Kipf, T., Fetaya, E., Wang, K.C., Welling, M., Zemel, R.: Neural relational inference for interacting systems. In: International Conference on Machine Learning, pp. 2688–2697. PMLR (2018)
9. Kreindler, D.M., Lumsden, C.J.: The effects of the irregular sample and missing data in time series analysis. Nonlinear Dyn. Psychol. Life Sci. (2006)
10. Li, M.M., Huang, K., Zitnik, M.: Representation learning for networks in biology and medicine: advancements, challenges, and opportunities. arXiv preprint arXiv:2104.04883 8 (2021)
11. Li, S.C.X., Marlin, B.: Learning from irregularly-sampled time series: a missing data perspective. In: International Conference on Machine Learning, pp. 5937–5946. PMLR (2020)
12. Luo, Y., Cai, X., Zhang, Y., Xu, J., et al.: Multivariate time series imputation with generative adversarial networks. In: Advances in Neural Information Processing Systems, vol. 31 (2018)
13. Mondal, D., Percival, D.B.: Wavelet variance analysis for gappy time series. Ann. Inst. Stat. Math. **62**(5), 943–966 (2010)
14. Rehfeld, K., Marwan, N., Heitzig, J., Kurths, J.: Comparison of correlation analysis techniques for irregularly sampled time series. Nonlinear Process. Geophys. **18**(3), 389–404 (2011)

15. Reiss, A., Stricker, D.: Introducing a new benchmarked dataset for activity monitoring. In: 2012 16th International Symposium on Wearable Computers, pp. 108–109. IEEE (2012)
16. Reyna, M.A., et al.: Early prediction of sepsis from clinical data: the physionet/computing in cardiology challenge 2019. In: 2019 Computing in Cardiology (CinC), p. 1. IEEE (2019)
17. Shang, C., Chen, J., Bi, J.: Discrete graph structure learning for forecasting multiple time series. arXiv preprint arXiv:2101.06861 (2021)
18. Shukla, S.N., Marlin, B.M.: Interpolation-prediction networks for irregularly sampled time series. arXiv preprint arXiv:1909.07782 (2019)
19. Shukla, S.N., Marlin, B.M.: Multi-time attention networks for irregularly sampled time series. arXiv preprint arXiv:2101.10318 (2021)
20. Suo, Q., Chou, J., Zhong, W., Zhang, A.: Tadanet: task-adaptive network for graph-enriched meta-learning. In: Proceedings of the 26th ACM SIGKDD International Conference on Knowledge Discovery & Data Mining, pp. 1789–1799 (2020)
21. Tipirneni, S., Reddy, C.K.: Self-supervised transformer for multivariate clinical time-series with missing values. arXiv preprint arXiv:2107.14293 (2021)
22. Tonekaboni, S., Li, C.L., Arik, S.O., Goldenberg, A., Pfister, T.: Decoupling local and global representations of time series. In: International Conference on Artificial Intelligence and Statistics, pp. 8700–8714. PMLR (2022)
23. Vaswani, A., et al.: Attention is all you need. In: Advances in Neural Information Processing Systems, vol. 30 (2017)
24. Wang, Y., Smola, A., Maddix, D., Gasthaus, J., Foster, D., Januschowski, T.: Deep factors for forecasting. In: International Conference on Machine Learning, pp. 6607–6617. PMLR (2019)
25. Wu, Y., et al.: Dynamic gaussian mixture based deep generative model for robust forecasting on sparse multivariate time series. In: Proceedings of the AAAI Conference on Artificial Intelligence, vol. 35, pp. 651–659 (2021)
26. Wu, Z., Pan, S., Chen, F., Long, G., Zhang, C., Philip, S.Y.: A comprehensive survey on graph neural networks. IEEE Trans. Neural Netw. Learn. Syst. **32**(1), 4–24 (2020)
27. Wu, Z., Pan, S., Long, G., Jiang, J., Chang, X., Zhang, C.: Connecting the dots: multivariate time series forecasting with graph neural networks. In: Proceedings of the 26th ACM SIGKDD International Conference on Knowledge Discovery & Data Mining, pp. 753–763 (2020)
28. Zerveas, G., Jayaraman, S., Patel, D., Bhamidipaty, A., Eickhoff, C.: A transformer-based framework for multivariate time series representation learning. In: Proceedings of the 27th ACM SIGKDD Conference on Knowledge Discovery & Data Mining, pp. 2114–2124 (2021)
29. Zhang, X., Zeman, M., Tsiligkaridis, T., Zitnik, M.: Graph-guided network for irregularly sampled multivariate time series. arXiv preprint arXiv:2110.05357 (2021)
30. Zhou, J., et al.: Graph neural networks: a review of methods and applications. AI Open **1**, 57–81 (2020)

Quasi-Periodicity Detection via Repetition Invariance of Path Signatures

Chenyang Wang(✉), Ling Luo, and Uwe Aickelin

University of Melbourne, Parkville, VIC 3010, Australia
chenyangw3@student.unimelb.edu.au,
{ling.luo,uwe.aickelin}@unimelb.edu.au

Abstract. Periodicity or repetition detection has a wide varieties of use cases in human activity tracking, music pattern discovery, physiological signal monitoring and more. While there exists a broad range of research, often the most practical approaches are those based on simple quantities that are conserved over periodic repetition, such as auto-correlation or Fourier transform. Unfortunately, these periodicity-based approaches do not generalise well to quasi-periodic (variable period) scenarios. In this research, we exploit the time warping invariance of path signatures to find linearly accumulating quantities with respect to quasi-periodic repetition, and propose a novel repetition detection algorithm Recurrence Point Signed Area Persistence. We show that our approach can effectively deal with repetition detection with period variations, which similar unsupervised methods tend to struggle with.

Keywords: Repetition detection · Path signature · Quasi-periodicity

1 Introduction

Repetition detection, or repeated pattern detection, is a common task in applications such as human activity tracking [11,17], music pattern discovery [5] or physiological signal monitoring [7]. The goal is to find intervals within a signal where a pattern is consecutively repeated multiple times. This task is often necessary for objectives such as forecasting, segment-wise behaviour classification or anomaly detection.

Detecting exact temporarily repeated patterns is not difficult, however in practice the problem is often complicated by noise, drifting baseline, variable scales, or variable period lengths. In particular, the variable period length case (which we call *quasi-periodicity*) is more challenging to detect because it is not compatible with conventional tools used for repetition detection, such as pattern matching or frequency analysis. While plenty of prior works exist for "rigid" repetition detection [7,11,17,18], many require supervised learning, and even

Supplementary Information The online version contains supplementary material available at https://doi.org/10.1007/978-3-031-33383-5_24.

© The Author(s), under exclusive license to Springer Nature Switzerland AG 2023
H. Kashima et al. (Eds.): PAKDD 2023, LNAI 13938, pp. 301–313, 2023.
https://doi.org/10.1007/978-3-031-33383-5_24

then they often do not innately account for potentially variable period lengths. Most existing approaches can either only work with quasi-periodic patterns with small period variations, or they must learn to recognise an unbounded number of possible variations for each repeated pattern, which is a much more difficult task. Ideally, we would like to identify repeated patterns with quasi-periodicity, rather than relying on "rigid" periodicity-based approaches.

In this research, we propose what is to our knowledge the first accurately specified unsupervised quasi-periodicity detection method, which relies on the warping invariance and repetition equivariance properties of the log signature [9]. By doing so, we are able to extend existing detection methods based on rigid-period conserved quantity to the variable-period case. We make the following contributions:

1. We show that log signatures are *equivariant with respect to time-warped repetitions.* Additionally, certain lower order signatures and log signatures satisfy the linear accumulation rule with additional prefix and suffix invariances, which allows us to detect repetition over arbitrary expanding windows.
2. We propose the algorithm *Recurrence Point Signed Area Persistence (RPSAP)* to detect quasi-periodic repetitions by exploiting the linear accumulation rule (a form of repetition invariance) of signed areas, and demonstrate that RPSAP not only performs on par or better than existing approaches, it is also robust under reparameterisation and uneven sampling.

Theorem proofs, experiment code and other supplementary materials for this paper can be found at https://github.com/Mithrillion/rpsap/.

2 Related Works

Existing methods typically approach this problem through either pattern matching, such as with template/motif-based methods [6,7] or recurrence/periodicity-based methods such as those using the auto-correlation function (ACF) [14,17], recurrence counting [11,17] or recurrence plot [10,13]. However, these approaches do not directly address the issue of variable period length. For example, periodicity-based approaches such as ACF often require that the periodicity of patterns does not drift too much. Supervised, template-based methods can learn variations of repeated patterns, but often must do so case by case. They also cannot deal with unknown patterns. Recently, deep learning-based approaches have also gained much traction [18,20], including methods which combine deep learning with recurrence-based representations [10]. However, ultimately they still rely on learning period variations instance by instance, and are often data-specific and are therefore not generally-applicable, unsupervised approaches.

Motif-based approaches try to convert repetition detection to a most frequent subsequence detection problem. Radius Profile [4] finds the most repeated patterns in a time series and is capable of detecting repetition, however it has $\mathcal{O}(N^2)$ complexity and must operate on the whole sequence at once. R-SIMPAD [7] is an approach based on range-restricted similarity matching, and can perform online unsupervised repetition detection. However, converting repetition detection directly to subsequence matching exploits the translation invariance of *rigid* periodic sequences, which does not take into account potential time warping equivalence between patterns. This approach also needs to perform multiple similarity searches for different pattern lengths.

Of all existing approaches, the simplicity and general applicability of periodicity-based approaches such as ACF or Fourier Transform make them attractive in use cases where run-time performance or data efficiency is needed. The main limitation is that these methods rely on quantities which are only conserved under strict periodicity. This suggests that if we could effectively extend such approaches to quasi-periodic time series data, we could greatly expand the range of real-world problems solvable with lightweight, unsupervised algorithms.

3 Definitions

In this section, we introduce the necessary definitions that lead to a conserved quantity under quasi-periodic repetition.

Definition 1 (Repeating and Periodic Sequences). *A repeating sequence* $\{\mathbf{X}_i\}$ *is the concatenation of multiple copies of a sequence* $\mathbf{U}$ *(called the* ***repeating unit****), i.e.* $\mathbf{X} = \mathbf{U}|\mathbf{U}|...|\mathbf{U}$. *In particular, A periodic sequence* $\{\mathbf{X}_i\}$ *of period* τ *is a time series satisfying* $\mathbf{X}_{i+k\tau} = \mathbf{X}_i$ *for all* i, k *that are well-defined for the sequence. Generalising to continuous time, a periodic curve* $\mathbf{C}$ *is a curve satisfying* $\mathbf{C}_{t+k\tau} = \mathbf{C}_t$.

Note that all periodic sequences or curves are repeating, but a repeating sequence or curve need not return to the same value after each repetition.

Definition 2 (Time Warping and Time Reparameterisation). *A (discrete) time warping of a sequence* $\{\mathbf{X}_i\}$ *is a transformation of* $\mathbf{X}$ *which consists of two actions: repeating a point multiple times, or merging repeated points. This notion of time warping is what the Dynamic Time Warping (DTW) algorithm [12] uses for alignment.*

A time reparameterisation of a curve $\mathbf{C}$ *is a mapping* $\varphi : [t_a, t_b] \rightarrow [t_a, t_b]$ *where* $\frac{d\varphi}{dt} > 0$. *In other words, it is a mapping of the time domain to the time domain itself that preserves the order of time points. It is analogous to time warping for discrete sequences, and may also be referred to as* ***continuous time warping****.*

Definition 3 (Quasi-periodicity). *A quasi-periodic sequence is a sequence that can be turned into a periodic sequence through a discrete time warping. A quasi-periodic curve is a curve that can be turned into a periodic curve under time reparameterisation.*

Definition 4 (Path Signature). *The signature of a curve*[1] $\mathbf{X}$ *on* $[a, b]$ *is defined as [8]:*

$$Sig(\mathbf{X})_{[a,b]} = \sum_{k=0}^{\infty} \int_a^b \int_a^{t_1} ... \int_a^{t_{k-1}} d\mathbf{X}_{t_1} \otimes d\mathbf{X}_{t_2} ... \otimes d\mathbf{X}_{t_k}$$

where $\otimes$ *is the tensor product.*

[1] or the linear interpolation of a discrete sequence.

The signature of a d-dimensional curve consists of multiple terms distinguished by a multi-index $i_1 i_2 ... i_m$, where each $i_k \in [1, d] \cap \mathbb{N}$.

$$Sig(\mathbf{X})^{i_1 i_2 ... i_k}_{[a,b]} = \int_a^b \int_a^{t1} ... \int_a^{t_{k-1}} d\mathbf{X}^{i_1}_{t_1} d\mathbf{X}^{i_2}_{t_2} ... d\mathbf{X}^{i_k}_{t_k}$$

The length of the multi-index is called the *depth* or *order* of the signature term. The 0th depth term of the signature is the identity element $\mathbf{1}$. The *truncated signature* $\Pi_m Sig(\mathbf{X})$ is defined as the signature up to depth m.

The path signature is invariant to time-reparameterisation [2,9]. Its invariance under time-reparameterisation can be proven with the change of variables formula for either the Riemann or Riemann-Stieljes integral.[2]

For a discrete sequence, the convention is to define its path signature as the path signature of its piecewise linear interpolation, which is a continuous curve.

A powerful result from the theory of path signatures is that they follow Chen's equality [2], which allows us to represent the concatenation of two curves as the tensor product of their signatures:

$$Sig(\mathbf{X}|\mathbf{Y}) = Sig(\mathbf{X}) \otimes Sig(\mathbf{Y}) \tag{1}$$

$$Sig(\mathbf{X}|\mathbf{X}|...|\mathbf{X}) = Sig(\mathbf{X})^{\otimes n} \tag{2}$$

Remark 1. Because path signatures are invariant to reparameterisation, each repetition in the above equality only has to be equivalent up to time warping.

We now introduce the concept of the log-signature [8,9]. Given the definition of integer power under the tensor product, $\mathbf{X}^{\otimes n}$, we can also define the exponential map and the logarithm map using their infinite series expansions, with $\otimes$ in place of the regular multiplication. Note that $\otimes$ is not commutative.

Definition 5 (Log Signature). *We define the log-signature as the logarithm map of the signature [8]:*

$$LogSig(\mathbf{X}) := Log(Sig(\mathbf{X}) - \mathbf{1}) = \bigoplus_{n=1}^{\infty} \frac{(-1)^{n-1}}{n} (Sig(\mathbf{X}) - \mathbf{1})^{\otimes n} \tag{3}$$

Similarly, we define the *truncated log signature*, $\Pi_m LogSig(\mathbf{X})$ as log signature up to depth m. Unlike signatures, log signatures have $\mathbf{0}$ as their 0th depth term.

4 Motivation

We would like to detect and quantify quasi-periodic repetitions by finding a quantity that is conserved under time-warped repetitions. In particular, we seek something that accumulates linearly with each repetition. We first define a notion of equivariance under periodic and quasi-periodic repetition.

[2] The reader may refer to [3] for more in-depth information regarding path signature.

Definition 6 (Linear Accumulation Rule). *A function over sequences (or curves) $\phi : \mathbf{X} \to \mathbb{R}^m$ satisfies the linear accumulation rule (LAR) if:*

$$\phi(\mathbf{U}|\mathbf{U}|...|\mathbf{U}) = n \cdot \phi(\mathbf{U}) \tag{4}$$

*Additionally, it is **prefix and suffix-invariant** if:*

$$\phi(\mathbf{P}|\mathbf{U}|\mathbf{U}|...|\mathbf{U}|\mathbf{Q}) = n \cdot \phi(\mathbf{U}) + \gamma(\mathbf{P}, \mathbf{Q}) \tag{5}$$

The motivation for the linear accumulation rule is that if $\phi(\mathbf{X})$ satisfies such a rule, then the stream of ϕ values evaluated on an expanding window of a repeating sequence $\mathbf{X}$ will exhibit a linear trend, which can be exploited to detect repeating patterns. Indeed, this is the approach used by many existing repetition and periodicity detection methods.

The prefix and suffix-invariant linear accumulation rule is particularly useful, due to the following observation:

Theorem 1 (Linear Trend from Invariance). *If ϕ satisfies the linear accumulation rule and it is prefix and suffix-invariant, then $\{\phi(\mathbf{X}_{[0:t]})\}$, the trajectory of $\phi(\mathbf{X})$ over expanding windows follows a linear trend on intervals where $\mathbf{X}$ is quasi-periodic.*

Proof. For readability, we moved detailed proofs of theorems to the supplementary material. The general idea is that $\{\phi(\mathbf{X}_{[0:t]})\}$ will follow the trend of $\phi(\mathbf{U})$, and if the suffix $\mathbf{Q}$ represents a part of the next repetition $\mathbf{U}$, then the same phase generates the identical linear offset from the trend line.

Example 1. For rigid periodic sequences, the (unnormalised) Fourier Transform satisfies linear accumulation. Let T be the repetition period, then $\mathcal{F}[nT](f_i) \approx n\mathcal{F}[T](f_i)$ where f_i are the dominant frequencies (i.e. peaks of the FT).

Such relationships are agnostic to the value of the repetition period, but they no longer hold if the repetition period varies significantly. In existing literature, the path signature [2,9] is known as a representation of parametric curves that is invariant to time-reparameterisation. In the following section, we discuss how fundamental properties of the path signature lead to a conserved quantity we can exploit for quasi-periodic repetition detection.

5 Theoretical Analysis

Lemma 1 (Elastic Repetition Equivariance). *The log signature of a time-warped repetition of a curve $\mathbf{U}$ is an integer multiple of the log signature of $\mathbf{U}$, i.e.*

$$LogSig(\mathbf{U}|\mathbf{U}|...|\mathbf{U}) = n \cdot LogSig(\mathbf{U}) \tag{6}$$

Proof. This follows directly from applying the logarithm map to both sides of Eq. 2 and using the property of the logarithm map $log(x^\alpha) = \alpha log(x)$.

Since $n \cdot \mathbf{S}/\|n \cdot \mathbf{S}\| = \mathbf{S}$, the unit norm of the log signature a repeating sequence is conserved. We have therefore discovered that *the log signature is repetition invariant* for repeating sequences. Just like the ACF and FT with fixed-period repetitions, the log signature of repeating pattern sequences stay collinear in the vector space of log signatures. In practice, we may test whether $LogSig(\mathbf{X})_{[a,c]}$ and $LogSig(\mathbf{X})_{[a,b]}$ have high cosine similarity to determine if $X_{[a,c]}$ either repeats $X_{[a,b]}$, or shares a smaller unit of repetition with $X_{[a,b]}$.

However, although the log signature satisfies the linear accumulation rule, it is not prefix-invariant. Indeed, because $\otimes$ is not commutative, the logarithm map for the signature does not satisfy $log(a \otimes b) = log(a) + log(b)$, therefore we cannot isolate $log(P)$ in $log(P \otimes U^{\otimes n})$ from $\mathbf{U}$-dependent terms. The general formula for evaluating $log(P \otimes U^{\otimes \lambda})$ from $p = log(P)$ and $u = log(U)$ is given by the Baker-Campbell-Hausdorff formula [19]:

$$log(P \otimes U^{\otimes \lambda}) = p + \lambda u + \frac{\lambda}{2}[u,p] + \frac{\lambda^2}{12}[u,[u,p]] - \frac{\lambda}{12}[p,[u,p]] + ... \qquad (7)$$

where $[x,y] = x \otimes y - y \otimes x$ is the Lie bracket. Notice that $log(P \otimes U^{\otimes \lambda})$ not only has terms depending on p all the way to higher order terms, but these higher order terms also depend on higher powers of λ.

Despite this, if we restrict our attention to *quasi-periodic* curves and sequences, all points in a quasi-periodic cycle, and particularly the endpoints of each repeated $\mathbf{U}$ segment, are thus recurrent (i.e., returns to the same value). Given that depth 2 signatures represent signed areas enclosed by two of the curve dimensions [3] (for curves of at least 2D), we immediately have the following:

Theorem 2 (LAR of Signed Areas). *The signed area terms of* $\mathbf{X} = \mathbf{P}|\mathbf{U}|...|\mathbf{U}$, $Sig(\mathbf{X})^{ij}, i \neq j$ *satisfy the prefix-invariant linear accumulation rule if* $\mathbf{U}|\mathbf{U}|...|\mathbf{U}$ *is quasi-periodic.*

Corollary 1. *LAR for signed areas is prefix and suffix-invariant.*

There is an intuitive visual explanation for the prefix and suffix-invariance of signed areas. In Fig. 1a We see that a geometric interpretation of the order 2 terms of the log signatures is the Lévy area (signed area enclosed by the curve and the line connecting the end points) on each of the coordinate planes [3]. The (only) order 2 log signature term for the above 2D curve equals the (+) area minus the (−) area. In Fig. 1b we see that generally, the cumulative signed areas depend on the segment linking the start and the end of the curve, and therefore depend on the prefix OP_1. However, as in Fig. 1c, if the curve is (quasi-)periodic around the area A_2 and returns to the same point P after each cycle, then the increment of the signed area after each repetition is A_2, which is unaffected by OP. The cumulative signed area will simply be $(-A_1 + n \cdot A_2)$. We may also notice that the area A_2 is the same regardless of which point on the curve around A_2 we choose as the start point of the repetition unit, implying that this linear trend is the same for any phase choice of the repetition unit. *The Lévy area increments between any phase-matching points in a repeating interval are all multiples of the increment over one repetition.* Furthermore, if there is a suffix PQ, then its contribution (region OPQ) to the signed area is not affected by A_2.

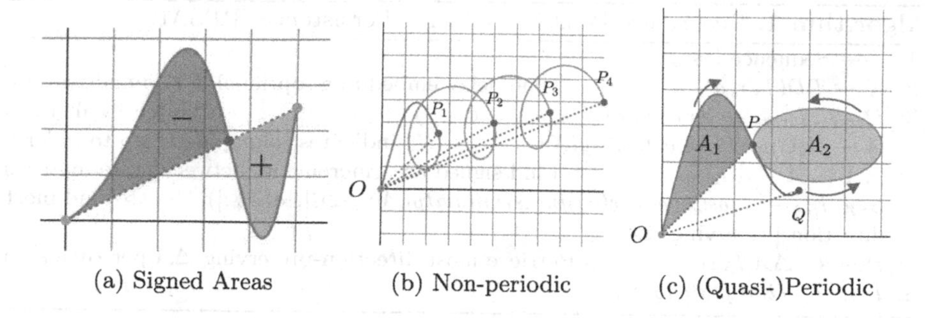

(a) Signed Areas (b) Non-periodic (c) (Quasi-)Periodic

Fig. 1. Prefix and suffix-invariance of Lévy areas.

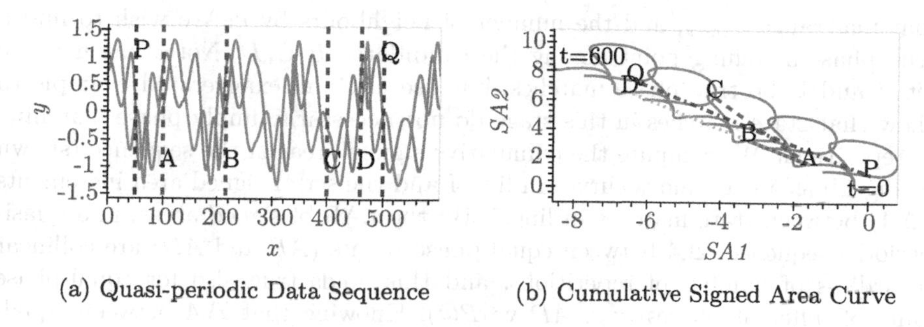

(a) Quasi-periodic Data Sequence (b) Cumulative Signed Area Curve

Fig. 2. RPSAP Example

6 Algorithmic Design

We propose Recurrence Point Signed Area Persistence (RPSAP). Given that the Lévy area increments between any equal-phase points must retain the same linear trend over repeating intervals, we can look for quasi-periodic repetition by testing whether such increments stay collinear. In principle, we may use any linear accumulation rule to detect quasi-periodic repetition by looking for linear trends over expanding windows (such as in the cumulative log signature of a sequence). However, only with the prefix and suffix-invariant rule do all phases of the repetition follow the same linear trend. Since higher order log signatures violate prefix and suffix invariance, we restrict our attention here to order 2 log signatures, i.e. signed (Lévy) areas.

We illustrate the workflow of Algorithm 1 (RPSAP) via an example in Fig. 2. We first embed the time series in a state space (if it is not already so) via time-delay embedding with embedding dimension d.[3] For each state vector A_t in the state sequence in Fig. 2a, we find its most similar state matches D_{state} and their indices I_{state} (e.g. the states B, C, D in Fig. 2a) via a range-restricted kNN search (band_knn), where the search range is defined by the minimal range w_{min} and

[3] We may automatically find the best parameters for time-delay embedding with well-established heuristics, such as False Nearest Neighbours [1].

Algorithm 1. Recurrence Point Signed Area Persistence (RPSAP)

1: $T \leftarrow$ Sequence Data
2: $X \leftarrow TD(T, \tau, d)$ ▷ Time-delay embedding, optional if T is multivariate
3: $D_{state}, I_{state} \leftarrow band\_knn(X, k, w_{min}, w_{max})$ ▷ Match local states
4: $\mathcal{A} \leftarrow \Pi_2 Cumulative\_LogSig(X)$ ▷ Π_2 indicates taking logsig up to order 2
5: $\Delta\mathcal{A} \leftarrow \mathcal{A} - \mathcal{A}[I_{state}]$ ▷ Find signed area increments between state matches
6: $S_{SA}, I_{SA} \leftarrow max\_argmax(cosine\_similarity(\mathcal{A}[: -\Delta t], \mathcal{A}[\Delta t :]))$ ▷ Find most direction-preserving $\Delta\mathcal{A}$
7: $\Delta\mathcal{A}^* \leftarrow \Delta\mathcal{A}[I_{SA}]$ ▷ Retrieve most direction-preserving $\Delta\mathcal{A}$ per time step
8: $I_{label} \leftarrow find\_plateau(S_{SA})$

maximal range w_{max}, and the number of neighbours by k. We wish to find a true phase-matching point among the candidates B, C, D. Note that here we let B and D be true phase matches, but use C as an exaggerated example to show that state matches in this stage do not necessarily imply phase matching of repetitions. We compute the cumulative signed areas of the sequence (shown in Fig. 2b as a continuous curve) in line 4 and the set of signed area increments ($\Delta\mathcal{A}$) between state matches in line 5. By the LAR of signed areas, in a quasi-periodic sequence, $\Delta\mathcal{A}$ between equal-phase points (AB and AD) are collinear regardless of number of repetitions, and this holds true also for equal-phase pairs of different phases (e.g. AD vs PQ). Knowing that $\Delta\mathcal{A}$ between equal-phase points must align with the trend of LAR, whereas $\Delta\mathcal{A}$ between spurious state matches can be in any direction (e.g. AC), it is reasonable to assume that the LAR trend will be the most shared direction for $\Delta\mathcal{A}$ for all A_t, and that if we take the nearest neighbour pair between the set of $\Delta\mathcal{A}$ at t and a nearby point $t + \Delta t$ as in line 6, they will most likely be aligned with this trend. We use this procedure to estimate a LAR trend $\Delta\mathcal{A}^*$ at each point (e.g. in the direction of AB or AD) in line 7, and label the segments with persistent $\Delta\mathcal{A}^*$ directions as possible intervals of repetition (line 8).

Time Complexity. Let N be the sequence length, then RPSAP has a time complexity of $\mathcal{O}(2Nw_{max}d + Nk^2d^2)$, where the significant terms come from the range-restricted kNN for recurrent states, and finding the maximum cosine similarity between sets of $\Delta\mathcal{A}$ for each pair of nearby points. Overall, it is linear in sequence length, and since the embedding dimension d and recurrence point candidate number k are usually well-constrained, RPSAP is comparable to other range-restricted kNN methods such as R-SIMPAD in terms of efficiency.

7 Experiments

Dataset and Baselines. As many related approaches for repetition detection were designed primarily for human activity recognition, we evaluate RPSAP on the widely-used PAMAP2 [16] and RecoFit [11] datasets against baseline methods. Both datasets contain accelerometer recordings of annotated human exercises.

Table 1. Rep Detection on PAMAP2 and RecoFit

	PAMAP2			RecoFit		
	NASC	RSIMPAD	RPSAP	NASC	RSIMPAD	RPSAP
accuracy	0.948	0.933	**0.949**	0.886 ± 0.049	0.939 ± 0.034	$\mathbf{0.946 \pm 0.032}$
precision	0.969	0.970	**0.994**	0.867 ± 0.077	0.902 ± 0.069	$\mathbf{0.945 \pm 0.054}$
recall	0.921	**0.923**	0.909	0.765 ± 0.160	$\mathbf{0.901 \pm 0.085}$	0.886 ± 0.097
f1 score	0.944	0.946	**0.949**	0.802 ± 0.111	0.899 ± 0.064	$\mathbf{0.911 \pm 0.064}$

We compare RPSAP with two baselines NASC [14] (an autocorrelation-based approach) and R-SIMPAD [7] (a subsequence similarity-based approach). They are the most widely reported traditional statistic-based method and the most recent unsupervised approach for repetition detection, respectively. On PAMAP2 we use pre-optimised hyperparameters for NASC and SIMPAD from [6] with added rolling window majority voting for smoothing the NASC results. On RecoFit, we use the first 25 observations to optimise hyperparameters for each method, then use the rest (101 observations) for evaluation. We illustrate the different characteristics of these approaches with a synthetic example.

Experiment Results. Figure 3 highlights the difference between the three approaches compared with a synthetic example. Here, we have generated a signal with a periodic segment, followed by an irregular segment, then a warped periodic segment. All methods can recognise the exact periodic segment, but NASC fails to find a suitable threshold value to accommodate both periodic segments while excluding the irregular part, whereas R-SIMPAD is only able to detect parts of the quasi-periodic segment with similar degrees of warping. On the other hand, our RPSAP is able to clearly find both periodic and quasi-periodic segments. In the last subplot, we included the first two principal component of the signed areas direction, which confirms that the direction of signed areas is indeed conserved over repeating intervals.

For repetition detection evaluation, we use the same ambiguity masking procedure as in [6,7], consistent with the baseline papers. In Table 1 we see that averaged across instances, RPSAP outperforms NASC and R-SIMPAD under most metrics across both datasets. Surprisingly, we found the auto-correlation baseline NASC performs much better than expected on PAMAP2 than [6] reported after using a simple sliding window label smoothing technique. Indeed, it is likely that we are already close to saturating the performance potential even with NASC. However, we do see a much larger performance difference on the much larger and diverse RecoFit dataset, indicating that RPSAP is able to deal with more varied types of repetitions more effectively.

To further investigate robustness under uneven sampling, we then randomly drop $\alpha\%$ of data points from RecoFit uniformly, to simulate uneven data sampling. We incrementally set α to be higher and see how robust R-SIMPAD and RPSAP are to this synthetically injected time warping. As we see from Fig. 4, RPSAP is significantly more robust than R-SIMPAD under such data defor-

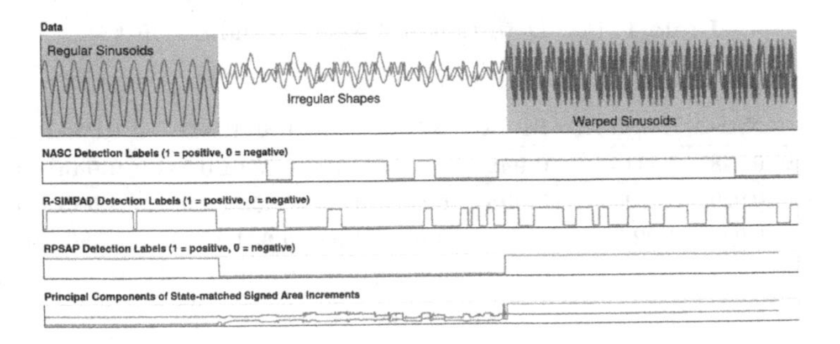

Fig. 3. Repetition Detection Results Comparison with a Synthetic Example

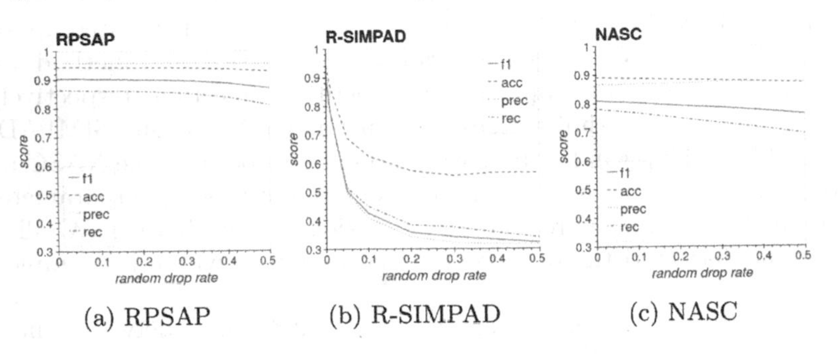

(a) RPSAP (b) R-SIMPAD (c) NASC

Fig. 4. Performance Comparison under Random Data Drop

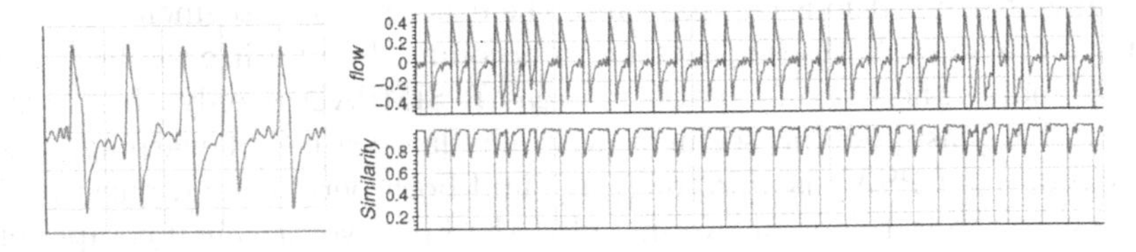

Fig. 5. Quasi-periodic segmentation with log signatures. Left: signal corresponding to the unit log signature mode. Right: segmentation result via similarity comparison between cumulative log signature and the log signature mode.

mation, retaining much of the performance even after 30%+ of data points are dropped. Surprisingly, its performance even improves under 0–20% random dropping, suggesting removing some high frequency noise in the data actually helps with RPSAP. In contrast, the performance of R-SIMPAD degrades significantly even with only 1% of data points being dropped, and NASC exhibits a more linear performance falloff with random data point drop. Here, we truly see the advantage of having innate time warping invariance built into the algorithm, as RPSAP is the only algorithm without performance loss when subjected to moderate amounts of random drop corruption.

8 Discussion and Conclusion

Result Analysis. Based on the results in the previous section, we see that RPSAP performs favourably compared to a subsequence similarity-based approach, and is much more resistant to uneven sampling or local time warping, demonstrating the importance of designing the algorithm around *quasi-periodicity* rather than rigid repetition detection. Considering that we do not require more complexity than rigid subsequence similarity methods such as R-SIMPAD, and that all operations used are stream-friendly, RPSAP is well-suited for use cases where we need online detection of repetition with timing or sampling irregularities.

Applications. While repetition detection in itself is often only the initial step in sequence data analysis, the repetition equi-/invariance property of log signatures that we exploited have greater implications. Once we identify a sufficiently consistent interval of quasi-periodic repetition with RPSAP, we immediately know that the log signature over such an interval is a summary feature which is invariant to the number of repetitions, as well as the particular time warping each repetition is subject to. Such a feat is impossible to achieve with either alignment-based approaches such as DTW (which require equal number of cycles) or conventional summary features (which are often not warp-invariant). This can be particularly attractive in use cases where we wish to obtain a description of the "normal behaviour" without pre-defined cycle boundaries and compare it with a new behaviour containing an indefinite number of cycles. In Fig. 5 we demonstrate an example where we take a respiratory signal [15], find a repeating interval, extract the "mode" of unit log signatures over said interval, then use the unit log signature to identify cycle end points for the airflow signal, without prior knowledge about breathing waveforms. Because log signatures are fixed-length vectors, we may also use RPSAP to break a time series into homogeneous segments, and feed the signature features over each segment into algorithms which require fixed-length inputs, such as deep learning classifiers, achieving both semantic segmentation and segment classification.

Conclusion. In this research, we showed that the log signature, and in particular signed areas, are equivariant under quasi-periodic repetition, and designed the RPSAP algorithm to detect such repetitions using the invariance of signed area trends over repeating intervals. We demonstrated that by accounting for time warping invariance within our algorithm, we achieved higher performance on human activity recognition tasks and significantly better robustness under uneven sampling. Furthermore, we proposed that such repetition invariance approach can be the basis for further pattern extraction and feature representation tasks.

References

1. Cao, L.: Practical method for determining the minimum embedding dimension of a scalar time series. Physica D **110**(1–2), 43–50 (1997). https://doi.org/10.1016/S0167-2789(97)00118-8

2. Chen, K.T.: Integration of paths, geometric invariants and a generalized baker-hausdorff formula. Ann. Math. 163–178 (1957). https://doi.org/10.2307/1969671
3. Chevyrev, I., Kormilitzin, A.: A primer on the signature method in machine learning. arXiv preprint arXiv:1603.03788 (2016)
4. De Paepe, D., Van Hoecke, S.: Mining recurring patterns in real-valued time series using the radius profile. In: 2020 IEEE International Conference on Data Mining (ICDM), pp. 984–989. IEEE (2020). https://doi.org/10.1109/ICDM50108.2020.00113
5. Hsu, J.L., Chen, A.L., Chen, H.C.: Finding approximate repeating patterns from sequence data. In: ISMIR, p. 2004 (2004)
6. Li, C.T., Cao, J., Liu, X., Stojmenovic, M.: mSIMPAD: efficient and robust mining of successive similar patterns of multiple lengths in time series. ACM Trans. Comput. Healthcare **1**(4), 1–19 (2020). https://doi.org/10.1145/3396250
7. Li, C.T., Shen, J., Yang, Y., Cao, J., Stojmenovic, M.: Repetitive activity monitoring from multivariate time series: a generic and efficient approach. In: 2021 IEEE 18th International Conference on Mobile Ad Hoc and Smart Systems (MASS), pp. 36–45. IEEE (2021). https://doi.org/10.1109/MASS52906.2021.00014
8. Lyons, T., McLeod, A.D.: Signature methods in machine learning. arXiv preprint arXiv:2206.14674 (2022)
9. Lyons, T.: Rough paths, signatures and the modelling of functions on streams. In: Proceedings of the International Congress of Mathematicians Seoul 2014 (2014)
10. Mirmomeni, M., Kulik, L., Bailey, J.: A transferable technique for detecting and localising segments of repeating patterns in time series. In: 2021 International Joint Conference on Neural Networks (IJCNN), pp. 1–10. IEEE (2021). https://doi.org/10.1109/IJCNN52387.2021.9534157
11. Morris, D., Saponas, T.S., Guillory, A., Kelner, I.: Recofit: using a wearable sensor to find, recognize, and count repetitive exercises. In: Proceedings of the SIGCHI Conference on Human Factors in Computing Systems, pp. 3225–3234 (2014). https://doi.org/10.1145/2556288.2557116
12. Müller, M.: Dynamic time warping. In: Müller, M. (ed.) Information Retrieval for Music and Motion, pp. 69–84. Springer, Heidelberg (2007). https://doi.org/10.1007/978-3-540-74048-3_4
13. Panagiotakis, C., Karvounas, G., Argyros, A.: Unsupervised detection of periodic segments in videos. In: 2018 25th IEEE International Conference on Image Processing (ICIP), pp. 923–927. IEEE (2018). https://doi.org/10.1109/ICIP.2018.8451336
14. Rai, A., Chintalapudi, K.K., Padmanabhan, V.N., Sen, R.: Zee: zero-effort crowdsourcing for indoor localization. In: Proceedings of the 18th Annual International Conference on Mobile Computing and Networking, pp. 293–304 (2012). https://doi.org/10.1145/2348543.2348580
15. Rehm, G.B., Kuhn, B.T., Nguyen, J., Anderson, N.R., Chuah, C.N., Adams, J.Y.: Improving mechanical ventilator clinical decision support systems with a machine learning classifier for determining ventilator mode. Stud. Health Technol. Inform. **264**, 318–322 (2019). https://doi.org/10.3233/SHTI190235
16. Reiss, A., Stricker, D.: Introducing a new benchmarked dataset for activity monitoring. In: 2012 16th International Symposium on Wearable Computers, pp. 108–109. IEEE (2012). https://doi.org/10.1109/ISWC.2012.13
17. Shen, C., Ho, B.J., Srivastava, M.: MiLift: efficient smartwatch-based workout tracking using automatic segmentation. IEEE Trans. Mob. Comput. **17**(7), 1609–1622 (2017). https://doi.org/10.1109/TMC.2017.2775641

18. Soro, A., Brunner, G., Tanner, S., Wattenhofer, R.: Recognition and repetition counting for complex physical exercises with deep learning. Sensors **19**(3), 714 (2019). https://doi.org/10.3390/s19030714
19. Stillwell, J.: Naive Lie Theory. Springer, New York (2008). https://doi.org/10.1007/978-0-387-78214-0
20. Torres-Soto, J., Ashley, E.A.: Multi-task deep learning for cardiac rhythm detection in wearable devices. NPJ Digit. Med. **3**(1), 1–8 (2020). https://doi.org/10.1038/s41746-020-00320-4

Targeted Attacks on Time Series Forecasting

Zeyu Chen(✉), Katharina Dost, Xuan Zhu, Xinglong Chang, Gillian Dobbie, and Jörg Wicker

The University of Auckland, Auckland, New Zealand
{zche677,zxua238,xcha011}@aucklanduni.ac.nz,
{katharina.dost,g.dobbie,j.wicker}@auckland.ac.nz

Abstract. Time Series Forecasting (TSF) is well established in domains dealing with temporal data to predict future events yielding the basis for strategic decision-making. Previous research indicated that forecasting models are vulnerable to adversarial attacks, that is, maliciously crafted perturbations of the original data with the goal of altering the model's predictions. However, attackers targeting specific outcomes pose a substantially more severe threat as they could manipulate the model and bend it to their needs. Regardless, there is no systematic approach for targeted adversarial learning in the TSF domain yet. In this paper, we introduce targeted attacks on TSF in a systematic manner. We establish a new experimental design standard regarding attack goals and perturbation control for targeted adversarial learning on TSF. For this purpose, we present a novel indirect sparse black-box evasion attack on TSF, *n*VITA. Additionally, we adapt the popular white-box attacks Fast Gradient Sign Method (FGSM) and Basic Iterative Method (BIM). Our experiments confirm not only that all three methods are effective but also that current state-of-the-art TSF models are indeed susceptible to attacks. These results motivate future research in this area to achieve higher reliability of forecasting models.

Keywords: Adversarial Learning · Time Series · Targeted Attack · Forecasting

1 Introduction

Time Series Forecasting (TSF) has received much attention recently in numerous high-stake applications such as stock market prediction [17,24], energy consumption estimation [8], traffic flow forecasting [18,26], and climate change investigation [27]. Those applications use a time series of historical data to predict a few timesteps in the future." With the recent advances in machine learning, *Neural Networks (NNs)* have been increasingly established to fulfill these tasks [25]. Despite their impressive performance, NNs are vulnerable to explicitly designed malicious attacks.

Supplementary Information The online version contains supplementary material available at https://doi.org/10.1007/978-3-031-33383-5_25.

© The Author(s), under exclusive license to Springer Nature Switzerland AG 2023
H. Kashima et al. (Eds.): PAKDD 2023, LNAI 13938, pp. 314–327, 2023.
https://doi.org/10.1007/978-3-031-33383-5_25

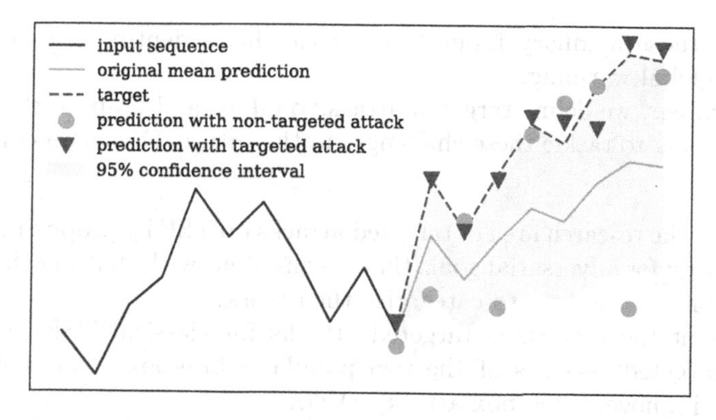

Fig. 1. Comparison of targeted and non-targeted attacks on TSF. While non-targeted attacks alter the forecast in an uncontrolled way, targeted attacks control the direction and the magnitude of the prediction error.

Studying a model's weaknesses not only helps the researchers understand more about the model itself but also allows them to develop defense mechanisms against malicious attacks, which can prevent potentially fatal consequences after model deployment. Adversarial learning, first proposed for image classification, is a field of machine learning designed to study weaknesses of prediction models and has revealed major vulnerabilities even for state-of-the-art NNs [6]. However, only a few studies [7,14,16,22,23] have been dedicated to the TSF problem.

We categorize our research as evasion attacks since our attack aims to manipulate the input data at the model test time, and we divide adversarial attacks on TSF further into two types: (i) *Non-Targeted Forecasting Attacks (for short here: Non-Targeted Attacks)* where the goal is to damage the model's performance by decreasing the metrics generally, and (ii) *Targeted Forecasting Attacks (for short here: Targeted Attacks)* where the attacker exploits the model's auto-correlative nature to identify the action needed to achieve certain false predictions. As evident from the comparison in Fig. 1, targeted attacks are more powerful than non-targeted ones since they grant an attacker control over the direction and magnitude of the forecast's error. Investigating how to tweak the input time series to push the prediction to a specified target value would not only help a researcher understand their trained model in depth and expose vulnerabilities early on. It is also crucial to ensure that the model behaves in a fair way. Nevertheless, most current studies [14,16,22,23] of adversarial attacks focus on non-targeted attacks, which cannot provide insights and explainability of the model as the targeted attacks do. Two examples that illustrate targeted attacks are given below.

Credit Score: Suppose a bank uses a model to calculate a client's credit score based on their account history, future balance forecast, etc. This score determines the amount of an allowed loan. Assessing which changes need to be made to the input time series to achieve a certain credit score can help the bank to improve the model and prevent malicious attacks in advance.

Climate Change: Assuming we have a reliable model to forecast the future global temperature based on a diversity of features including greenhouse gases emissions, electricity generation, deforestation, etc. Understanding how to manipulate those auxiliary input features can help identify new strategies to slow down global warming.

In this paper, we define targeted attacks on TSF and propose three suitable attacks allowing to tackle these challenges in the future. Our main contributions are:

1. We define the research area of targeted attacks on TSF by proposing a suitable terminology for adversarial goals that is consistent with that in other domains and enables researchers to categorize their work.
2. We present the first three targeted attacks for classical TSF: FGSMt and BIMt, targeted versions of the two popular white-box attacks FGSM and BIM, and a novel black-box attack, nVITA.
3. We establish a new standard experimental design for adversarial learning on TSF that allows for a fair comparison of our as well as future targeted attacks.

The remainder of this paper is organized as follows: Sect. 2 states the problem and reviews related research. Our proposed targeted attacks are introduced in Sect. 3 and evaluated under the experimental setup introduced in Sect. 4. Sections 5 and 6 discuss our findings with a real-world application and conclude the paper, respectively.

2 Background

Due to the different nature of classification and regression problems, attack methods and evaluation metrics based on classification problems cannot be directly transferred to TSF. Aiming to propose a system tailored to regression problems, in this section, we define the problem and present related research on existing evasion attacks on TSF and related algorithms.

Problem Statement and Notation. Let T_X be a multivariate input time series with time index $t \in \mathcal{T}$ and feature index $f \in \mathcal{F}$. A TSF model M with parameters θ is fit to the historic data of T_X and makes predictions/forecasts $M(T_X) = T'_y$ approximating the actual value T_y. An adversary can attack the model by carefully crafting ϵ-bounded, additive perturbations $\eta = \{[\eta_{t,f}]\}$ with $\eta_{t,f} \leq \epsilon \ \forall t, f$, that alter T_X to an ideally hardly distinguishable perturbed time series $T_X^{\text{adv}} := T_X + \eta$ causing a different model output.

Depending on the adversary's goal, the desired attacked prediction T_y^{adv} can differ. We formulate two novel attack types in the context of TSF: non-targeted and targeted attacks. In a *non-targeted attack*, the adversary's goal is to damage the model's performance while remaining undetected. The attack success can be measured in terms of any statistic representing the model's performance, such as absolute error. In a *targeted attack*, the adversary aims to force the model M to make specific predictions g at a particular time index i. A targeted attack goal G can be defined as the tuple (i, g). Although it is potentially possible to achieve multiple targeted attack goals with a single attack, for the sake of simplicity, we only consider one goal per attack in this paper.

Table 1. Adversarial attacks on TSF

Attack	Reference	TSF Type	Attack Scenario	Attack Goal	Indirect	Sparsity
FGSM	[16,22]	Non-prob.	White-box	Non-targeted	False	Dense
BIM	[16,22,23]	Non-prob.	White-box	Non-targeted	False	Dense
Variants of BIM	[22]	Non-prob.	White-box	Non-targeted	False	Dense
Liu *et al.*	[14]	Prob.	White-box	Non-targeted	True	Medium
Dang-Nhu *et al.*	[7]	Prob.	White-box	Targeted	False	Dense
BIMt, FGSMt	Our work	Non-prob.	White-box	Targeted	False	Dense
*n*VITA	Our work	Non-prob.	Black-box	Both	True	Sparse

In a *white-box* scenario, an attacker has full access to and knowledge of the model (i.e., its architecture, parameters, and hyperparameters). A *black-box* scenario is more restrictive, and the attacker does not have knowledge of the model.
Time Series Forecasting Models. In this paper, we focus on adversarial attacks on four *Neural Network (NN)* models that are well established and commonly adopted in TSF: (i) *Convolutional Neural Network (CNN)* [13] which, originally designed for images, uses convolution layers that exploit spatial relations in the input image/data to abstract it to an activation map, (ii) *Long Short-Term Memory (LSTM)* [10], a variant of *Recurrent Neural Network (RNN)* [19] that contains a memory component allowing the learner to access past events, and (iii) *Gated Recurrent Unit (GRU)* [4], a simplified version of LSTM. LSTM and GRU models are considered to be the state-of-the-art NN models for TSF [11]. Moreover, we include *Random Forest (RF)* [3] to represent the non-neural network machine learning model for our novel white-box attack exclusively.

Existing Attacks on TSF. To the best of our knowledge, there is no general targeted attack on TSF. However, there are non-targeted attacks and a targeted attack limited to probabilistic TSF [7] outputting probability distributions rather than values.

Table 1 gives an overview of existing attacks on TSF. Beyond the TSF Type (probabilistic/non-probabilistic), the attack scenario, and the attack goal, we provide two additional characteristics: An *indirect* [14] attack does not perturb the feature to be forecast, i.e., y. We define the level of the sparsity of attacks as follows. If the attack is limited to k feature time combinations, it is a sparse attack with k levels. If the attack is limited to k features, it is a medium sparse attack with k levels. If the attack perturbs all the input time series, it is a dense attack (see Fig. 2 for a comparison).

Initially proposed for image data, *Fast Gradient Sign Method (FGSM)*, *Basic Iterative Method (BIM)*, and variants of it are gradient-based non-targeted attacks that have been directly adapted to TSF [16,22,23]. Unfortunately, this adaptation results in inaccuracies with respect to the perturbation measurement and limited applicability of the crafted adversarial sample in the real world (see Sect. 4 – "Perturbation Measurement"). We propose targeted versions of these attacks in our paper.

Dang-Nhu *et al.* [7] and Liu *et al.* [14] proposed attacks on probabilistic TSF attacks. However, as a result of their different outputs, the attacks designed for

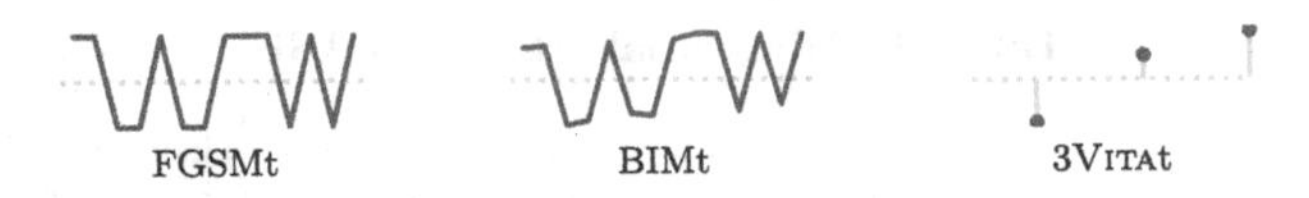

Fig. 2. Sparsity comparison for our proposed methods' perturbations.

probabilistic models cannot be directly applied to non-probabilistic models and vice versa.

Differential Evolution and One Pixel Attack. *Differential Evolution (DE)* [20] is a population-based meta-heuristic search algorithm that aims to solve optimization problems. By starting with a list of candidate solutions as the parent population, DE iteratively creates children through cross-over with randomly selected parents. If a child holds a larger fitness score than its parent according to a pre-defined fitness function, it takes the parent's place in the next generation. Based on DE, the *One Pixel Attack* [21] has been proposed for image classification perturbing only a single pixel per image to fool a classifier. Similarly, we use DE to introduce a sparse attack on TSF.

3 Proposed Method

In this section, we introduce three new proposed attacks for the targeted scenario. First, we adjust the popular white-box attacks FGSM and BIM to optimize towards a specified target instead of maximal chaos and annotate them with the suffix "t" to indicate the targeted versions. In the spirit of the traditional FGSM and BIM attacks, both add perturbations to the entire input time series requiring an attacker to manipulate the data in every timestep and feature. While in real-world applications, it is very often that it is infeasible to change all of the input, and some features simply cannot be altered Second, we introduce nVITA, a novel black-box attack capable of targeted and non-targeted attacks while overcoming the above limitations: nVITA only requires an attacker to manipulate exactly n feature/timestep pairs, and it does not require access to the trained model.

3.1 Fast Gradient Sign Method (FGSM) for TSF

First introduced for the image classification domain, FGSM was adapted to TSF by perturbing each feature of a multivariate time series at each timestep instead of each of an image's pixels by ϵ [16,22]. Although the perturbation's magnitude is fixed, it can be either additive or subtractive. FGSM chooses the sign based on the model's loss function's gradient. The existing non-targeted FGSM aims to maximize the error between model prediction T'_y and ground truth T_y. The goal of targeted attacks, however, is to approximate a target prediction g at a target time index i as closely as possible. Hence, we alter the perturbation aiming to achieve the goal within magnitude ϵ to

$$\eta_{\mathrm{FGSMt}} := -\epsilon \cdot \operatorname{sign}\left(\nabla_{T_X} J_M\left(\theta_M, T_X, g\right)\right).$$

where J_M and θ_M denote the loss function of the attacked model M and its parameters, respectively. Note that $(\eta_{\text{FGSMt}})_{t,f} \in \{-\epsilon, +\epsilon\}$ for every timestep t and feature f leading to an overall perturbation of $\epsilon \times |\mathcal{T}| \times |\mathcal{F}|$. Figure 2 contains an adversarial example crafted by FGSMt in TSF.

3.2 Basic Iterative Method (BIM) for TSF

The basic iterative method (BIM) is an extended version of FGSM. Starting from the initial input $T_{X_0}^{\text{adv}} := T_X$, BIM iteratively applies FGSM with α as the perturbation bound and clips the crafted adversarial example to ensure that the overall perturbation lies within the ϵ-bounds. The adjustments required for BIM to perform targeted attacks are similar to those for FGSMt. We find the perturbed time series for BIMt as follows:

$$T_{X_{N+1}}^{\text{adv}} := \text{Clip}_{T_X,\epsilon}\left\{T_{X_N}^{\text{adv}} - \alpha \cdot \text{sign}\left(\nabla_x J\left(T_{X_N}^{\text{adv}}, g\right)\right)\right\},$$

for each iteration $N \in \mathbb{N}_0$. BIMt stops iterating after a fixed number of iterations. Since FGSM is a white-box attack, BIMt also requires full access to the model. Similar to FGSMt, BIMt perturbs every timestep and feature; however, the magnitude lies in the interval $[-\epsilon, +\epsilon]$, and the overall perturbation can be smaller than that of FGSMt. An adversarial example generated by BIMt in TSF is shown in Fig. 2. Extended versions of BIM omit the clipping constraint [23] or refrain from perturbing all features but select based on their importance [22].

3.3 *n*VITA (*n*-Values Time Series Attack)

We propose *n*VITA, a novel sparse indirect black-box attack on TSF that can tackle non-targeted and targeted scenarios. *n*VITA uses the population-based meta-heuristic search algorithm DE to find the small specified number $n \geq 1$ of timestamps to mislead a forecasting model. It bettered FGSMt and BIMt algorithms as it requires no knowledge of the model to be attacked and can achieve comparable attacking outcomes whilst altering much fewer timesteps and features.

Parent Generation. Given a time series T_X, *n*VITA aims to perturb exactly n timestep-feature combinations. Hence, each of these n perturbations can be represented as a triple (t, f, p) of timestamp $t \in \mathcal{T}$, feature $f \in \mathcal{F}$, and perturbation $p \in [-\epsilon, +\epsilon]$ that will be added to the input time series. Combining these n triples into a vector of length $3n$, we obtain the overall perturbation representation

$$\eta_{n\text{VITA}} = [t_1, f_1, p_1, t_2, f_2, p_2, \ldots, t_n, f_n, p_n].$$

For simplicity, we omit the index for η in this section.

We start the search for the optimal combination of n perturbations with a set of s random parent samples, the parent generation $\text{Gen}_1 = \{\eta_1^1, \ldots, \eta_s^1\}$. Each

η follows the representation introduced above and is randomly generated such that every $t \in \mathcal{T}$, $f \in \mathcal{F}$, and $p \in [-\epsilon, +\epsilon]$. Using cross-overs and mutations, DE iteratively creates new generations, continuously improving qualitatively with respect to a fitness function.

Fitness Function. We consult a *Fitness Function (FF)* to decide if a solution should be replaced with a mutated one in the next generation. Depending on the type of attack, we suggest two different fitness functions.

The goal of a *non-targeted* attack is to damage the model's performance. In our implementation, we use the MSE between the actual test data T_y and the faulty model prediction T_y^{adv} caused by the attack, i.e., with t as the length of T_y,

$$\text{FF}_{\text{non-targeted}} = \frac{1}{t} \sum_{i=1}^{t} \left(T_{y_i} - T_{y_i}^{\text{adv}}\right)^2 .$$

A *targeted* attack aims to cause a specified target prediction. Choices for the FF are similar to those in the non-targeted case; however, since we restricted our scope to only one attack goal per attack, the FF will be evaluated on only one point. Hence, minimizing Absolute Error (AE) and Squared Error (SE) leads to the same result, and we choose AE to avoid the unnecessary square in our implementation, i.e.,

$$\text{FF}_{\text{targeted}} = \left|G - T_{y_i}^{\text{adv}}\right| \quad \text{for} \quad G = (i, g).$$

Note that the non-targeted FF evaluates the model's performance over the entire test length, whereas the targeted attack optimizes only for its goal G.

Customization. Both perturbation representation and fitness function can be easily customized to handle different complex real-world scenarios. One example is the FULLVITA adaptation that perturbs not only n but all timestamp-feature combinations in the test data. In this case, the perturbation representation can be reduced to

$$\eta_{\text{FULLVITA}} = \left[p_1, p_2, \ldots, p_{|\mathcal{T}|\cdot|\mathcal{F}|}\right]$$

to optimize the search space size since perturbation positions are now fixed.

Another customization could include restricting the features to be perturbed to those that can be influenced in practice, as in the use-case we discuss in Sect. 5. This customization makes nVITA an indirect attack.

4 Experimental Setup

This paper proposes three novel targeted attacks on TSF, nVITA, FGSMt, and BIMt. To demonstrate their efficiency, we compare them in an experimental setup designed for targeted attacks. As nVITA can operate in a non-targeted scenario, we provide additional benchmarks there. To guarantee a fair evaluation, we had to adjust the typical setup. This section discusses these adjustments and provides detailed and reproducible information regarding our experimental setup[1] .

[1] All materials, including Python implementation, supplementary materials, and additional results, are available at https://github.com/ProfiterolePuff/nvita.

Datasets. We use the following real-world dataset: (i) *Electricity* (German Electricity Consumption)[2] with $n_f = 3$ features, (ii) *NZTemp* (New Zealand Land Temperatures; subset of Earth Surface Temperature Data [5]) with $n_f = 8$, (iii) *CNYExch* (USD/CNY (CNY=X) Exchange Rate)[3] with $n_f = 5$, and (iv) *Oil* (iPath Pure Beta Crude Oil ETN)(See footnote 3) with $n_f = 6$. We preprocess the datasets by filling in missing values with linear interpolation between the previous and the following available value if possible and cut otherwise. All datasets use the sliding window approach with window size n_w tuned using Autocorrelation Function plots for each dataset individually. We randomly select 10% as our validation data and 100 observations as test data to evaluate the attack performance. The remaining data is our training data and is normalized into $[0, 1]$. The test and validation data are scaled accordingly. This procedure deviates from that in the existing literature [16,22,23] as the normalization is typically carried out before the training/validation/test split leading to faulty perturbation measurements and information leakage. We discuss this problem under "Perturbation Measurement".

Models and Parameters. We evaluate our method on the NN models introduced in Sect. 2 (see our supplementary materials for the architectures). Since nVITA can attack any machine learning models, we include RF for nVITA exclusively. We tune all hyperparameters, i.e., the learning rate, the number of epochs, and the number of trees for RF, for each dataset by grid search on the validation set.

Baselines and Competitors. To benchmark FGSMt and BIMt with $\alpha = \epsilon/200$, we provide the *Baseline Random Sign attack (BRS)* which randomly selects 1 or -1 as the sign of perturbation. nVITA is tested with $n \in \{1, 3, 5\}$ with 200 maximum generations and a convergence tolerance of 0.01, and all other parameters as specified in Sect. 3.3. We benchmark nVITA with the *Baseline Random n Value attack (BRnV)* which randomly selects n values to attack with perturbation ϵ and a random sign.

Perturbation Measurement. Existing studies [16,22,23] borrow the fixed perturbation measurement from the time series classification domain, which considers a time series as a whole. After normalizing the entire time series into $[0, 1]$, a fixed perturbation threshold ϵ is granted, and, for example, $\epsilon = 0.2$ directly corresponds to 20% perturbation. While this procedure is reasonable for time series classification, TSF uses the sliding window method, where a prediction is carried out based on the partial time series cut by a single window. As illustrated in Fig. 3 for three different sample windows, the range of values can differ strongly between windows and will not be normalized anymore. It can even exceed 1 as the overall normalization takes place on the training set only to avoid information leakage. Thus, we propose a new dynamic perturbation control system for adversarial machine learning on TSF. For a univariate time series, we scale the perturbation threshold with the window range, i.e., our new $\epsilon' = \epsilon \cdot W_i$, where

[2] Source: https://open-power-system-data.org/.

[3] Source: https://finance.yahoo.com/.

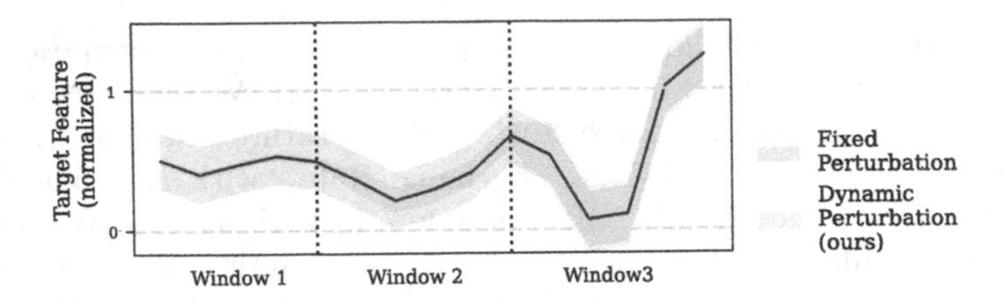

Fig. 3. Comparison of fixed (blue) and dynamic (orange) perturbation limitations for $\epsilon = 0.2$ (Color figure online)

W_i denotes the range of values within window i. Figure 3 shows a comparison of the fixed perturbation control (blue) and our new dynamic approach (orange). Intuitively, we allow only small perturbations on windows with a small range where a large jump might be easily detectable and loosen the restrictions on windows with a large range where large perturbations might be concealed better. Note that our perturbation bound is generally tighter than the fixed one for all windows with a range below 1. This concept extends to multivariate time series by assigning individual ϵ'-bounds for each feature time series depending on its window range. In our experiments, we control the perturbation as introduced here and test $\epsilon \in \{0.05, 0.1, 0.15, 0.2\}$.

Evaluation. To quantify the effectiveness of our attacks, we evaluate for each window the Absolute Error (AE) as defined in Sect. 3. Mean Absolute Error (MAE) is then calculated to represent the attack performance. We also measure the attacks' success using the confidence intervals of the models in their prediction. Intuitively, assessing if an adversarial attack can impact a model to push its prediction out of its, for example, 90% confidence interval measures the attacks' power, and it enables fair comparison among several test windows, datasets, and models. In order to calculate the confidence interval for NNs, we use BLiTZ [2] to create uncertainty in our NNs' model prediction and calculate the standard deviation among 100 predictions. For random forests, the standard deviation for all trees' predictions is used instead. The targets for a targeted attack in a particular window are set to the upper bound and lower bounds of the 99% confidence interval ± 0.1, respectively, to make it particularly challenging to achieve. A targeted attack is considered *successful* if the attacked model's prediction is altered beyond the confidence interval percentiles.

We repeat each experiment 5 times to combat the model's randomness and average the results for statistical stability. Note that we largely include 95% confidence intervals in our plots but omit to provide explicit variances among runs since we observed that they are negligible. See our supplementary materials for detailed results.

5 Results

This section summarizes the results for the non-targeted attack scenario and provides detailed results for targeted attacks. For non-targeted attacks, AE is

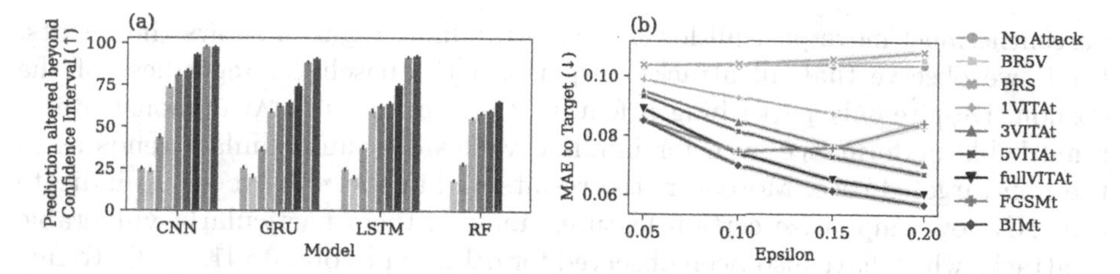

Fig. 4. Comparison of the impact of different models (a) and different ϵ values (b) on the attack performance. For the models, we investigate how much the prediction can be altered in terms of the models' confidence intervals. For ϵ, we measure the MAE to the target value.

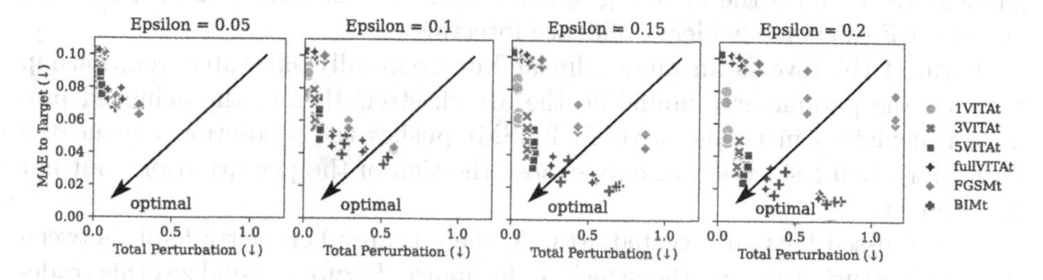

Fig. 5. The Trade-Off between Attack Performance and Total Perturbation for targeted attacks under varying ϵ values.

measured between the model prediction and ground truth values. A larger MAE indicates worse model performance and hence a better attack result. In contrast, for targeted attacks, AE is measured between the model prediction and the target instead of the ground truth values. A lower error here indicates that the model prediction is closer to our target and hence a better attack result. To avoid confusion, we annotate each measure with an arrow indicating if a higher (↑) or a lower (↓) value indicates a better result. Note that while nVITA is flexible enough to attack non-NN models like Random Forests, the other attack cannot. To enable a fair comparison, whenever we average over different models, we omit RF and include it only in Fig. 4 (a).

Non-targeted Attack Results. As a sanity check, we benchmarked nVITA against FGSM and BIM (see our supplementary materials for the results). We observe that FULLVITA achieves a competitive attack performance while using substantially less perturbation and operating under a black-box scenario. For lower n, the attack performance gradually decreases. However, the total perturbation drops by a larger margin making it a strong attack in applications where no model or training data access can be obtained, and the attacker's opportunities to alter the data are limited.

Targeted Attack Results. For the targeted attacks, we investigate by how much we can alter the model's prediction and measure the distance in terms of the models' confidence percentiles. Note that a prediction moving past the p%

percentile must move past all lower percentiles first. Figure 4 shows the results. First, we observe that all attacks outperform the baselines, regardless of the epsilon. Despite only perturbing n feature/time tuples, nVITAt demonstrates a remarkable performance even for minimal values of n and similar trends as in the non-targeted case. Moreover, the results within the nVITAt group indicate that NNs overemphasize certain features, making them particularly vulnerable to attacks which have also been observed for other types of data [1,9,12]. To further investigate the differences between models, Fig. 4 (a) measures the average distance (in terms of confidence intervals) by which a prediction could be altered. We observe that CNN is the most vulnerable. GRU and LSTM are less vulnerable without a substantial difference. RF seems more robust than NN models, particularly because the strong gradient-based attacks cannot attack RF and because RF calculates wider confidence intervals.

Figure 4 (b) reveals an almost linear but eventually saturated relationship between the parameter ϵ bounding the attack strength and the achieved prediction deviation in terms of MAE. FGSMt pushes the prediction beyond the target for $\epsilon > 0.1$ since it can only choose the sign of the perturbations but not the amount.

As discussed for non-targeted attacks, there is an inherent trade-off between the total perturbation and the attack performance. Figure 5 visualizes this trade-off for different ϵ values where a perfect attack would be located in $(0, 0)$. We observe similar trends to the non-targeted results: The gradient-based methods are more effective, but the nVITAt family perturbs the input time series substantially less. An in-between solution is FULLVITAt combining the nVITAt approach with the concept of FGSMt/BIMt that perturb the entire time series. In addition, the figure shows that larger ϵ enables smaller MAE differences to the target value while increasing the total perturbation.

Case Study. In this subsection, we apply the nVITAt on the USA covid prediction model to explore the insights we can gain in real-life applications.

We use the CovidUSA dataset [15] containing daily updates on the number of COVID-19 cases, deaths, ICU and hospital patients, tests, vaccinations, stringency index, population, and GDP per capita in the US. On this dataset, we train an LSTM model that takes all features from the previous seven days and outputs the next day's number of new cases. The experiments are set up as in Sect. 4 with minor changes. We attack a 10% test set using 1VITA, 3VITA, 5VITA with maximally positive and negative targets. To avoid perturbations in features we cannot control in real life, the attack is prohibited from altering "New Cases", "Population", and "GDP per capita".

Assuming the model is highly accurate, it can act as a proxy for reality: Learning how to alter the model prediction can provide insights on how to change reality and, in this case, how to better manage the next pandemic. Figure 6 presents the frequencies of (time index, feature, direction) triples, averaged over all test windows. The results suggest that if there is an increasing trend in "new vaccinations" as well as a decreasing trend in "ICU patients", the number of new COVID-19 cases will decrease and vice versa. The targeted attacks also give us

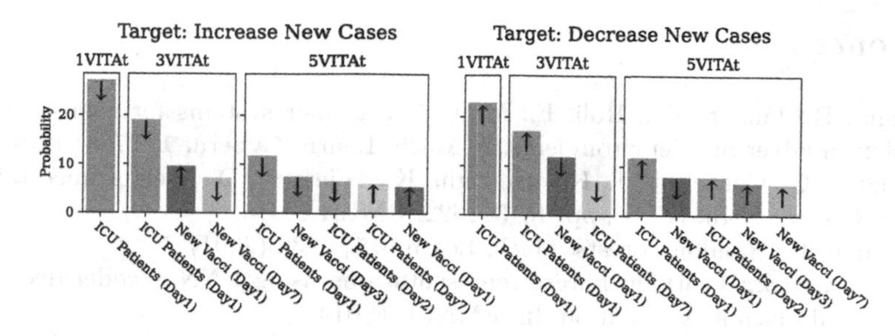

Fig. 6. CovidUSA Case Study Result, Arrows on the bar indicate the direction of perturbation, (↑) for positive and (↓) for negative.

the most influential levers in the COVID-19 defense: "new vaccinations", followed by "ICU patients".

These results seem reasonable. Vaccinating the population has proven to be the key ingredient in protecting against COVID-19. The number of ICU patients is likely correlated with the number of vaccinations as the course of the disease is typically less severe for vaccinated patients. Overall, we conclude that targeted TSF attacks can provide meaningful and explainable insights into the model.

6 Conclusion

In this paper, we introduced a targeted adversarial attack setting on Time Series Forecasting and proposed three attacks for this purpose, FGSMt, BIMt, and nVITAt. Targeted attacks on TSF are particularly threatening as they not only cause a machine learning model to fail but also grant adversaries the power to bend the model predictions to their needs. We establish a new standard experimental design correcting flaws in previous designs by focusing on a fair comparison of different attacks among different datasets. Our results demonstrate that targeted adversarial attacks are not only a theoretical threat but also quite effective in practice, and even state-of-the-art models expose serious vulnerabilities. In future research, we will strengthen our proposed attacks, investigate how nVITAt, in particular, could exploit dataset and model characteristics for heightened performance, and move beyond the evasion strategy towards poisoning for TSF. Simultaneously, we will explore how to respond to this threat by developing adequate methods to detect and defend against such attacks. nVITA attacks might show more outliers than true data samples and can potentially be detected accordingly. For the attacks perturbing the entire time series, we plan to develop a time-sensitive discriminative model or explore adversarial training.

Acknowledgements. The authors wish to acknowledge the use of New Zealand eScience Infrastructure (NeSI) national facilities - https://www.nesi.org.nz.

References

1. Biggio, B., Fumera, G., Roli, F.: Multiple classifier systems for robust classifier design in adversarial environments. J. Mach. Learn. Cybern. **1**, 27–41 (2010)
2. Blundell, C., Cornebise, J., Kavukcuoglu, K., Wierstra, D.: Weight uncertainty in neural network. In: ICML, pp. 1613–1622. PMLR (2015)
3. Breiman, L.: Random forests. Mach. Learn. **45**, 5–32 (2001)
4. Cho, K., et al.: Learning phrase representations using RNN encoder-decoder for statistical machine translation. In: EMNLP (2014)
5. Cowtan, K.: The climate data guide: Global surface temperatures: berkeley earth surface temperatures (2019). https://bit.ly/3fAqtVg Accessed 18 Feb 2022
6. Dalvi, N., Domingos, P., Sanghai, S., Verma, D.: Adversarial classification. In: The tenth ACM SIGKDD International Conference on Knowledge Discovery and Data Mining, pp. 99–108 (2004)
7. Dang-Nhu, R., Singh, G., Bielik, P., Vechev, M.: Adversarial attacks on probabilistic autoregressive forecasting models. In: III, H.D., Singh, A. (eds.) The 37th ICML. vol. 119, pp. 2356–2365. PMLR, 13–18 Jul 2020
8. Deb, C., Zhang, F., Yang, J., Lee, S.E., Shah, K.W.: A review on time series forecasting techniques for building energy consumption. Renew. Sustain. Energy Rev. **74**, 902–924 (2017)
9. Demontis, A., et al.: Yes, machine learning can be more secure! A case study on android malware detection. IEEE Trans. Dependable Secure Comput. **16**(4), 711–724 (2017)
10. Hochreiter, S., Schmidhuber, J.: Long short-term memory. Neural Comput. **9**(8), 1735–1780 (1997)
11. Van Houdt, G., Mosquera, C., Nápoles, G.: A review on the long short-term memory model. Artif. Intell. Rev. **53**(8), 5929–5955 (2020). https://doi.org/10.1007/s10462-020-09838-1
12. Kołcz, A., Teo, C.H.: Feature weighting for improved classifier robustness. In: CEAS '09, Mountain View, CA, USA (2009)
13. Lecun, Y., Bottou, L., Bengio, Y., Haffner, P.: Gradient-based learning applied to document recognition. IEEE **86**(11), 2278–2324 (1998)
14. Liu, L., Park, Y., Hoang, T.N., Hasson, H., Huan, J.: Towards robust multivariate time-series forecasting: adversarial attacks and defense mechanisms. In: KDD 2022 Workshop on Mining and Learning from Time Series - Deep Forecasting: Models, Interpretability, and Applications (2022)
15. Mathieu, E., et al.: Coronavirus pandemic (covid-19). Our World in Data (2020). https://ourworldindata.org/coronavirus
16. Mode, G.R., Hoque, K.A.: Adversarial examples in deep learning for multivariate time series regression. In: 2020 IEEE Applied Imagery Pattern Recognition Workshop (AIPR), pp. 1–10. IEEE (2020)
17. Mondal, P., Shit, L., Goswami, S.: Study of effectiveness of time series modeling (arima) in forecasting stock prices. IJCSEA **4**(2), 13 (2014)
18. Razvan-Gabriel Cirstea, Chenjuan Guo, B.Y.: Graph attention recurrent neural networks for correlated time series forecasting. In: KDD MiLeTS19 (2019)
19. Rumelhart, D.E., Hinton, G.E., Williams, R.J.: Learning representations by back-propagating errors. Nature **323**, 533–536 (1986)
20. Storn, R., Price, K.: Differential evolution: A simple and efficient adaptive scheme for global optimization over continuous spaces. J. Global Optim. **23** (1995)

21. Su, J., Vargas, D.V., Sakurai, K.: One pixel attack for fooling deep neural networks. IEEE Trans. Evol. Comput. **23**(5), 828–841 (2019)
22. Wu, T., Wang, X., Qiao, S., Xian, X., Liu, Y., Zhang, L.: Small perturbations are enough: Adversarial attacks on time series prediction. Inf. Sci. **587**, 794–812 (2022)
23. Xu, A., Wang, X., Zhang, Y., Wu, T., Xian, X.: Adversarial attacks on deep neural networks for time series prediction. In: 2021 10th ICICSE, pp. 8–14 (2021)
24. Yoon, Y., Swales, G.: Predicting stock price performance: a neural network approach. In: The Twenty-Fourth Annual Hawaii International Conference on System Sciences, vol. 4, pp. 156–162 (1991)
25. Zhang, G., Patuwo, B.E., Hu, M.Y.: Forecasting with artificial neural networks: the state of the art. Int. J. Forecast. **14**(1), 35–62 (1998)
26. Zhang, X., et al.: Traffic flow forecasting with spatial-temporal graph diffusion network. In: The AAAI Conference on Artificial Intelligence, vol. 35, pp. 15008–15015 (2021)
27. Zhang, Z.: Multivariate Time Series Analysis in Climate and Environmental Research. Springer, Cham (2018). https://doi.org/10.1007/978-3-319-67340-0

cPNN: Continuous Progressive Neural Networks for Evolving Streaming Time Series

Federico Giannini(✉), Giacomo Ziffer, and Emanuele Della Valle

DEIB - Politecnico di Milano, Via Ponzio 34/5, 20133 Milano, Italy
{federico.giannini,giacomo.ziffer,emanuele.dellavalle}@polimi.it

Abstract. Dealing with an unbounded data stream involves overcoming the assumption that data is identically distributed and independent. A data stream can, in fact, exhibit temporal dependencies (i.e., be a time series), and data can change distribution over time (concept drift). The two problems are deeply discussed, and existing solutions address them separately: a joint solution is absent. In addition, learning multiple concepts implies remembering the past (a.k.a. avoiding catastrophic forgetting in Neural Networks' terminology). This work proposes Continuous Progressive Neural Networks (cPNN), a solution that tames concept drifts, handles temporal dependencies, and bypasses catastrophic forgetting. cPNN is a continuous version of Progressive Neural Networks, a methodology for remembering old concepts and transferring past knowledge to fit the new concepts quickly. We base our method on Recurrent Neural Networks and exploit the Stochastic Gradient Descent applied to data streams with temporal dependencies. Results of an ablation study show a quick adaptation of cPNN to new concepts and robustness to drifts.

Keywords: Data streams · Catastrophic forgetting · Concept drift

1 Introduction

In a context where data comes as an unbounded data stream and is continually evolving, we must overcome the central hypothesis of Machine Learning (ML): the assumption according to which data is independent and identically distributed (shortly, i.i.d). It does not hold for any data stream where data could suffer from changes in its distribution (the so-called "concept drift") and shows temporal dependencies. While the literature has deeply investigated the two situations separately, few works deal with the joint problem. The need to find a combined solution is, thus, increasingly emerging. We formalize the mentioned problem by calling it **Evolving Streaming Time Series (ESTS)**. "Evolving" indicates the possibility of concept drift, while "Streaming" refers to data points arriving continually from an unbounded data stream. We use "Time Series" to stress the presence of temporal dependencies. Working with concept drifts and multiple concepts makes it necessary to consider the well-known stability-plasticity dilemma [14], according

© The Author(s) 2023
H. Kashima et al. (Eds.): PAKDD 2023, LNAI 13938, pp. 328–340, 2023.
https://doi.org/10.1007/978-3-031-33383-5_26

to which too much plasticity results in forgetting past knowledge. This problem is known as **catastrophic forgetting (CF)** [10]. Too much stability leads, instead, to difficulties in learning new knowledge.

Among the models for dealing with time series, sequential models based on Recurrent Neural Networks (RNN) are widely used in the literature [7]. Applying Neural Networks (NN) to the streaming scenario allows it to exploit its learning algorithm's (Stochastic Gradient Descent, SGD) adaptability. In contrast, SGD can suffer when the new concept differs significantly from the previous one, and a NN forgets the last concept when it learns a new one. To resolve these issues, Progressive Neural Networks (PNN) [19] consist of NN architectures to jointly remember the previously learned knowledge and use transfer learning to recycle the knowledge gained from old concepts [16]. However, this methodology is not meant to deal with an ESTS.

Our work, thus, aims to investigate the following **research question**: *in the context of an Evolving Streaming Time Series, is there a solution to jointly manage concept drifts, temporal dependencies, and catastrophic forgetting?* In this paper, we positively answer this question by contributing **Continuous PNN (cPNN)**, a novel continuous version of PNNs that extends them to an ESTS scenario. We first propose a strategy to exploit SGD in a streaming scenario to tame temporal dependencies. Secondly, our approach utilizes PNN-based architectures to efficiently address both concept drifts and CF, using transfer learning to enable rapid adaptation to new concepts while maintaining the predictive ability of previously learned ones. A crucial feature of cPNN is that the architecture can be potentially applied to each type of RNN model. We conduct an ablation study on a binary classification problem during the experiment phase to test cPNN on synthetically generated data streams. We compare cPNN with two ablated architectures: cLSTM and mcLSTM. After a concept drift, cLSTM continues training on the new concept. It, thus, does not avoid CF and is not concept drift aware. mcLSTM avoids CF, but it does not use transfer learning. Temporal dependencies are tamed by using RNN models. Results show that cPNN performs better after concept drifts than ablated architectures.

The rest of the paper is organized as follows. Firstly, Sect. 2 analyzes the already present ideas in literature. Section 3 exposes our method and contributions. Then, Sect. 4 discusses the settings of our experiments, while Sect. 5 exhibits the results. Finally, Sect. 6 discusses conclusions and future works.

2 Related Works

Continual Learning (CL) thoroughly investigated methods to learn and avoid CF continually [12]. The Task Incremental Learning scenario assumes that data is split into batches of samples (named experiences) provided over time. Each of them represents a task. The data distribution and objective function are normally fixed within a task. In this paper, we refer to this scenario whenever we use CL.

In this context (shown in Fig. 1.a), **PNNs** [19] are NN architectures that use transfer learning to recycle knowledge gained from previous tasks. Furthermore,

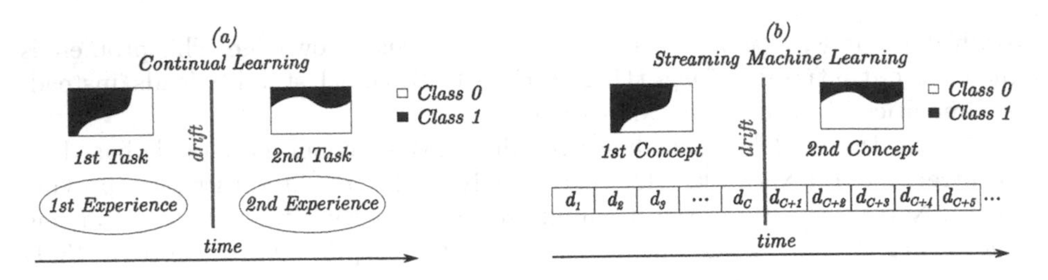

Fig. 1. Comparison of CL and SML scenarios.

the parameters associated with the old tasks are frozen to avoid CF. PNNs, thus, can learn a new task while keeping the predictive ability on the earlier tasks. The architecture is built dynamically and starts with a single NN (named **column**). Equation 1 shows that, for each new task k, a column is added whose *i-th* layer receives the *(i-1)-th* layer's output $h_{i-1}^{(k)}$ of column k and the *(i-1)-th* layers' outputs $h_{i-1}^{(j)}$ of all the earlier columns. W_i and U_i are the weight matrices to be learned. U_i are called **lateral connections** and implement transfer learning.

$$h_i^{(k)} = f\left(W_i^{(k)} h_{i-1}^{(k)} + \sum_{j<k} U_i^{(k:j)} h_{i-1}^{(j)})\right) \tag{1}$$

PNNs, on their own, do not tame temporal dependencies. To do so, they must use **RNNs** as columns, which operate on fixed-size sequences of items. RNNs recursively express the *i-th* hidden layer's output h_i as a function of the *i-th* item's features X_i and the output h_{i-1} of the *(i-1)-th* hidden layer. Due to the vanishing gradient, such an architecture cannot tame long temporal dependencies [7]. **Long short-term memory (LSTM)** [8] resolves this issue by memorizing only the helpful information and introducing the memory cell representing past cumulated knowledge. **Gated Incremental Memory (GIM)** [4] develops a recurrent version of PNNs using LSTM as columns. Column k receives lateral connections only from column $k-1$ to decrease the number of parameters. The *i-th* item's hidden layer $h_i^{(k)}$ of column k is computed as expressed by Eq. 2. For each item i, its features X_i and the previous column's *i-th* hidden layer output are concatenated. Lateral connections are represented by the weights applied to the output of the previous column's hidden layer. The model's output is computed for each sequence element i by applying a further layer after h_i.

$$h_i^{(k)} = LSTM([X_i, h_i^{(k-1)}]) \tag{2}$$

The works mentioned above assume all data in each experience to be accessible at once. The specific paradigm called **Streaming Machine Learning (SML)** [3], instead, was introduced to learn continually from a data point (or mini-batch) at a time (see Fig. 1.b). **Concept drift**, that is a phenomenon in which the statistical properties of a target domain change over time in an arbitrary way [13], is a crucial issue that SML tames. We can distinguish two types of concept drift: virtual and real. It is easy to take them apart in the context of streaming classification. Virtual concept drifts do not affect the decision boundary, while real ones do. Additionally, in an abrupt drift, the new concept replaces the old one in a short period or in an exact instant, while in gradual and incremental drifts, the new concept gradually or incrementally replaces the old one. Finally, the concepts could reoccur over time. Concept drift detectors can detect all the mentioned types of concept drift [13].

Most SML methods assume that the data stream's points are independent. In the real world, this assumption is unrealistic since they can exhibit temporal dependencies. Despite many works raising the issue that ignoring this situation can cause problems in the learning and evaluation processes [18,23,24], taming of temporal dependencies in an evolving data stream is still an open issue.

3 Proposed Method

This work proposes cPNN, a novel methodology for applying NNs to perform binary classification of an ESTS' data points. In Sect. 3.1, we analyze SGD behavior on data streams containing concept drifts. Section 3.2 proposes a method to exploit SGD in an ESTS scenario. Finally, Sect. 3.3 presents cPNN.

3.1 Stochasting Gradient Descent for Evolving Data Streams

The SGD's iterative nature makes it possible to apply it on data streams by buffering the data points in fixed-size batches [7].[1] Figure 2 illustrates this idea by analyzing a NN composed of a single linear neuron with two weights and no bias. Let's assume that the NN at d_{C1}, when a first abrupt drift occurs, has learned the decision boundary illustrated in Fig. 2.a. Notice that the second concept only marginally modifies the boundary between classes. Thus, SGD can quickly adapt to the drift since the minimum of the new concept's loss function is close to the previous one. On the contrary, the third concept swaps the classes when it occurs at d_{C2}. In this case, the new minimum is distant, and the SGD algorithm requires more iterations to reach it. Furthermore, the performance initially collapses since the starting configuration optimizes the inverted problem. In any case, when the model adapts to the new concept, it forgets the previous one since SGD has reached the new minimum. The more the new decision boundary changes, the lower the performance. Thus, a simple NN cannot deal with CF.

3.2 Stochasting Gradient Descent for Streaming Time Series

As already stressed, although the i.i.d. assumption is usually made for each concept, data can show time dependence that requires RNN models like LSTM. To ensure that SGD is an unbiased gradient, we cannot sample an entire i.i.d.

[1] See MOA's Perceptron application: https://www.cs.waikato.ac.nz/~abifet/MOA/API/classmoa_1_1classifiers_1_1functions_1_1perceptron.html.

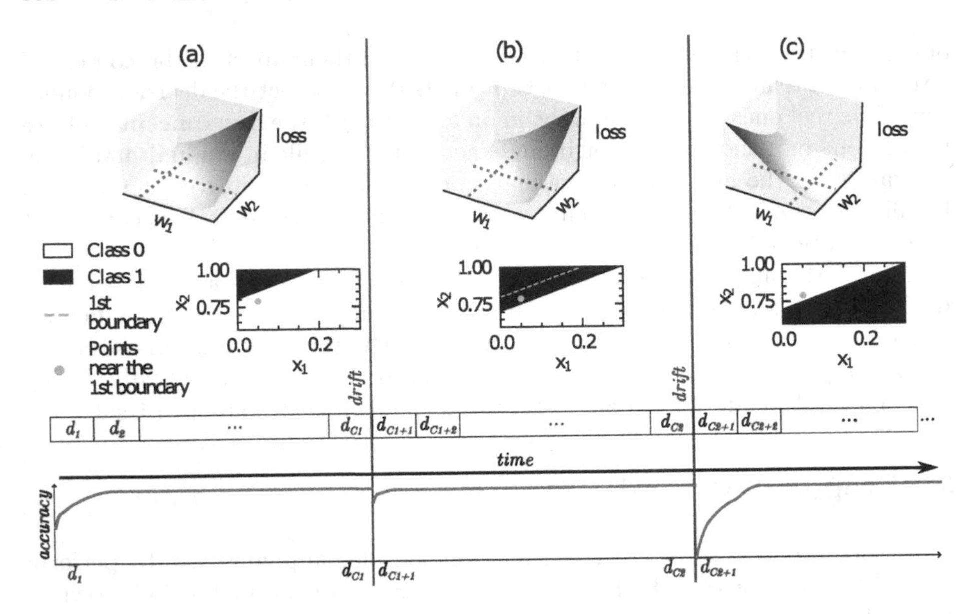

Fig. 2. Loss functions' minimum and accuracy trend of a single linear neuron associated with the following classification functions: (a) $-x_1+x_2-0.8 \geq 0$ (b) $-x_1+x_2-0.7 \geq 0$ (c) $-x_1 + x_2 - 0.7 < 0$.

training set [15]. The data points are, in fact, not available at once, and data has autocorrelations. We, therefore, input the data points in chronological order. Notice that, in this way, we are not minimizing the loss function to all the data but only to the most recently seen data points [1]. Indeed, the literature on data streams [2] commonly assigns greater weight to recent data points because we expect that future data points related to the current concept will bear greater similarity to recent data. In particular, we adopt windowing from Data Stream Management Systems to propose (see Fig. 3) to buffer data points in a batch with size B and build the sequences using a sliding window (with size W) once the batch is complete. In this way, we produce B-W+1 sequences for each batch. Notably, the windowing approach permits us to keep the temporal order.

3.3 Continuous PNN (cPNN)

To better adapt to the concept drift, we propose a methodology to combine the knowledge gained from previous concepts with that learned from the current one. At the same time, we deal with catastrophic forgetting and, thus, provide accurate predictions for all the concepts. Moreover, we handle data points arriving continually from an unbounded data stream and tame temporal dependencies.

PNNs and GIM can recycle old knowledge and avoid CF but are meant to be applied to CL experiences. We, thus, combine SML and CL techniques to build **Continuous PNN (cPNN)**: a continuous version of PNNs. We first define **Continuous LSTM (cLSTM)**, a continuous version of LSTM whose input is

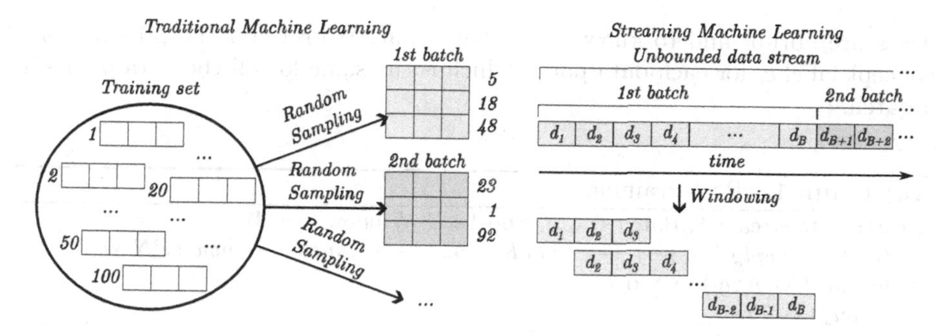

Fig. 3. Data processing in cases of Traditional Machine Learning and SML.

built as explained in Sect. 3.2. cLSTM outputs a probability distribution for each sequence item. Each data point's probability distribution on the target classes is computed by averaging its probability distributions associated with all the sequences to which it belongs. We then consider each concept as a task of CL. We use cLSTM as the base model (column) of cPNN to learn continually from an unbounded data stream's data points and tame temporal dependencies. Lateral connections are implemented as suggested by GIM to reduce the parameters. The architecture (shown in Fig. 4) can be edited by changing the column's type from cLSTM to any RNN model.

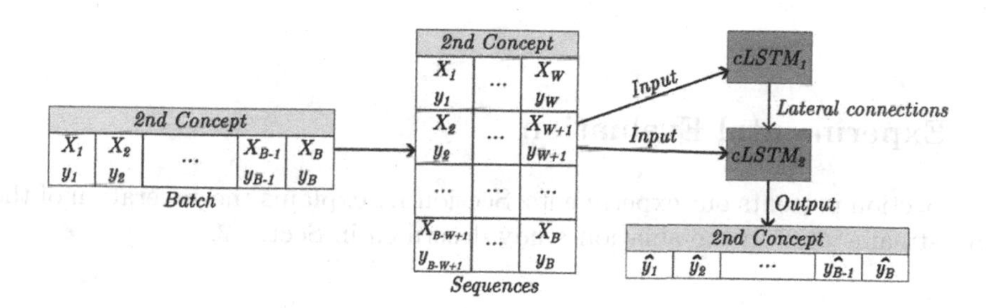

Fig. 4. The cPNN architecture during the second concept training.

Algorithm 1 details the cPNN's lifecycle. The architecture initially has a single column. We buffer the data stream in a batch with size B (Line 5) and create the model's input (Line 9) when the batch is complete. We, then, apply Prequential evaluation [6] (Lines 10-12), taking first the model's predictions and evaluating the performance on the entire batch. Finally, we train the model on the batch for several epochs. After a concept drift, the model receives the batch accumulated up to that time (Line 7). Then, we add a new column to the architecture, building lateral connections and freezing the weights of the previous column (Line 15). Since CL assumes that the label associated with each experience is known and experiences are not mixed, we also assume that drifts are abrupt and to know when they occur. We rely on the presence of a concept label c_t for each data point, which is the same for all the data points in a batch.

Algorithm 1. cPNN training

Input: Data stream S, Batch size B, Epochs E, Window Size W.

1: $batch \leftarrow$ *empty list*, $perf \leftarrow$ *empty list*, $c_{t-1} \leftarrow -1$, $model \leftarrow new\ cPNN()$
2: **for all** (X_t, y_t, c_t) in S **do**
3: $\quad drift \leftarrow True$
4: $\quad$ **if** $c_t = c_{t-1}$ **then**
5: $\quad\quad$ Append (X_t, y_t) to $batch$
6: $\quad\quad drift \leftarrow False$
7: $\quad$ **if** $|batch| = B$ OR $drift = True$ **then**
8: $\quad\quad$ **if** $|batch| \geq W$ **then**
9: $\quad\quad\quad X, Y \leftarrow BuildSequences(batch, W)$
10: $\quad\quad\quad pred \leftarrow model.predict(X)$
11: $\quad\quad\quad$ Append $Evaluate(pred, Y)$ to $perf$
12: $\quad\quad\quad model.fit(X, Y, E)$
13: $\quad\quad batch \leftarrow$ *empty list*
14: $\quad$ **if** $drift = True$ **then**
15: $\quad\quad model.addColumn()$
16: $\quad\quad$ Append (X_t, y_t) to $batch$
17: $\quad c_{t-1} \leftarrow c_t$

4 Experimental Evaluation

This Section presents our experiments. Section 4.1 explains the generation of the data streams used in the ablation study described in Sect. 4.2.

4.1 Generated Data Streams

As detailed in [21], the most commonly used SML benchmarks containing temporal dependencies (Electricity and CoverType) are unsuitable for our purpose. The most known synthetic data stream generators (SINE [5], SEA [22], Hyperplane [9], and STAGGER [20]) do not, instead, introduce temporal dependencies in the data. We, thus, propose the construction of a synthetic generator to have a simple and controlled case study to apply the models and analyze their behaviors. We start from SINE and produce a variant whose generated points have temporal dependencies. We begin from a randomly generated two-dimension point in (0,1). Each coordinate of the following points is generated by summing a random value (random walk [17]) to the previously generated point's value. Every random walk's sign is generated to prevent exceeding the range (0,1). After quantifying the autocorrelation between data points using the

Partial Autocorrelation Function plot, we set the maximum size of data points having the same label as ten. To identify the boundaries of the classes, we utilize the two SINE generator's **boundary functions** defined in Eqs. refeqspssine1 and refeqspssine2.

$$S1 : x_1 - sin(x_2) = 0 \tag{3}$$

$$S2 : x_1 - 0.5 - 0.3\ sin(3\ \pi\ x_2) = 0 \tag{4}$$

We denote by and the **classification functions** that classify with "1" the points above, respectively, the S1 and S2 curves, while with "0" the remaining ones. and invert the labels of and respectively. We generate one data stream for every classification function, each representing one concept and containing 50k data points. Let us introduce the term **sign drift** as the drift where a new concept reverses the labels while maintaining the boundary function (e.g. from to) or changing it (e.g., from to). We combine the data streams in two ways. Firstly, **classification inversion drift** produces a single sign drift that keeps the boundary function unchanged (Fig. 2.c). Secondly, **boundary function drift** combines all four concepts' data streams by alternating the boundary functions (Fig. 2.b) and producing one or two sign drifts. By design, more than 50% of the sequences with the same label have a maximum length of five, and labels are balanced. When we change the boundary function without a sign drift (e.g., from to), 65% of the points keep the same label. If we combine a sign drift with a boundary function drift (e.g., from to), the percentage drops to 35%. Finally, all the points change their labels after a classification inversion drift (e.g., from to).

4.2 Experimental Setting

We conduct an ablation study for our hypothesis formulation and compare cPNN with two alternative architectures. **mcLSTM** (Multiple cLSTMs) remove the lateral connections so that each new column does not consider the previous column's hidden layer output. Direct application of the base model **cLSTM** (see Sect. 3.3) removes, instead, the creation of different columns, resulting in a cPNN with only one column that ignores drifts. Hyperparameter values are chosen as follows after executing the preliminary experiments. Epochs number: 10, window size: 10, batch size: 128, learning rate: 0.1, hidden's layer size: 50.[2] The final performance is computed by averaging the batches. Since the labels are balanced and we do not focus on a particular class, we evaluate the accuracy.

Our hypothesis is that cPNN can adapt to new concepts in an ESTS more quickly than the other two architectures. Additionally, we expect that models can quickly adapt to a new concept if it is similar to the previous one. A sign drift would be more complicated if the model does not learn to invert its past knowledge. We evaluate the final accuracy in four **cases** to verify the two hypotheses for each concept. The first two cases (**[1,50]** and **[1,100]**) analyze how models adapt to the new concept by considering the accuracy at the end of the first 50

[2] Complete source code available at https://github.com/federicogiannini13/cpnn.

Table 1. Accuracies on classification inversion drift. cPNN outperforms the ablated versions in all cases.

concept	case	cPNN	cLSTM	mcLSTM	cPNN	cLSTM	mcLSTM
2nd	[1,50]	**.96, .004**	*.872, .026*	.903, .008	**.864, .017**	.742, .024	.75, .006
	[1,100]	**.972, .002**	*.921, .014*	.933, .004	**.888, .02**	.79, .024	*.774, .006*
	(100,)	**.989, .001**	.98, .002	*.975, .001*	**.927, .018**	.883, .029	*.834, .009*
	[1,)	**.984, .001**	.965, .004	.964, .001	**.917, .018**	.859, .026	*.818, .007*

and 100 batches after the drift.[3] A reasonably accurate model in the first part of the concept is robust to concept drifts. The third case (**[100,)**), which covers the batch range from 100 onwards, assesses the accuracy of models in response to the newly introduced concept once they have adapted to it. Finally, the fourth case (**[1,)**) monitors the entire concept by investigating accuracy from the first batch onwards. Each experiment is repeated ten times, and their average accuracy is analyzed. Tables 1, 2 and 3 of Sect. 5 report results using the mentioned notations.

5 Results

This section analyzes the results of the different experiments described in Sect. 4.2. Tables 1, 2 and 3 report the ten executions' average accuracies and standard deviations. Since the first concept's architectures are the same, we make comparisons from the second concept onwards. Architectures are compared in pairs. We report in bold the statistically best-performing architecture (if it is statistically better performing than the remaining two) and in italics the less-performing one. We, thus, first conduct a Shapiro-Wilk test to check for normality. If we cannot reject the null hypothesis for both distributions, we conduct a Welch's t-test. Otherwise, we run a Wilcoxon signed-rank test. We perform a one-sided test in both cases. We underline the not normally distributed samples. All the tests are conducted with a significance level of 0.05.

5.1 Classification Inversion Drift

Results in Table 1 show that after the concept drift, cLSTM performance collapse. Since they are similar, we only report results for two data streams. For the S1 classification function, the mcLSTM's random initialization of the parameters works better than the cLSTM one (which is the inverse concept's optimal one), but from the 100th till the end of the concept, cLSTM outperforms mcLSTM. cPNN can adapt quickly to the new concept. It results in being the best-performing model in all the experiments. In the case of S2, the gap between

[3] Thresholds represent the number of batches during preliminary experiments, after which the more robust and less robust models achieved good performance.

cPNN and the other models is more significant. These experiments suggest that cPNN could learn to invert past knowledge. cLSTM requires more iterations to reach the new optimal setting since it starts from the inverse concept one. At the end of each concept, cLSTM's new optimal configuration is still worse than cPNN's one.

Table 2. Accuracies on data streams , , , and , , , . cPNN always recovers faster from concept drifts than the ablated versions. In some cases, a single cLSTM performs better in the long run, but in the end, it only remembers that last concept since it does not manage CF. mcLSTM that does not use transfer learning and resets the parameter configuration performs worse in almost all situations.

concept	case	cPNN	cLSTM	mcLSTM	cPNN	cLSTM	mcLSTM
2nd	[1,50]	**.759, .012**	.746, .011	*.738, .006*	**.781, .006**	*.743, .015*	.753, .006
	[1,100]	**.803, .007**	.791, .012	*.762, .005*	**.808, .004**	.772, .008	.775, .004
	(100,)	.877, .006	**.897, .013**	*.829, .011*	**.876, .005**	.851, .011	*.823, .009*
	[1,)	.858, .005	**.87, .012**	*.812, .009*	**.859, .003**	.831, .009	*.811, .007*
3rd	[1,50]	**.951, .004**	.893, .022	.905, .008	**.947, .004**	.94, .009	*.908, .004*
	[1,100]	**.964, .004**	.932, .013	.935, .004	**.961, .003**	.955, .006	*.936, .003*
	(100,)	.982, .002	.982, .001	*.975, .001*	.981, .002	.982, .001	*.975, .001*
	[1,)	**.977, .002**	.969, .004	*.965, .001*	.976, .002	.975, .002	*.965, .001*
4th	[1,50]	**.855, .012**	.778, .016	*.754, .005*	**.862, .008**	.777, .025	*.738, .005*
	[1,100]	**.88, .011**	.822, .012	*.776, .004*	**.88, .006**	.831, .019	*.76, .003*
	(100,)	.907, .007	.91, .009	*.826, .01*	.909, .009	.915, .013	*.827, .009*
	[1,)	**.9, .008**	.888, .009	*.813, .008*	.902, .008	.894, .014	*.81, .007*

Table 3. Accuracies on data streams , , , and , , , .

concept	case	cPNN	cLSTM	mcLSTM	cPNN	cLSTM	mcLSTM
2nd	[1,50]	.931, .014	.933, .012	*.915, .004*	**.921, .008**	.892, .022	.903, .003
	[1,100]	.943, .01	**.951, .008**	.941, .003	.939, .009	.934, .012	.934, .003
	(100,)	*.974, .003*	**.983, .001**	.976, .001	*.971, .004*	**.983, .001**	.975, .001
	[1,)	.966, .005	**.974, .003**	.967, .001	.963, .005	**.97, .003**	.964, .001
3rd	[1,50]	**.824, .013**	.768, .023	.754, .005	**.835, .013**	.779, .016	*.755, .007*
	[1,100]	**.851, .015**	.802, .024	*.774, .004*	**.863, .01**	.812, .018	*.776, .003*
	(100,)	.896, .01	.892, .018	*.835, .011*	.903, .009	.893, .017	*.831, .01*
	[1,)	**.885, .011**	.869, .02	*.819, .009*	**.893, .009**	.872, .017	*.817, .008*
4th	[1,50]	**.952, .006**	.942, .007	*.907, .008*	**.953, .007**	.923, .017	.92, .004
	[1,100]	**.963, .004**	.957, .003	*.936, .005*	**.962, .004**	.945, .01	.944, .002
	(100,)	.98, .004	**.983, .002**	*.975, .001*	.979, .003	**.982, .002**	*.976, .001*
	[1,)	.975, .004	.976, .001	*.965, .002*	.975, .003	.973, .004	*.967, .0*

5.2 Boundary Function Drift

Results regarding boundary function drift (shown in Tables 2 and 3) indicate that cPNN adapts more quickly to a new concept after a sign drift and when the new boundary function is more complex than the previous (a drift from S1 to S2). In this case, cPNN outperforms the other architectures in the first 50 and 100 batches. From the 100th batch, cLSTM and cPNN have similar performance. cLSTM outperforms cPNN in the first batches only after the first drift from S2 to S1, with no sign drift. mcLSTM performs worse in almost all the experiments.

6 Conclusion

This paper pioneers a novel continuous version of PNNs for Evolving Streaming Time Series. We proposed CPNN to deal simultaneously with concept drifts and temporal dependencies while avoiding catastrophic forgetting. To do so, we presented a continuous adaptation of LSTM (namely cLSTM) that exploits the SGD algorithm to tame temporal dependencies in a data stream. A similar method was used by [11] on a complex architecture and real datasets. Instead, our goal was to analyze the models' behaviors using a simplified scenario. cPNN's architecture is based on PNNs to tame CF and use transfer learning to fit new concepts quickly. To investigate cPNN behavior, we generated synthetic data streams and conducted an ablation study. cPNN performance highlighted a quicker adaptation to new concepts. Its average accuracy after each concept drift is, in fact, statistically greater than the ablated ones. cPNN resulted, thus, in being more robust to concept drifts, especially in the case of sign drift.

One of the main limitations of cPNN is that its complexity increases linearly with the number of concepts. We, thus, imagine that this architecture could be applied in the case of reoccurrent drifts where we would need to check whether the new concept has been seen before. Additionally, when dealing with data streams, the selection of hyperparameters can become challenging, and the resulting outcomes may be highly sensitive to these choices. Moreover, we only studied the models in a simplified scenario with abrupt concept drifts and synthetic data streams containing only two features. In our future works, we intend to explore more types of drift in a higher dimensional space and complex classification functions. Finally, as in many CL experiments, we assumed to have an "oracle" that knows the concept associated with each data point. In our future works, we will apply concept drift detection methods. cPNN performance results suggested that it could automatically learn to invert past knowledge when there is a sign drift. We also think its quicker adaptation to the new concept is due to past recycling ability. We will analyze the model's parameters in future works to verify it. In the long term, we intend to investigate how cPNN learns in contexts where real data evolves via gradual or incremental concept drifts. We will most likely need to examine other types of columns, like Gated Recurrent Units or Transformers.

References

1. Anagnostopoulos, C., Tasoulis, D.K., Adams, N.M., Pavlidis, N.G., Hand, D.J.: Online linear and quadratic discriminant analysis with adaptive forgetting for streaming classification. Stat. Anal. Data Min. **5**(2), 139–166 (2012)

2. Babcock, B., Babu, S., Datar, M., Motwani, R., Widom, J.: Models and issues in data stream systems. In: PODS, pp. 1–16. ACM (2002)
3. Bifet, A., Gavaldà, R., Holmes, G., Pfahringer, B.: Machine learning for data streams: with practical examples in MOA. MIT press (2018)
4. Cossu, A., Carta, A., Bacciu, D.: Continual learning with gated incremental memories for sequential data processing. In: IJCNN, pp. 1–8. IEEE (2020)
5. Gama, J., Medas, P., Castillo, G., Rodrigues, P.: Learning with drift detection. In: Bazzan, A.L.C., Labidi, S. (eds.) SBIA 2004. LNCS (LNAI), vol. 3171, pp. 286–295. Springer, Heidelberg (2004). https://doi.org/10.1007/978-3-540-28645-5_29
6. Gama, J., Sebastião, R., Rodrigues, P.P.: Issues in evaluation of stream learning algorithms. In: KDD, pp. 329–338. ACM (2009)
7. Goodfellow, I.J., Bengio, Y., Courville, A.C.: Deep Learning. Adaptive Computation and Machine Learning, MIT Press, Cambridge (2016)
8. Hochreiter, S., Schmidhuber, J.: Long short-term memory. Neural Comput. **9**(8), 1735–1780 (1997)
9. Hulten, G., Spencer, L., Domingos, P.M.: Mining time-changing data streams. In: KDD, pp. 97–106. ACM (2001)
10. Lange, M.D., et al.: A continual learning survey: defying forgetting in classification tasks. IEEE Trans. Pattern Anal. Mach. Intell. **44**(7), 3366–3385 (2022)
11. Lemos Neto, Á.C., Coelho, R.A., Castro, C.L.: An incremental learning approach using long short-term memory neural networks. J. Control Autom. Electr. Syst. 1–9 (2022). https://doi.org/10.1007/s40313-021-00882-y
12. Lesort, T., Lomonaco, V., Stoian, A., Maltoni, D., Filliat, D., Rodríguez, N.D.: Continual learning for robotics: definition, framework, learning strategies, opportunities and challenges. Inf. Fusion **58**, 52–68 (2020)
13. Lu, J., Liu, A., Dong, F., Gu, F., Gama, J., Zhang, G.: Learning under concept drift: a review. IEEE Trans. Knowl. Data Eng. **31**(12), 2346–2363 (2019)
14. McCloskey, M., Cohen, N.J.: Catastrophic interference in connectionist networks: the sequential learning problem. In: Psychology of Learning and Motivation, vol. 24, pp. 109–165. Elsevier (1989)
15. Meng, Q., Chen, W., Wang, Y., Ma, Z., Liu, T.: Convergence analysis of distributed stochastic gradient descent with shuffling. Neurocomputing **337**, 46–57 (2019)
16. Pan, S.J., Yang, Q.: A survey on transfer learning. IEEE Trans. Knowl. Data Eng. **22**(10), 1345–1359 (2010)
17. Pearson, K.: The problem of the random walk. Nature **72**(1865), 294–294 (1905)
18. Read, J., Rios, R.A., Nogueira, T., de Mello, R.F.: Data streams are time series: challenging assumptions. In: Cerri, R., Prati, R.C. (eds.) BRACIS 2020. LNCS (LNAI), vol. 12320, pp. 529–543. Springer, Cham (2020). https://doi.org/10.1007/978-3-030-61380-8_36
19. Rusu, A.A., et al.: Progressive neural networks. CoRR abs/1606.04671 (2016)
20. Schlimmer, J.C., Granger, R.H.: Incremental learning from noisy data. Mach. Learn. **1**(3), 317–354 (1986)
21. de Souza, V.M.A., dos Reis, D.M., Maletzke, A.G., Batista, G.E.A.P.A.: Challenges in benchmarking stream learning algorithms with real-world data. Data Min. Knowl. Discov. **34**(6), 1805–1858 (2020)
22. Street, W.N., Kim, Y.: A streaming ensemble algorithm (SEA) for large-scale classification. In: KDD, pp. 377–382. ACM (2001)

23. Ziffer, G., Bernardo, A., Della Valle, E., Cerqueira, V., Bifet, A.: Towards time-evolving analytics: Online learning for time-dependent evolving data streams. Data Sci. 1–16 (in press)
24. Zliobaite, I., Bifet, A., Read, J., Pfahringer, B., Holmes, G.: Evaluation methods and decision theory for classification of streaming data with temporal dependence. Mach. Learn. **98**(3), 455–482 (2015)

Open Access This chapter is licensed under the terms of the Creative Commons Attribution 4.0 International License (http://creativecommons.org/licenses/by/4.0/), which permits use, sharing, adaptation, distribution and reproduction in any medium or format, as long as you give appropriate credit to the original author(s) and the source, provide a link to the Creative Commons license and indicate if changes were made.

The images or other third party material in this chapter are included in the chapter's Creative Commons license, unless indicated otherwise in a credit line to the material. If material is not included in the chapter's Creative Commons license and your intended use is not permitted by statutory regulation or exceeds the permitted use, you will need to obtain permission directly from the copyright holder.

Dynamic Variable Dependency Encoding and Its Application on Change Point Detection

Hao Huang[1(✉)] and Shinjae Yoo[2]

[1] GE Research, Niskayuna, NY 12309, USA
hao.huang1@ge.com
[2] Brookhaven National Lab, Upton, NY 11973, USA
sjyoo@bnl.gov

Abstract. Multivariate time series usually have complex and time-varying dependencies among variables. In order to spot changes and interpret temporal dynamics, it is essential to understand these dependencies and how they evolve over time. However, the problem of acquiring and monitoring them is extremely challenging due to the dynamic and non-linear interactions among time series. In this paper, we propose a dynamic dependency learning method, which learns dependency latent space with a two-level attention model. The first level is a bi-sided attention module to learn the short-term dependencies. Once the sequence of short-term dependencies is collected over a certain period of time, a temporal self-attention module is applied to obtain the actual dependencies for the current timestamp. The coordinates in latent space is descriptor of the temporal dynamic. We apply this descriptor to change point detection, and experiments show that our proposed method outperforms popular baselines.

Keywords: Change point detection · Dynamic variable dependency

1 Introduction

Among all the time series modeling tasks, learning variable dependencies across time is of critical importance in various domains such as biological networks, activity recognition and industrial system monitoring. For example, learning interactions between molecules (e.g. proteins or genes) can help us better understand the temporal interaction between molecules. More importantly, it can provide hints to discover core mechanisms in time domain, which can be used as marker to indicate the stage of specific diseases such as cancer.

One way of representing these dependencies is to construct dynamic networks (graphs) where each entity is a node associated with a variable (i.e. a univariate time series), while an edge at each timestamp represents the current relationship between the two connected nodes. Nevertheless, the underlying dependencies of variables are usually **not known a priori**. An important problem that arises in many applications is using observational time series alone to infer these relationships and their evolution over time.

© The Author(s), under exclusive license to Springer Nature Switzerland AG 2023
H. Kashima et al. (Eds.): PAKDD 2023, LNAI 13938, pp. 341–352, 2023.
https://doi.org/10.1007/978-3-031-33383-5_27

In many applications, such dependencies can be **non-linear and time-variant**, which makes it very challenging to extract them through traditional methods such as vector autoregressive model (VAR) [11], Granger causality [17], network inference [8] or even recent deep learning based methods [12,24].

Given multivariate time series as input, we propose *Dynamic Variable Dependency Encoding*, or ***DVDE***, which learns dynamic dependencies among variables across time. It consists of two levels of attention modules. It first learns short-term dependencies via a bi-sided attention module in a certain period of time before current timestamp. Once it collected a sequence of short-term dependencies, it uses a temporal self-attention module to enhance some parts of the sequence while diminishing other parts in order to acquire the actual dependency for current timestamp. The dependencies are encoded in a latent space, which is used as a descriptor to represent the current temporal pattern. In the experimental study, we test change point detection upon the descriptor output and show that our model outperforms state-of-the-art methods.

2 Related Works

In multivariate time series modeling, an important problem that arises in many applications is using observational multivariate time series to infer the dependency among variables in order to understand how the structure of these complex systems changes over a period of interest [10].

To learn these dynamic dependencies, state-of-the-art studies can generally be divided into three categories:

1. *Classic stochastic process models.* They include models such as latent space based Kalman Filters [5] and VAR [11] that are extended to deal with multivariate time series. Although they have been popularly used in many applications throughout the years, their success highly rely on the assumption that the dependencies among time series variables are linear. When used in real world problems, their approximation to nonlinear interactions does not lead to satisfactory results [7].
2. *Graphical lasso relevant methods.* Inferring dynamic networks via graphical lasso have been explored by [10,21,22]. In their work [10], Hallac et.al proposed time-varying graphical lasso model that can infer dynamic networks from raw time series data. Compared against the previous works, it is able to model different types of network evolutionary patterns. However, the selection of the proper evolutionary penalty is very important for obtaining accurate results [10]. In real world applications, it is extremely challenging to set this penalty parameter without prior knowledge, which makes their method less practical. Tomasi et.al. in [22] presented a method that models dynamic networks via latent variable space of observed time series. It is an attempt to generalise both latent variable and dynamic network inference.
3. *Deep-learning based methods.* Applying deep learning models to multivariate time series data has recently gained growing interest in a variety of applications. Among all these efforts, Huang et.al inferred static nonlinear networks via deep learning in [12,24]. To study dynamic networks, Tank et al.

[20] proposed a class of nonlinear architectures with Multi Layer Perceptron (MLP) or Recurrent Neural Network (RNN). Dang et al. [7] proposed a deep learning framework that consists of multiple customized gated recurrent units (GRUs) to discover nonlinear and inter-time-series dependencies. Specifically, it uses GRUs to represent the dynamic dependencies as high-dimensional vectors. Such vectors collected from all component time series form the informative output to discover inter-dependencies. IMV-LSTM [9] is a multi-variable attention-based LSTM model that makes use of the hidden state matrix to construct an associated update scheme, such that each row of the hidden matrix encapsulates information exclusively from a specific variable of the input. Temporal Causal Discovery Framework (TCDF) was proposed in [18], which is based on generic Temporal Convolutional Network (TCN) architecture and uses multiple convolutional layers with dilatation to increase the receptive field.

Table 1. List of key notations used in this paper.

X	an observed multivariate time series
m	number of variables
a	length of receptive window to learn global dependency
$Y, \hat{Y}$	the true and predicted next timestamp
$x^i_{t-\ell+1:t}$	the i-th variable's values of the latest ℓ timestamps
G_t	the short-term features at timestamp t
$\tilde{G}_t$	the global features at timestamp t
U, V	the left and right basis of dependency
L_t, R_t	the left and right attention matrix at timestamp t
$\mathcal{A}$	the graph seed collection
P_t	the short-term weights of dependency basis
$\tilde{P}_t$	the (global) weights of dependency basis
A_t	the dependency graph at timestamp t

3 Problem Setting and Notations

Table 1 lists the key notations used in this paper. Given multivariate time series as input, our *DVDE* aims at learning the dependency graph $A_t \in \mathbb{R}^{m \times m}$ at each timestamp t where m is the number of nodes. Each node is associated with a variable (i.e. a univariate time series). Each edge is weighted with continuous value that represents the current dependency between the two connected nodes. While the number of nodes m is fixed, the weights of edges are time-varying and can be either positive or negative (or even zero).

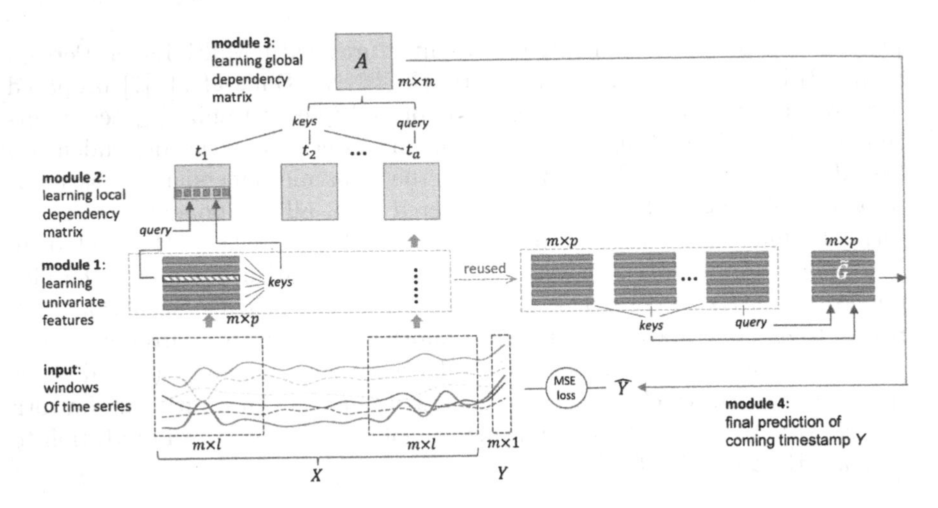

Fig. 1. *DVDE* consists of four modules. It takes multivariate time series as input and predicts the coming timestamp ($\hat{Y}$). Dependency graph A is intermediate output.

4 The Proposed DVDE Model

Instead of learning A_t directly in the format of $\mathbb{R}^{m \times m}$, we learn its coordinates in the dependency latent space, i.e. weights of dependency basis. The reason is that low-rank approximation can lead to less noise influence, better accuracy and higher scalability [24]. Our dependency basis are learnable network parameters that shared by the whole time series dataset. They consist of two parts: left and right basis, noted as U and $V \in \mathbb{R}^{m \times k}$ where k is rank size. Each column in U and V will be multiplied (vector product) to obtain a seed of dependency graph $\in \mathbb{R}^{m \times m}$, therefore in total there are k^2 such graph seeds. If U and V contains sufficient information, any A_t can be represented by a weighted sum of all these graph seeds, that is,

$$\mathcal{A} = \| \ (U(:,i)V(:,j)^T)_{i \in \{1,k\}, j \in \{1,k\}},$$
$$A_t = \sum_{\iota=1}^{k^2} (f(\tilde{P}_t))_\iota \mathcal{A}_\iota, \tag{1}$$

where $\| \ (*)$ and $(*)^T$ denotes the matrix concatenation and transpose operation respectively, $\mathcal{A} \in \mathbb{R}^{k^2 \times m \times m}$ are the collection of k^2 graph seeds, and $\tilde{P}_t \in \mathbb{R}^{k \times k}$ are the (global) dependency weights at timestamp t with $f(*)$ a flatten function to convert square matrix to vector $\in \mathbb{R}^{1 \times k^2}$. In our work, $\tilde{P}_t$ is then fed into regular change point detection methods such as Bottom-up [14] to detect change points.

Figure 1 illustrates the framework of our *DVDE*. Given input time series $X \in \mathbb{R}^{m \times n}$, *DVDE* outputs the predicted value of the coming timestamp $\hat{Y} \in \mathbb{R}^{m \times 1}$.

A_t (and the hidden dependency weights $\tilde{P}_t$) are intermediate output during the regression process. The whole framework consists of four modules:

1) Module 1 applies temporal convolutional networks (TCN) on each variable respectively on a sliding window basis before timestamp t, to acquire a sequence of univariate features. Then an attention model is used to learn the global nonlinear features $\tilde{G}_t$ at timestamp t.
2) Module 2 is a bi-sided attention module, to learn the sequence of short-term dependency weights at $t_1, ..., t_a$.
3) Given the sequence of short-term dependency, module 3 obtains the global dependency by using a temporal attention module. Then the module follows Equation (1) to obtain dependency graph A_t.
4) Module 4 performs the final prediction.

The following subsections will describe them in details.

4.1 Module 1: Learning Nonlinear Univariate Features

Module 1 is to acquire nonlinearly transformed features from each input variable respectively, without convolving intervariable information. It first applies a TCN on each variable to acquire their nonlinear features. The output of TCN are a sequence of univariate features at each timestamp. The output from all variables are concatenated and noted as $G_t \in \mathbb{R}^{m \times p}$. The second part of module 1 is a temporal attention model that takes the sequence of concatenated output to acquire the global nonlinear features $\tilde{G}_t$, which will be used in module 4 later.

4.2 Module 2: Learning Short-term Dependency

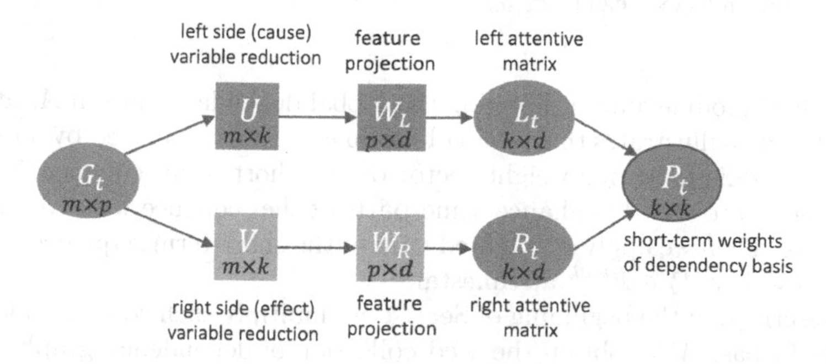

Fig. 2. Module 2 is a bi-sided attention model to learn short-term dependency weights P_t. Here, circles denote network outputs and squares denote weight parameters.

The second module is to learn short-term dependencies at each timestamp. The short term dependency sequence will be fed into module 3 later to learn the global dependency.

As we described in the beginning of Sect. 4, we learn dependency graph through weights of dependency basis. The process is illustrated in Fig. 2. It learns left and right attentive matrix respectively, which are then multiplied together to obtain the short term weights P_t. First of all, we define and initialize dependency basis U and V (as network parameters). Then to learn the left attentive matrix, we take input G_t from module 1 and projects the original variable space (m) to the space of left dependency basis (k) using U. Next, a second projection is applied to the feature space that projects p to attention space with d dimensions using another network parameter $W_L \in \mathbb{R}^{p \times d}$. The output is the left attentive matrix noted as $L_t \in \mathbb{R}^{k \times d}$. Similarly, we can obtain right attentive matrix $R_t \in \mathbb{R}^{k \times d}$ from the lower path in Fig. 2. The short-term weight P_t is obtained by:

$$P_t = tanh(L_t R_t^T), \tag{2}$$

where $(*)^T$ denotes the matrix transpose operation.

4.3 Module 3: Learning Actual Dependency

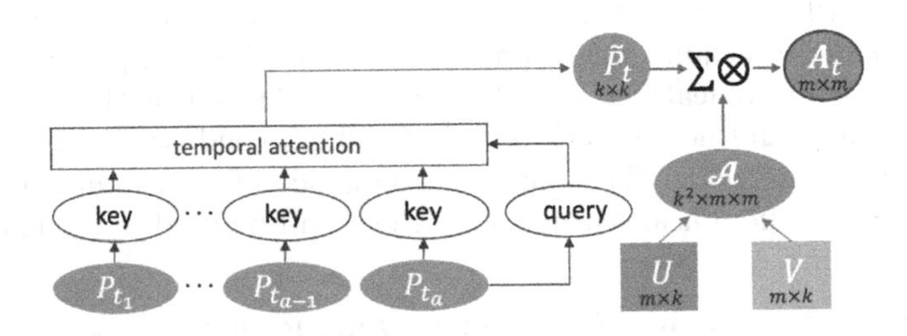

Fig. 3. Module 3 is to learn the global dependency graph A_t from the sequence of short-term dependency weights $P_{t_1:t_a}$.

The third module aims at learning the global dependency graph A_t at timestamp t. Figure 3 illustrates the design of module 3. This is achieved by a temporal attention module to learn a weight vector on the short-term sequence. The temporal weights are used to enhance some parts of the sequence while diminishing other parts. Then we apply a weighted sum of the short-term sequence to obtain the global weight $\tilde{P}_t \in \mathbb{R}^{k \times k}$ at timestamp t.

As described in the beginning of Sect. 4, we multiply each column in left basis U and right basis V to obtain the seed collection of dependency graph. At last, the dependency graph A_t is obtained by Eq. (1).

4.4 Module 4: Final Prediction and Application

Module 4 takes input from $\tilde{G}_t$ (from module 1) and A_t (from module 3), and outputs the predicted values of the coming timestamp $\hat{Y}$. Here A_t is used to enroll intervariable information by $D_t = (\tilde{G}_t^T A_t)^T$ where $D_t \in \mathbb{R}^{m \times p}$. Specifically, $\tilde{G}_t$ represents the collection of univariate features that is then convoluted by the dependency matrix A_t. At the last step, the model learns a weight parameter matrix $W_Y \in \mathbb{R}^{m \times p}$ and applies it on D_t with a row-wise dot product to output $\hat{Y} \in \mathbb{R}^{m \times 1}$. Residual is calculated by mean squared error between $\hat{Y}$ and actual Y.

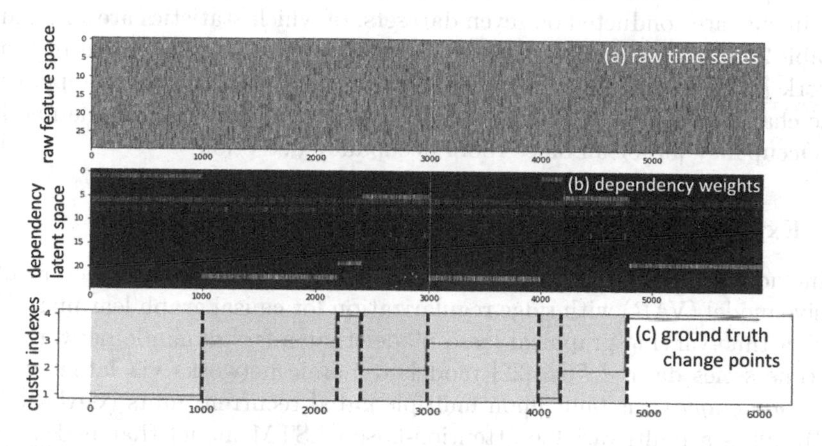

Fig. 4. Change point detection example. The x axis of the three subfigures are aligned time axis. Figure (a) shows a synthetic time series (plot as heatmap). Figure (b) shows the (flattened) dependency weights encoded by our model that are consistent with ground truth change points (Figure (c)).

To examine the learning quality of dependency we test our method on change point detection. For each timestamp, we flatten the weight vector $\tilde{P}_t$ and treat it as the pattern descriptor at timestamp t. After collecting all $\|_{t=1}^{n} (\tilde{P}_t) \in \mathbb{R}^{n \times k^2}$, we feed it to a popular signal segmentation method, Bottom-up change point detection [14], to obtain change points. Figure 4 illustrates an example of our method on a synthetic time series (Syn1 in Table 2).

Table 2. Dataset statistics and description.

Dataset	#time series	#dim	length	#change points	description
Syn1	1	30	6,000	7	bio synthetic
Syn2	10	65	10,000	100	bio synthetic
Flights1	20	47	4,025-7,084	60	flight sensors
Flights2	20	47	10,045-20,122	60	flight sensors
Beedance	6	3	608-1,124	117	bee activity
HASC	18	3	11,738-12,000	196	human activity
Occupancy	1	4	8143	12	room occupancy

5 Experiments

Our code and datasets are available in shorturl.at/sESZ4. We focus on two perspectives of experiments: the quality of the learned dependencies and the performance of change point detection. All experiments are carried out in *unsupervised manner*: other than raw multivariate time series given as input, there is no prior knowledge such as the number of change points or graph sparsity degree. The experiments are conducted on seven datasets, of which statistics are summarized in Table 2. They include two synthetic datasets that simulate gene regulatory network [25], two flight datasets [13] where change points indicate operational phase change, Beedance [19] and HASC [1] that record bee and human activity, and Occupancy [3] for an office room occupancy detection.

5.1 Experiments on Learning Dynamic Dependencies

We include the following popular baselines. *VAR* [11] applies the vector autoregressive model (VAR) with ridge regularization for causal graph learning. *TVGL* [10] is a time-varying graphical lasso model that infers dynamic networks from raw time series data. *LTGL* [22] models dynamic networks via latent variable space. *seq2graph* [7] is built upon multiple gated recurrent units (GRUs). *IMV-LSTM* [9] is a multi-variable attention-based LSTM model that makes use of the hidden state matrix and develops an associated update scheme. *TCDF* [18] is based on generic Temporal Convolutional Network (TCN) and uses multiple convolutional layers with dilatation to increase the receptive field.

Two simulated data sets (Syn1 and Syn2) are created where we can quantify the graph learning quality with ground truths. The ground truth are binary where 1 indicates directed dependency and 0 means no dependency. All learned dependencies are converted to absolute value for evaluation.

For multivariate time series, our model infers a sequence of graphs, which are evaluated against ground truth graphs one by one respectively with two metrics: Area Under Precision and Recall curve ($AUPR$) and Area Under Receiver Operating Characteristics curve ($AUROC$). For both metrics, higher score (up to 1) means better accuracy. The average result across all graphs from each dataset is shown in Fig. 5. Generally speaking, deep learning based methods, including seq2graph, TCDF, IMV-LSTM and our $DVDE$, outperform the other three methods. It shows that deep neural networks are more capable to capture nonlinear relationships among variables. Particularly, our proposed $DVDE$ outperforms the other algorithms over 18% on $AUROC$ and 30% on $AUPR$. It proves that by projecting time series into dependency latent space using our two-level attention model, we can learn dynamic dependencies among variables more effectively.

5.2 Experiments on Change Point Detection

Change point detection methods are usually considered as unsupervised learning due to the scarcity of labels. Therefore, a common challenge that arises in practice is to design a fair comparison with or without algorithm parameters tuning. To

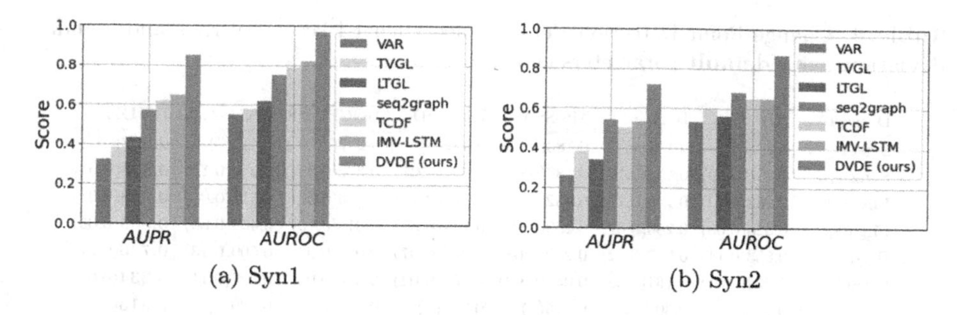

(a) Syn1 (b) Syn2

Fig. 5. Comparison of dynamic dependency (graph) learning.

Table 3. Change Point Detection Performance with F1 score (average and standard deviation) with **best** parameters choice.

Dataset	NB	BinSeg	MS-SSA	KL-CPD	BOCPDMS	mSSA-MW	**DVDE**
Syn1	0.349(–)	0.320(–)	0.362(–)	0.733(–)	0.398(–)	0.623(–)	**1.000**(–)
Syn2	0.321(0.10)	0.267(0.05)	0.253(0.03)	0.683(0.09)	0.510(0.08)	0.453(0.05)	**0.910**(0.04)
Flights1	0.562(0.08)	0.682(0.06)	0.582(0.10)	0.755(0.08)	0.727(0.05)	0.742(0.13)	**0.831**(0.06)
Flights2	0.593(0.06)	0.651(0.05)	0.607(0.08)	0.781(0.10)	0.731(0.12)	0.721(0.15)	**0.865**(0.05)
Beedance	0.612(0.07)	0.597(0.10)	0.583(0.06)	0.401(0.05)	0.167(0.07)	0.659(0.12)	**0.752**(0.07)
HASC	0.028(0.01)	0.304(0.08)	0.265(0.07)	0.156(0.01)	0.204(0.06)	0.327(0.10)	**0.356**(0.06)
Occupancy	0.340(–)	0.308(–)	0.462(–)	0.341(–)	0.474(–)	0.783(–)	**0.963**(–)
average	0.401	0.447	0.445	0.550	0.459	0.615	**0.811**
p-value	< 0.01	< 0.01	< 0.01	< 0.01	< 0.01	< 0.01	

overcome this, we follow the evaluation setup in [4,23], where for each dataset and algorithm we report (1) the F1-score using default parameters, and (2) the best F1-score over a grid search of parameter configurations. The default parameter configuration details can be found in our source code shorturl.at/sESZ4.

We include the following six baselines for comparison. *NB* [14] is a naive Bottom-up change point detection. *BinSeg* [15] is a sequential detection approach. *MS-SSA* [2] detects change points by calculating the martingale score. *KL-CPD* [6] is a deep learning based kernel learning framework. *BOCPDMS* [16] extends Bayesian online change point detection to online model selection and non-stationary spatio-temporal processes. *mSSA-MW* [4] models the underlying dynamics through a moving window based spatio-temporal model.

To evaluate the detected change points, we employ the same strategy that is used in [4]. Here the evaluation metric is **F1-score**, which is defined as: $F1 = \frac{2 \times TP}{2 \times TP + FP + FN}$, where TP, FP, and FN denote the number of true positives, false positives, and false negatives, respectively. Particularly, the following soft true positive (TP) rule is used: for a constant $\eta > 0$, a TP is recorded if the algorithm detects a change point $\hat{\tau}$ such that $|\hat{\tau} - \tau| \leq \eta$ where τ is a true change point; otherwise a FP is recorded. Here we set $\eta = 10$.

In Tables 3 and 4, the best and default performance of each algorithm are shown as the average and standard deviation of the F1-scores for each dataset. For the dataset with only one time series (Syn1 and Occupancy), standard deviation can not be reported. Again deep-learning based algorithms (KL-CPD and

Table 4. Change Point Detection Performance with F1 score (average and standard deviation) with **default** parameters choice.

Dataset	NB	BinSeg	MS-SSA	KL-CPD	BOCPDMS	mSSA-MW	**DVDE**
Syn1	0.278(–)	0.160(–)	0.183(–)	0.542(–)	0.287(–)	0.412(–)	**0.857**(–)
Syn2	0.236(0.09)	0.134(0.10)	0.187(0.08)	0.454(0.07)	0.322(0.08)	0.256(0.11)	**0.838**(0.05)
Flights1	0.233(0.13)	0.146(0.12)	0.352(0.08)	0.320(0.12)	0.456(0.08)	0.517(0.06)	**0.763**(0.07)
Flights2	0.328(0.09)	0.280(0.07)	0.230(0.09)	0.296(0.08)	0.318(0.10)	0.408(0.05)	**0.805**(0.05)
Beedance	0.553(0.11)	0.097(0.02)	0.279(0.08)	0.252(0.07)	0.092(0.02)	0.500(0.13)	**0.746**(0.07)
HASC	0.052(0.01)	0.161(0.05)	0.049(0.02)	0.155(0.01)	0.078(0.04)	0.177(0.11)	**0.334**(0.08)
Occupancy	0.293(–)	0.308(–)	0.375(–)	0.302(–)	0.198(–)	0.480(–)	**0.813**(–)
average	0.282	0.184	0.236	0.332	0.250	0.393	**0.737**
p-value	< 0.01	< 0.01	< 0.01	< 0.01	< 0.01	< 0.01	

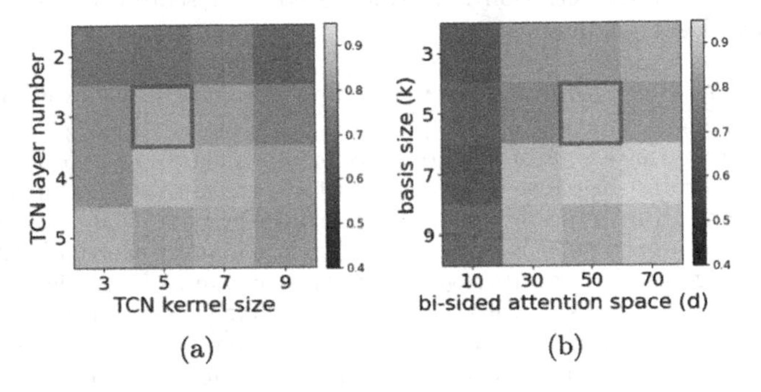

Fig. 6. Parameter sensitivity test of *DVDE* on graph learning (AUPR).

our *DVDE*) have better performance on the datasets with high nonlinearity (Syn1 and Syn2). Our proposed *DVDE* shows consistently superior performance over the other methods across all datasets. Particularly, our proposed *DVDE* is over 30% better than the second best method with best parameter, and over 90% better with default parameters. Moreover, all the p-values are lower than 0.01, which means that *DVDE*'s superior performance is highly statistically significant.

Figure 6 shows the performance robustness of our *DVDE* method under four key parameter tuning, with the red surrounded cells indicate the performance with default parameter setting. We can observe that even with sub-optimal parameter configuration, our *DVDE* still performs reasonably and stably well. However, it is worth to notice that the bi-sided attention space has to be large enough (e.g. ≥ 30 in this example) to capture dynamic dependency.

To understand the contribution by each part of our model, an ablation study is shown in Fig. 7. In short, the full model achieves the best performance, which shows that all parts are integrated systematically and each plays an indispensable role. The version without module 2 performs the worst. It shows that module 2 is the key component to acquire dynamic dependency. Besides, module 1 is also critical because of its ability to extract nonlinear univariate features. Additionally, the attentions in module 1 and 3 also contribute to higher performance.

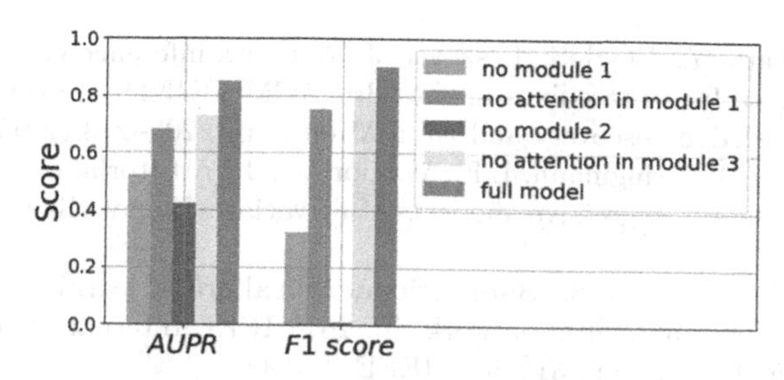

Fig. 7. Ablation study on Syn1 dataset.

6 Conclusion

This work presents a dynamic dependency learning model on multivariate time series. It learns the dependency graph through the latent space of dependency basis. The latent space coordinates can be used as descriptors of the temporal dynamic at every timestamp. We provide comprehensive experiments on both dynamic graph learning and change point detection to show the outperformance of our model.

Acknowledgment. This work was supported in part by United State, Department of Energy, under Contract Number DE-SC0012704.

References

1. Human activity sensing consortium challenge. http://hasc.jp/hc2011/download.html (2012)
2. Microsoft nimbusml package, ssa class. https://docs.microsoft.com/enus/python/api/nimbusml/nimbusml.timeseries.ssachangepointdetector. (2020)
3. Occupancy detection data set. https://archive.ics.uci.edu/ml/datasets/Occupancy+Detection+/. (2016)
4. Alanqary, A., Alomar, A., Shah, D.: Change point detection via multivariate singular spectrum analysis. In: Advances in Neural Information Processing Systems, vol. 34 (2021)
5. Artico, I., Wit, E.C.: Dynamic latent space relational event model. arXiv preprint arXiv:2204.04753 (2022)
6. Chang, W-C., Li, C-L., Yang, Y., Póczos, B.: Kernel change-point detection with auxiliary deep generative models. arXiv preprint arXiv:1901.06077 (2019)
7. Dang, X-H., Shah, S.Y., Zerfos, P.: Seq2graph: discovering dynamic non-linear dependencies from multivariate time series. In: 2019 IEEE International Conference on Big Data (Big Data), pp. 1774–1783. IEEE (2019)
8. Friedman, J., Hastie, T., Tibshirani, R.: Sparse inverse covariance estimation with the graphical lasso. Biostatistics **9**(3), 432–441 (2008)
9. Guo, T., Lin, T., Antulov-Fantulin, N.: Exploring interpretable LSTM neural networks over multi-variable data. In: International Conference on Machine Learning, pp. 2494–2504. PMLR (2019)

10. Hallac, D., Park, Y., Boyd, S., Leskovec, J.: Network inference via the time-varying graphical lasso. In: Proceedings of the 23rd ACM SIGKDD International Conference on Knowledge Discovery and Data Mining, pp. 205–213 (2017)
11. Haslbeck, J.M.B., Bringmann, L.F., Waldorp, L.J.: A tutorial on estimating time-varying vector autoregressive models. Multivariate Behav. Res. **56**(1), 120–149 (2021)
12. Huang, H., Xu, C., Yoo, S.: Bi-directional causal graph learning through weight-sharing and low-rank neural network. In: 2019 IEEE International Conference on Data Mining (ICDM), pp. 319–328. IEEE (2019)
13. Huang, H., Xu, C., Yoo, S., Yan, W., Wang, T., Xue, F.: Imbalanced time series classification for flight data analyzing with nonlinear granger causality learning. In: Proceedings of the 29th ACM International Conference on Information & Knowledge Management, pp. 2533–2540 (2020)
14. Keogh, E., Chu, S., Hart, D., Pazzani, M.: An online algorithm for segmenting time series. In: Proceedings 2001 IEEE International Conference on Data Mining, pp. 289–296. IEEE (2001)
15. Killick, R., Eckley, I.: changepoint: an r package for changepoint analysis. J. Stat. Softw. **58**(3), 1–19 (2014)
16. Knoblauch, J., Damoulas, T.: Spatio-temporal bayesian on-line changepoint detection with model selection. In: International Conference on Machine Learning, PMLR (2018)
17. Lütkepohl, H.: New Introduction to Multiple Time Series Analysis. Springer, Cham (2005)
18. Nauta, M., Bucur, D., Seifert, C.: Causal discovery with attention-based convolutional neural networks. Mach. Learn. Knowl. Extract. **1**(1), 312–340 (2019)
19. Oh, S.M., Rehg, J.M., Balch, T., Dellaert, F.: Learning and inferring motion patterns using parametric segmental switching linear dynamic systems. Int. J. Comput. Vis. **77**(1), 103–124 (2008)
20. Tank, A., Covert, I., Foti, N., Shojaie, A., Fox, E.: Neural granger causality. arXiv preprint arXiv:1802.05842 (2018)
21. Tomasi, F., Tozzo, V., Barla, A.: Temporal pattern detection in time-varying graphical models. In: 2020 25th International Conference on Pattern Recognition (ICPR). IEEE (2021)
22. Tomasi, F., Tozzo, V., Salzo, S., Verri, A.: Latent variable time-varying network inference. In: Proceedings of the 24th ACM SIGKDD International Conference on Knowledge Discovery & Data Mining (2018)
23. van den Burg, G.J.J., KI Williams, C.: An evaluation of change point detection algorithms. arXiv preprint arXiv:2003.06222 (2020)
24. Xu, C., Huang, H., Yoo, S.: Scalable causal graph learning through a deep neural network. In: Proceedings of the 28th ACM International Conference on Information and Knowledge Management, pp. 1853–1862. ACM (2019)
25. Yao, S., Yoo, S., Dantong, Yu.: Prior knowledge driven granger causality analysis on gene regulatory network discovery. BMC Bioinf. **16**(1), 1–18 (2015)

Author Index

© The Editor(s) (if applicable) and The Author(s), under exclusive license
to Springer Nature Switzerland AG 2023

H. Kashima et al. (Eds.): PAKDD 2023, LNAI 13938, pp. 353–354, 2023.
https://doi.org/10.1007/978-3-031-33383-5

Printed in the United States
by Baker & Taylor Publisher Services

Printed in the United States
by Baker & Taylor Publisher Services